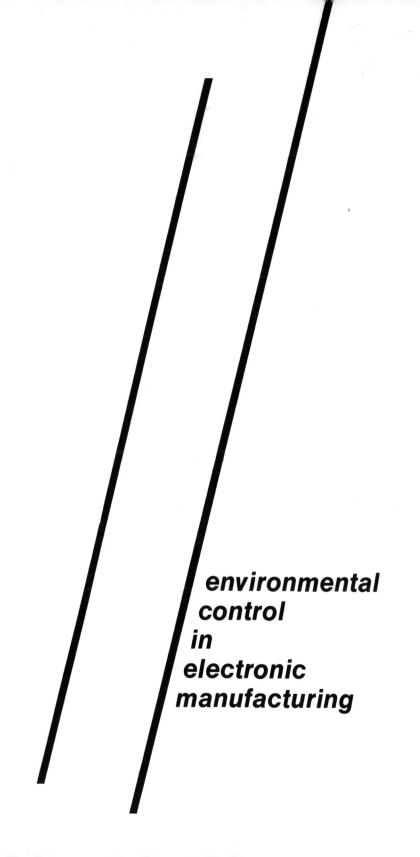

# environmental
# control
# in
# electronic
# manufacturing

## WESTERN ELECTRIC SERIES

*lasers in industry, edited by* **S. S. Charschan**
*environmental control in electronic manufacturing,*
  *edited by* **P. W. Morrison**

# environmental
# control
# in
# electronic
# manufacturing

*edited by **Philip W. Morrison***
*Member of Technical Staff*
*Western Electric*

**Van Nostrand Reinhold Company**

*New York / Cincinnati / Toronto / London / Melbourne*

Van Nostrand Reinhold Company Regional Offices:
New York  Cincinnati  Chicago  Millbrae  Dallas

Van Nostrand Reinhold Company International Offices:
London  Toronto  Melbourne

Library of Congress Catalog Card Number: 73-9518
ISBN: 0-442-25564-2

Manufactured in the United States of America

Published by Van Nostrand Reinhold Company
450 West 33rd Street, New York, N.Y. 10001

Published simultaneously in Canada by Van Nostrand Reinhold Ltd.

15  14  13  12  11  10  9  8  7  6  5  4  3  2  1

**Library of Congress Cataloging in Publication Data**

Morrison, P      W
    Environmental control in electronic manufacturing.

    (Western Electric series)
    Includes bibliographical references.
    1. Electronic apparatus and appliances—Design
and construction.  2. Contamination (Technology)
I. Title.  II. Series: Western Electric Company,
Inc. Western Electric series.
TK7870.M587      621.381'7      73-9518
ISBN 0-442-25564-2

## contributors

L. K. Barnes

C. Y. Bartholomew

T. H. Briggs

D. K. Conley

R. V. Day

J. E. Dennison

C. L. Fraust

D. K. Green

S. K. Kempner

A. S. Lakatos

H. J. Litsch

P. W. Morrison

W. W. Pcihoda

A. Pfahnl

W. V. Pink

J. F. Pudvin

G. L. Schadler

E. J. Sowinski

A. H. Szkudlapski

R. A. Whitner

All members of Engineering Staff, Western Electric or Bell Telephone Laboratories.

# foreword

The history of clean rooms is long and traditionally consisted of doing the best one knew at the time and trying to inculcate a sense of cleanliness into the habits and actions of the necessary workers. In the past few years substantial advances have been made in methods and measurement, some denying and some confirming prior art knowledge and techniques. The principle ingredient added has been an engineering, as opposed to a craft or technique, approach to the design of clean rooms and providing clean materials, water and other reagents.

This present volume presents the current status, methods, knowledge and standards as we know them in our own work in microelectronics. The material is presented on the basis of principles, to the extent possible, with examples based entirely on our own experience in microelectronics. The knowledge and data are incomplete, but we believe represent a sufficient step forward to justify collection and publication to other users. It is hoped that with the intent of writing in terms of principles, these principles can be extended to other fields as well as providing the basis for further understanding in the microelectronics field. Commonology, up to the point of stifling innovation, works to the advantage of all.

Our own interests in clean and controlled environment and materials go back long before the current era of microelectronics into electron tube development and manufacture. Both fields are characterized as being critical to sensitive levels of additives for performance and contaminants as "poisons." Generally, the levels of trace materials which can beneficially or adversely affect the devices are lower than existing measurement capabilities. The employment of clean environments and cleaning methods is thus not a cure-all, but a minimizer in the sense that gross effects are reduced or eliminated, but not assuring perfection.

We are most indebted to Willis Whitfield of our Sandia colleagues for his concept of the laminar flow approach to providing clean environment, to the several instrument companies for their activities in providing relatively rapid mensuration and to the Atomic Energy Commission for the development of the high efficiency particulate air filter. Our work really stems, in a meaningful engineering sense, from these advances.

Most microelectronic manufacture requires a succession of very dirty operations followed by requirements of very pure or clean environment. Hence, cleaning, cleaning fluids, and cleaning methods are no small part of the practical aspect of controlling environmental factors in practice. The control of contaminating gases in the sense of separation or filtering is a real need, but is, unhappily, without a general practical solution at this time.

A major indeterminate is an accurate answer to the simple question of how clean does one need to be. The environment desired is often easy to define—perfection. Economics and capability may not permit such a solution.

Hence economic justification sometimes becomes essentially an article of faith based on prior experience matched to the quality of product required. In a personal opinion we have never been clean enough. Yet the same opinion views in reflection many operations and practices, the best known at the time, which were probably ineffectual and hence costly. Further, it seems clear that the trend in the microelectronics field can be only in one direction—toward ever-growing higher standards of environment control. Hence, it follows that the quicker this field can be reduced to good, knowledgeable engineering practices and principles, the more likely we can provide required environments, prove by measurement that what we have is what is wanted, know how to employ the facility as to not destroy its effectiveness—and all with the best economy. It is hoped this book presents a step in this direction.

A. E. ANDERSON

# *preface*

Contaminants, which can be defined as unwanted matter or energy, play an especially important role in the manufacture or repair of such items as semiconductors, space vehicles, human organs, ball bearings and antibiotic drugs. One is struck by the dissimilarity of these subjects, which are examples from the microelectronic, aerospace, medical, precision equipment and pharmaceutical fields. Is there a common approach to achieve control of the contaminants affecting these diverse industries? Can a hospital specializing in organ transplants benefit from the experiences of a company making silicon integrated circuits? The premise of this book is that the principles of environmental or contamination control, as used by the microelectronic industries, do apply broadly to any series of precision operations where unwanted materials or influences affect their successful completion. For example, while it is unusual to think of the human body as a product or an operating room as a factory, the engineering principles of manufacturing environmental control are applicable. The control of infectious microorganisms in hospital operating rooms can be achieved in the identical manner used to protect against airborne particles during the pattern generation of semiconductor devices. In certain instances it can be said that the semiconductor device is more sensitive to such particles in the air. The human body will become infected due only to a living particle while the semiconductor device can be "killed" by viable, inert or chemically active particles depositing on its surfaces at different stages of manufacture. The risks of "manufacture" are different since the loss of human life has consequences beyond economics, but the technical approaches to environmental problem solving are similar. The same logic can be applied to examples from the other precision industries where trace quantities of unwanted materials or influences affect the cost and reliability of their quality products.

This book is written by the Western Electric engineering staff, with contributions from members of Bell Telephone Laboratories, to accelerate the understanding and acceptance of the new engineering discipline of environmental control. It is written for the bachelor degree level of engineer or science major who is interested in the technical challenge of manufacturing a precision device. The examples used throughout are taken from industrial experience in making microelectronic devices. However, they have been selected to illustrate the principles of environmental control engineering as well as some of the "how to do it" aspects of fabricating electron devices.

This volume follows a systematic engineering approach to making a product where contaminants play a significant role. Chapter 1 defines the basic elements of product manufacture, the contaminants and their carriers which influence engineering decisions. Semiconductor and thin film processes are outlined to acquaint readers with the rudiments of microelectronic processing. Chapter 2 delves into the product-contaminant relationships exhibited in the failure mechanisms of electron devices. The intent of this chapter is to take environmental problems beyond the empirical treatment of symptoms into a deeper understanding of the effects of contaminants. Chapters 3 through 5 describe the methods of contaminant measurements as both a diagnostic tool and a means of production quality control. Individual chapters are devoted to microscopy and particle measurement systems because of their broad role in contaminant data collection and analysis. Chapters 6 through 15 describe the principles and applications of control techniques for the various sources and carriers of contaminants in a manufacturing situation. These chapters will illustrate the interdependence of product requirements, the contamination control capabilities of environmental facility and plant service designs, and the cleaning techniques required to remove any remaining surface deposits. Chapter 16 completes the systems analysis of the manufacturing routine in the discussion of monitoring checkpoints for contaminants during production for feedback analysis. The insight gained by periodic analyses of actual conditions relative to engineering planning criteria leads back to the earlier steps of product-contaminant diagnosis and improved control of contaminants during manufacture.

The contributing authors of these chapters have day-to-day association with their specific subject which provide the basis for their written work. In certain instances a chapter represents the first grouping of heretofore discrete subjects into a cohesive philosophy. The ability to accomplish this task comes from the authors' understanding of environmental control engineering and the role that the otherwise diverse facets play within it.

We must sadly relate that one of the authors, Wilson V. Pink, died during the time lapse between completion of the manuscript and publication. An innovator in clean environment control, Wilson is missed by all who knew him. As editor I have assumed the responsibility for the final review of his chapters.

In any work of this kind there are many contributors who merit acknowledgement but remain anonymous because of the large numbers involved. This book would not have been possible without the collective knowledge of our colleagues in Western Electric Company and the Bell Telephone Laboratories; we recognize the debt owed to all of them. Specific acknowledgment is given to A. E. Anderson of the Western Electric Company; his insight and full support literally made this text a reality instead of a dream.

P. W. MORRISON

# contents

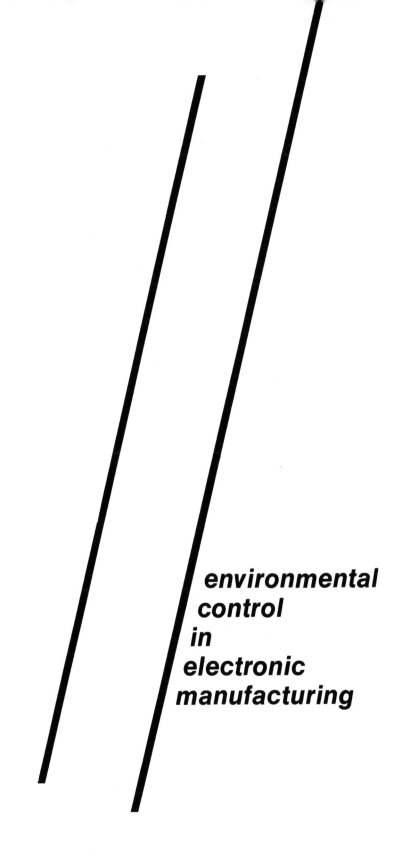

# environmental
## control
## in
## electronic
## manufacturing

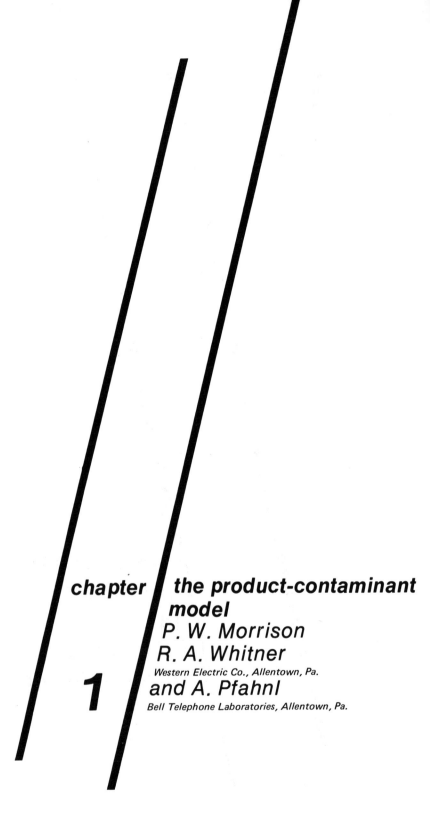

**chapter**

**the product-contaminant model**

*P. W. Morrison*
*R. A. Whitner*

*Western Electric Co., Allentown, Pa.*

**and A. Pfahnl**

*Bell Telephone Laboratories, Allentown, Pa.*

**1**

The term *microelectronics* is generally applied to the transistors, diodes, and integrated circuits of semiconductor technology and the resistors, capacitors, and interconnections of thin film technology which have revolutionized the world of electronic circuitry. These electron devices fit the definition of a precision product—any item exhibiting high extremes of quality performance which is manufactured on a volume basis to exacting tolerances. High quality, mass production techniques and tolerances held close to theoretically ideal conditions are what makes microelectronics so susceptible to the many forms of contaminants.

Precision product industries have a common concern for quality control in their manufacturing processes, and one group of variables in that routine is contamination. This chapter will establish what is meant by contamination and the relationship to one precision product industry—microelectronics. The intent is to emphasize the principles underlying this relationship so as to establish a general product-contaminant model applicable to other fields besides the electronic industry.

## the contaminant model

*Contaminants* or *contamination* is a broad term used in many diverse fields to denote an undesirable influence. When discussing precision product manufacture a contaminant is any substance or energy which adversely affects a product or process of interest. Then by extension, the control of contaminants is the planning, organization and implementation necessary to achieve the desired level of cleanness in, on, or around the product or process of interest. These definitions are admittedly broad, but it is important to recognize that there are innumerable mechanisms which can cause product failure. Figure 1-1 illustrates this point. Harmful influences may come from the impurities of raw materials, environmental deposits, or residues from necessary processes of fabrication. From the viewpoint of the finished product it is unimportant where the contaminant originates; however, in establish-

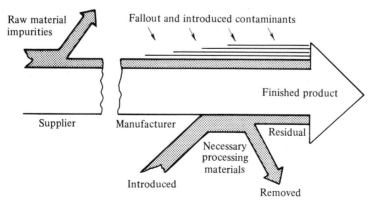

*Figure 1-1 Product contamination.*

ing control of contaminating influences an understanding as to the source of that influence is essential.

## classification of contaminants

It is difficult, if not impossible, to establish a contaminant classification system which has mutually exclusive subdivisions. Contaminants can be in any form, i.e., solid, liquid, gas, or energy. They can be classified by their effects on the product. Or they can be classified by the mechanisms of deposition—settlement, impingement, condensation, precipitation, etc. The selected contaminant category system of this text is based on the distinct properties of the contaminants themselves, which can then be related easily to the means of transport and the methods of control. Thus, contaminants are broken into (1) physical matter (particles); (2) organic chemicals; (3) inorganic chemicals; and (4) viable (microbial) organisms.

**Physical Matter (Particles).** The word *physical* is used in the sense that a material has body or form. Thus, we are concerned about solid substances and usually in particulate form. It is true that liquid mists can be considered particles and that film depositions are of a solid nature, but these exceptions can be grouped within this classification where their contaminating property is of a physical nature. Since in most cases physical contaminants are in particulate form the term *particle* will be used in this text to mean a physical contaminant.

For this type of contaminant the major concern is that the physical presence of the contaminating material will cause disruption to the product or process of interest. For example, a particle of sand can score a motor bearing or obstruct the passage of light in a photographic process. As silicon dioxide the particle may also have chemical properties, but for the contaminated processes under discussion its chemical nature is deemed inconsequential. Its physical nature is the property which makes the sand particle a contaminant.

Particulate contamination is almost a universal frustration to man. Housewives complain about dust; the miner must fear it as a cause of pneumoconiosis; and in such precision fields as microelectronics, aerospace, and the drug industries their products cannot be made in the presence of high dust concentrations.

The reason stems from three characteristics of particulate matter: (1) its universal presence, (2) the large number of available particles, and (3) the wide range of their size distribution. Figure 1-2 lists typical materials which may be found in a sample of particles taken at an industrial work position. Practically every solid material can degenerate into particulate form and be distributed everywhere by air currents. The many sources of particles create a fantastic number of airborne particles. The environment of a typical factory without special particle control facilities may contain 200,000–10,000,000 particles/cu. ft of air. For a factory which has about 1,000,000 airborne

| | |
|---|---|
| Cement dust | Sneeze and coughing discharges |
| Sand | Tobacco smoke |
| Lint | Algae, spores, bacteria, pollen |
| Clothing fibers | Metallic burr and particles |
| Hair | Wear particles |
| Epidermal scale | Discharges from furnaces and flames |

*Figure 1-2 Particulate contaminants.*

particles/cu. ft the removal of 99% of the particles by filtration would still leave a residue of 1,000 particles/cu. ft to cause product damage. However, the particle sizes of this residue usually will be in the submicron range in a practical system. The magnitude of the particle size variation problem can be seen in Fig. 1-3. A particle can be beach sand several millimeters in diameter or combustion nuclei with a 0.01 micron diameter.

**Organic Chemicals.** In the minds of many people the control of contaminants means particle control. However, as the ability to attain and verify particle-free environments is approached the influences of other contamination forms become more predominant. The chemical nature of materials is commonly divided according to organic or inorganic properties, and since all materials known to man can be a contaminant this subdivision is appropriate here. Organic chemistry is, by definition, the chemistry of carbon compounds where covalent bonds permit carbon atoms to attach themselves to each other in chains, rings, branches, and crosslinks in an almost infinite number of variations. The covalent nature of organic matter tends to classify into the following broad characteristics relative to ionic materials:
1. Low melting and boiling points.
2. Soluble in organic liquids and insoluble in water.
3. Burn readily.
4. Reactions are molecular and complex in their resultant products.
5. Solutions and melts do not conduct electricity.
6. Commonly exist as isomers, which are different compounds having the same molecular formula.
7. Mostly derived from ten elements, yet form combinations numbering about 1,000,000.

While there are many exceptions to the above, these general statements are helpful in differentiating between organic and inorganic matter.

The universal presence of organic material in any industry is testimony to their usefulness and their potential as a contaminant. Figure 1-4 lists some of the organic contaminants common to the electronics industry. The examples illustrate that organic contaminants may be deposited on product surfaces as particles from many different sources, surface residues from prior processes, vapor condensations from the environment, or as films left behind by lubrication oils or cleaning solvents. These deposits are differentiated from physical contaminants by the fact that the chemical properties of the

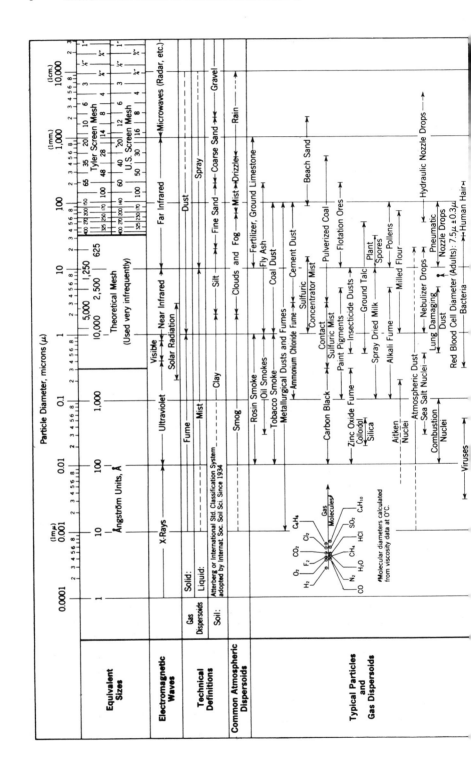

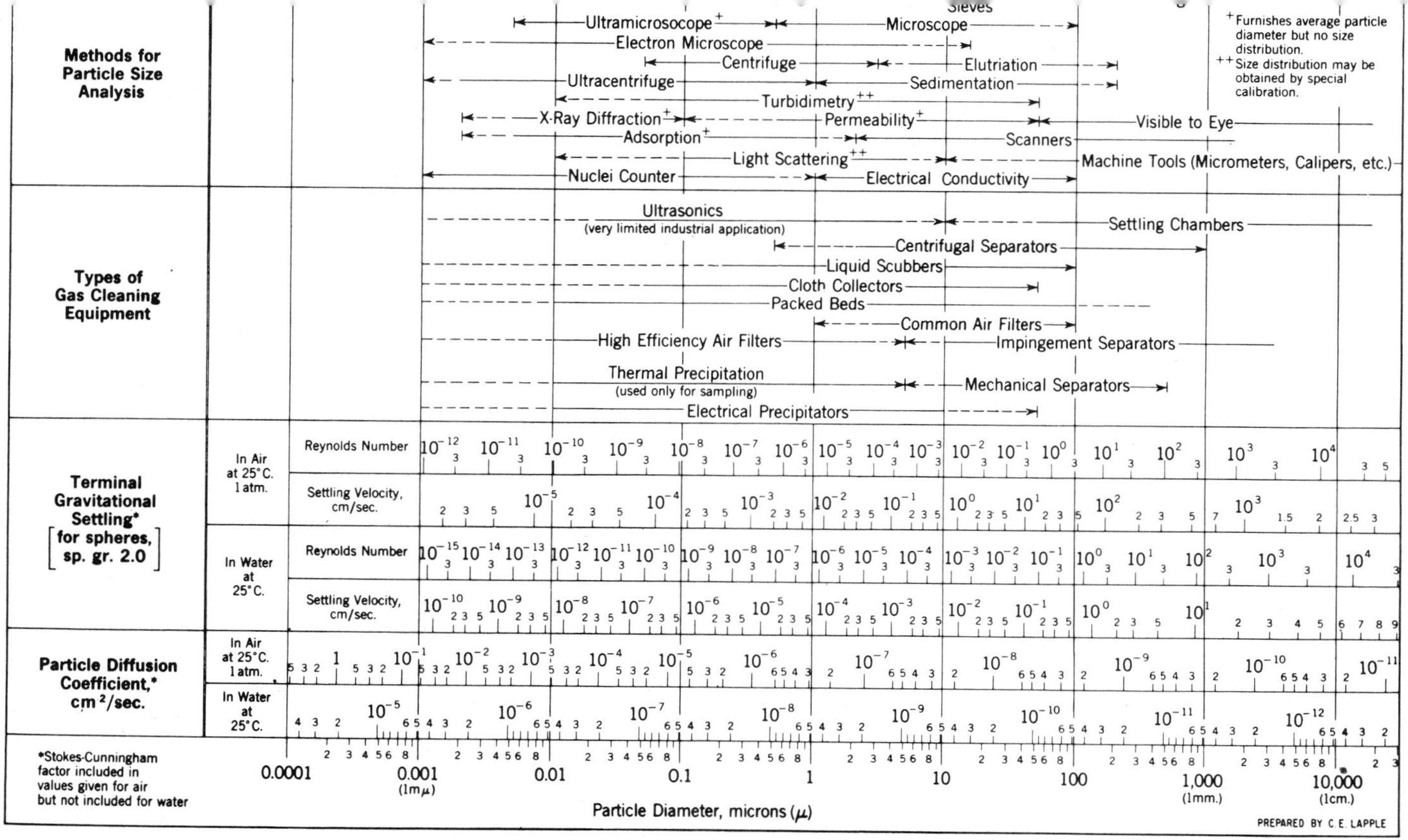

Figure 1-3  Characteristics of particles and particle dispersoids.  (Reprinted from the Stanford Research Institute Journal, Third Quarter, 1961.)

| Oils | Carbonaceous residues | Greases |
|------|----------------------|---------|
| Fats | Fluxes | Photoresists |
| Waxes | Paints | Solvents |
| Plastics | | Hydrocarbon flames |

*Figure 1-4  Organic contaminants.*

organic matter cause unique problems in contaminant measurement and removal.

**Inorganic Chemicals.** It is the ionic nature of most inorganic chemicals which is the major source of concern from a contamination viewpoint. Thus, the discussion of inorganic matter as a contaminant will emphasize those chemicals which display ionic characteristics. As in the case of organic (or covalent) matter some general tendencies can be assigned to inorganic (or ionic) materials:

1. High melting and boiling points.
2. Soluble in water and insoluble in organic liquids.
3. Solutions and melts conduct electric current.
4. Difficult to burn.
5. Reactions are ionic and simple in their resultant products.
6. Polymerization is extremely rare and isomerism is unusual.
7. Derived from any of 103 elements in combinations numbering about 100,000.

Again, there are many exceptions to these statements, but our purpose here is to illustrate central tendencies without belaboring the completeness of its validity.

The solubility of inorganics in water and their ability to conduct electric current are the major mechanisms by which electronic products are contaminated. Ionic material in the presence of even minute traces of water will establish an electrolytic cell. Current flows result which can short a semiconductor device, and galvanic corrosion occurs which can degrade product materials to the point of failure. Even without the presence of moisture as an electrolyte the charge of an ionic surface contaminant can cause redistribution of devices' bulk charges.

**Viable (Microbial) Organisms.** Microbial contaminants can be simply described as organic particles which can reproduce. It is the biological aspects of microorganisms which warrant their consideration as a separate contaminant category. In the performance of their living functions, microorganisms will multiply their numbers and cause either product degradation or disease in plants, animals, and human beings. The electronics industry usually is affected by microorganisms as particulate or organic deposits on product surfaces. Only in highly pure process water do the living aspects of microbes become a significant factor as will be shown in Chap. 8.

Even though there is little knowledge about the direct effects of microbes on electronic devices it is well to have an appreciation of the variety of organisms and their physical characteristics.   Thus, when speaking about microbes, they may be viruses, rickettsiae, fungi, protozoa, and algae. Figure 1-5 gives a size range comparison of these classifications. The number, variety and size range of these living, organic particles makes them a potential contamination threat to microelectronic products as well as people.

## carriers of contaminants

Experience has shown that contaminants usually originate from many different sources but are deposited onto a sensitive product by a limited number of contaminant carriers.  Generally, control is most effectively and economically achieved by the regulation of these contaminant carriers.

A contaminant deposit on the product surface can be an accumulation from the prior or immediate mechanical, electrical or chemical processes; the factory environment surrounding the process; the people handling the product; or any one of the factory water, reagent chemical, pressurized gas, and waste disposal systems supplied to the immediate process. The contribution of any one contaminant carrier may be paramount, or a specific condition might demand the control of all carriers.  Let us examine what each can contribute relative to each other.

**Fabricating Processes.**  Any process, be it mechanical, electrical, or chemical in nature, can be a carrier of contamination in the performance of a fabricating function for product assembly.   Thermocompression bonding tips or electrical test set probe contacts can accumulate particles and organic films on their surfaces and transfer the contaminants to a silicon wafer. Certain mechanical operations, such as welding, grinding, or lapping, generate particles and oil droplets which splatter onto other parts awaiting processing. Electroplating baths will degrade with use as environmental particles and chemical impurities are concentrated in the solution and are deposited on the plated surfaces.  The contaminant depositions by these processes are by-products or detracting influences on the prime function of bonding, testing, welding, electroplating, etc.  The significance of these by-products is determined by their effect on the yield of the prime process function.  Such contamination is intimately interwoven with specific products and processes and, therefore, cannot be discussed to advantage in this text.  However, their existence as a contaminant carrier should not be overlooked in a specific investigation into contaminants.

**Cleaning Processes.**  There are many processes, usually chemical in nature, whose prime purpose is cleaning rather than fabrication.  Cleaning processes are included within the manufacturing routine to remove the contaminant by-products of fabricating processes and the deposits from other contaminant carriers.  Cleaning may be achieved through material removal (acid etch),

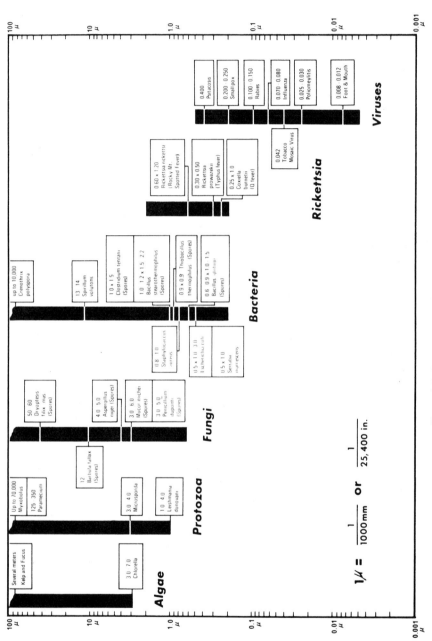

Figure 1-5 Relative sizes of microorganisms.

solvent action (trichloroethylene degreasing), dilution (cascade rinsing), saponification (detergent cleaning), or displacement (Freon* drying). The type of cleaning action will depend on the contaminant form, the type of material to be cleaned, and the degree of cleanness required.

As with processes with fabrication functions, the chemical cleaning processes can cause contaminant deposition as a by-product. For example, an immersion type solvent operation may well remove organic films from a product surface but redeposit particles which had collected in the bath during use. However, contamination can also result from the intrinsic inefficiency of the cleaning action, through improper application to a product material or through inadequate performance of the cleaning operation itself. Thus, cleaning processes can be both a cause and a cure of product contamination. If cleaning processes could remove all forms of harmful deposits at 100% efficiency on an economic basis, there would be little need for other techniques to control contaminants. Since this is not usually possible for precision products, the other carriers of contaminants must be controlled to a sufficient degree whereby an economic balance is achieved. Chapters 6 and 7 will discuss the theory and practice of surface cleaning via chemical processes.

**Factory Water Supply.** Water of many different qualities is used extensively in chemical cleaning processes as the solvent in aqueous solutions and for rinsing operations. The liquidborne matter contaminating the water supply may be particles with sizes ranging from colloids to sand grains, organic and inorganic salts from the raw water supplies, or viable algae and bacteria which may grow in distribution systems. The key factor affecting their measurement and control is the fact that the contaminants are waterborne. Water rinsing is so often used in microelectronics as the final surface cleaning process that "intrinsic" water has become the common request of the product engineer. Filtration, deionization, and sterilization, as described in Chap. 8, are common terms to the plant engineer supplying high quality water as well as the product engineer who specifies its use.

**Reagent Chemicals.** In the discussion on chemical cleaning processes the emphasis was placed on product contamination as the chemical solution degraded with use. However, any chemical bath can only be as good as the raw ingredients. Both the solvent and the solute must be sufficiently pure to insure the desired chemical action. In microelectronics many types of acids, alkalies, metallic salts, and organic solvents are used in large quantities. Generally, the highest attainable grades of chemicals are used to preclude contamination by chemical impurities. In some cases further purification is required when the commercial quality is inadequate. Process chemicals as another vehicle for product contamination are examined in Chap. 9 along with the necessary safety and handling criteria in manufacturing situations.

---

*Registered Trademark of E. I. du Pont de Nemours & Co., Wilmington, Del.

**Factory Environment.** The ambient of the factory contains a diverse collection of airborne particulates, vapors, and radiations which may cause adverse physical, chemical, biological, or energy influences on any product. Again, the emphasis is placed on the method of transport, i.e., airborne, rather than the contaminant form. The effect may be direct (a particle settling on a silicon wafer) or indirect (a particle settling into a cleaning solution bath where it is subsequently deposited on a silicon wafer). Figure 1-2 has shown that the sources of airborne contaminants may be from the outside environment, factory personnel, production facilities, or the wastes created by production. The factory environment integrates the effects of other contaminant sources into a highly mobile form. The factory environment is a major carrier of contaminants which can destroy electron devices. Its influence is such that many industries consider factory environment control and the control of contaminants to be synonymous terms. Chapters 10 and 11 are devoted to the theory and practice of factory environment control.

**Pressurized Gas Systems.** Another "airborne" carrier of contaminants in the microelectronics industry is the pressurized gas systems delivering compressed air, oxygen, hydrogen, nitrogen, helium, and argon to process environments associated with surface heat treatments, oxidation, diffusion, crystal growth, sputtering, inert storage, and many other microelectronic processes. While particle and residual gas impurities are the major forms of contaminants being distributed to the process point-of-use, the piping systems can be contaminated with such organic materials as fluxes, pipe dope, and detergent residues which result from improper pipe cleaning or assembly. These gases must be generated, stored, and distributed in such a manner that the point-of-use quality can be guaranteed. The difficulties in achieving this goal are explained in Chap. 12.

**People.** Human beings are an obvious source of contaminants in a factory and, at the same time, must be protected from those influences which affect health and safety. Product contamination by people may be felt at any point in the manufacturing routine, either through direct contact with product or by contributions to other contaminant carriers (cleaning processes, factory environment, etc.). Facetiously it has been said that if you want to keep a job clean, keep people away from it. However, people are essential; and it is important that health and safety planning be incorporated into any manufacturing situation. Product contamination by people can be minimized by the various forms of isolation described in Chap. 13; the protection of personnel health and safety is the subject of Chap. 14.

**Waste Disposal.** Until now the carriers of contamination were easily recognized as having intimate contact with the product. For the case of waste collection and treatment the contamination potential is considerable, but the emphasis shifts from the product to the community. Spent chemicals, hazardous vapors, and water wastes are by-products of chemical cleaning processes; and as a part of the manufacturing routine they are transported out of the factory environment to protect both the employees and the product. However, inadequate chemical exhaust or drain systems plus poor handling procedures for concentrated chemical waste will increase the vaporous contaminants in the factory air. Having been moved outside the factory, these chemical cleaning by-products require proper scrubbing, neutralization, precipitation or some other form of waste treatment to protect the community against pollution. The collection or transport of waste is the contaminant interface with the employees and the product, while the treatment or disposal of waste is the pollution interface with the community. Chapter 15 on the general subject of waste disposal is relevant to the control of contaminants since it is, in reality, the final step of chemical cleaning processes, and improper control can harm either the community, employees, or product.

## the product model

The manufacturing routine for precision products is often defined as the progression of events from the raw materials through the fabrication of piece parts, assembly operations, functional testing, and final inspection of the finished product. The details will vary radically for different precision products. However, a major portion of contaminants is generated or carried by facilities and processes which are common to most, if not all, precision industries. These are the so-called "common carriers" of contamination: the factory environment and the people within it, plant waters and gases, surface cleaning facilities and procedures, high purity production chemicals, and process effluent systems. Thus, the experience gained by one precision industry in the control of the common carriers of contamination has general application in other industries.

The remainder of this chapter discusses the major processes of semiconductor and thin film manufacture currently being utilized. They are grouped to illustrate the basics of the manufacturing routine, i.e., raw materials, fabrication processes, testing, and final inspection. In this manner the reader can draw analogies to other products, especially concerning the relative sensitivity to the various forms of product contaminants.

## semiconductor technology

A large number of different types of semiconductors and semiconductor devices are presently in manufacture or development. Regardless of the variations in material, size, or type, all are sensitive to chemical and particulate contamination. The degree to which a device is affected depends on the process, the device design, and the device requirements.

Semiconductor devices are manufactured from several different semiconductor materials, including germanium, silicon, gallium arsenide, and gallium phosphide. Initially, germanium was the prime material, although silicon quickly gained greater acceptance because it has better temperature characteristics. While silicon is the most useful semiconductor today, other materials are used in increasingly larger number to take advantage of their unique properties. Each material presents a different set of processing difficulties and requirements because of its particular chemistry and metallurgy; however, in many cases the equipment and environmental considerations are similar. For brevity and because of its commercial importance, this discussion will center on silicon technology and its use in integrated circuits.

### silicon integrated circuits

The IEEE's definition of an integrated circuit is, "A combination of interconnected circuit elements inseparately associated on or within a continuous substrate." Silicon integrated circuits (SIC) have been developed to reduce the cost, size, and power requirements and improve the electrical characteristics and reliability of circuits when compared to conventionally assembled circuits. Figure 1-6 shows a typical SIC package—a dual gate with 26 components on a square of silicon .060 in. on a side which employs the bipolar

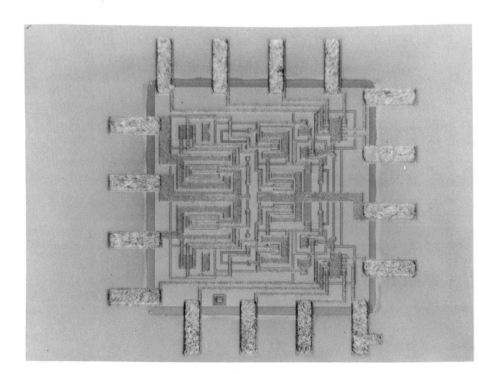

*Figure 1-6 Beam-lead, sealed-junction integrated circuit chip ready for bonding.*

junction isolation beam lead technology. Figure 1-7 shows the major steps currently used in the manufacture of such a SIC chip. Several new technologies are currently in development that will increase the component density at least an order of magnitude but, as usual, at the price of requiring more sensitive processes.

**TABLE 1-1**

| Region | Name | Silicon Type | Diffusion Source |
|--------|------|--------------|------------------|
| C | Buried Layer | N+[a] | Antimony |
| D | Isolation | P+ | Boron |
| E | Deep Collector | N+ | Phosphorus |
| F | Base | P | Boron |
| G | Emitter | N+ | Phosphorus |

[a]The + indicates a high concentration of impurities.

Figure 1-8 and Table 1-1 describe a schematic cross-section of a completed beam lead monolithic integrated circuit. In the illustration all $p$-type silicon is shaded and all $n$-type silicon clear.[*]

The type of silicon and the impurity concentration result either from the type of starting wafer or from subsequent treatment. Also shown are the main circuit elements of a bipolar SIC: a transistor, resistor, and diode. The overall manufacturing procedure for integrated circuits is concerned with how the different regions can be generated, the components interconnected, the circuit tested, and the chip packaged for mechanical and environmental protection.

### general processing—SIC

To generate the needed regions of $n$ or $p$ material, areas of the chip are selectively doped with the appropriate impurity by (1) growing a uniform oxide on the wafer; (2) cutting holes in the oxide with a photolithographic process; (3) diffusing the desired impurity in the exposed silicon (the oxide prevents doping of the silicon beneath it); and (4) growing a new oxide. This cycle is repeated up to 6 times in the production of an SIC circuit, as shown in Fig. 1-7. The patterns that are cut in the oxide become visible depressions

[*]Intrinsic (very pure) silicon has a relatively high resistance and few current conductors. There are two types of doped silicon.

1. $n$-type silicon in which an impurity (phosphorus, arsenic, antimony, etc.) has been added that provides excess electrons to conduct current.
2. $p$-type silicon in which an impurity (boron, gallium, etc.) has been added that lacks an electron (has a hole) and conducts current by movement of the positively charged hole.

The resistance of the silicon varies inversely to the amount of added impurity. Adjoining $p$ and $n$ regions form a rectifying junction which conducts current easily in one direction but not the other.

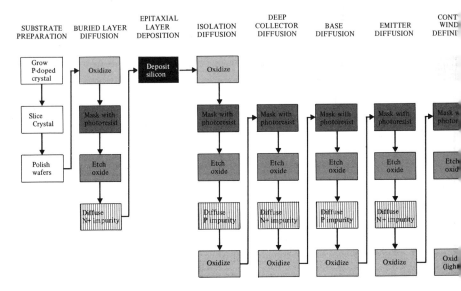

*Figure 1-7 Manufacturing process for a typical beam-lead sealed-junction silicon integrated circuit. Similar operations are indicated by equal shading.*

in the silicon after subsequent processing and are used to align the successive patterns. Appropriately doped layers of silicon can also be added to the silicon wafer by a process called epitaxy. Following is a discussion of each basic process step.

**Wafer Preparation.** The silicon wafer used in SIC manufacture may be 1.5-3.0 in. in diameter and .010 in. thick. It is an almost perfect single crystal oriented very carefully to a particular plane. To obtain this, silicon is refined by a gaseous reaction so that the impurity level is less than one part in $10^9$ parts silicon. The silicon is then melted and boron added to a level of one part in $10^8$. From this melt a single crystal ingot is pulled using a properly oriented seed crystal. The ingot is uniform in diameter and up to 20 in. long. This ingot is then sawed into wafers about .020 in. thick, and the wafers lapped and polished to a mirror finish on one side by a combined chemical and mechanical process to the final thickness of .010 in.

**Wafer Cleaning.** Achieving a clean wafer surface is one of the major concerns in integrated circuit manufacture. Any particles on the surface of a wafer prior to a high temperature operation will cause an oxide defect and a potential circuit failure. Chemical residues, which may not create a visible problem, may change the silicon impurity concentration causing either immediate circuit failure or even more important, poorer reliability. Wafer cleaning will be discussed in more detail in Chaps. 6 and 7.

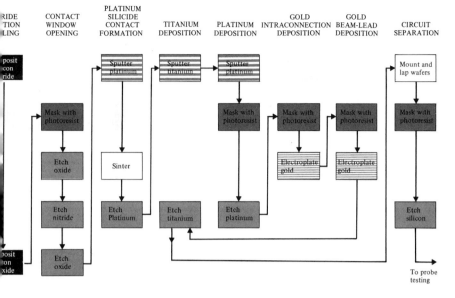

Figure 1.7 continued

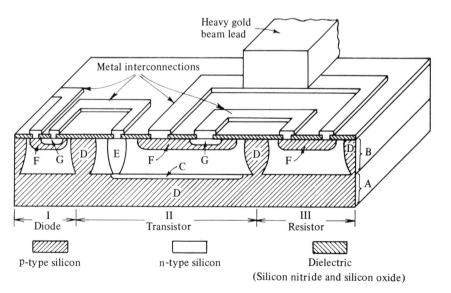

Figure 1-8 Schematic cross section of the major regions of a typical integrated circuit. A—Substrate; B—Epitaxial layer (n); C—Buried N$^+$ layer (n); D—Isolation diffusion (p); E—Deep colleter diffusion (n); F—Base diffusion (p); G—Emitter diffusion (n).

**Oxidation.** To grow the masking oxide, the polished silicon wafer is placed in an oxidizing ambient such as steam or wet oxygen at a temperature over 800°C. The oxide grows uniformly over the whole wafer, and the oxide appears colored due to the interference of light waves as they pass through the oxide and are reflected back from the silicon surface. Imperfections in the oxide appear as a change in the color and may be caused by the contamination of the silicon surface prior to oxidation. Some common contaminants are organic or inorganic residues that are not removed in wafer cleaning, airborne particles that fall on the wafer prior to oxidation, and particles generated in the furnace. Since any oxide defect can potentially lead to a circuit failure, contamination must be minimized.

**Photolithography.** Patterns are etched in the silicon oxide by photolithography to expose the silicon regions that are to be doped at a particular step. In this process an emulsion that is sensitive to a particular ultraviolet light band is whirled onto the wafer and then exposed through a precision mask which bears the desired pattern. This process is performed in special clean rooms with yellow light that does not affect the emulsion. In general, a mask has several hundred identical patterns in an $X$-$Y$ array, each pattern being capable of forming one circuit. The oxide is removed in the exposed areas by etching with a diluted hydrofluoric acid which dissolves the silicon oxide but does not etch the areas protected by the emulsion.

The emulsion on the wafer must be uniform and within certain thickness ranges for proper results. Very small images are more easily defined in thin films; however, thinner films tend to have pinholes that lead to unwanted holes in the oxide. The film thickness also determines the time required for exposure and excessive variation can cause improper exposure on a part of the wafer. To minimize variation both the viscosity and the application procedure are carefully controlled, and the emulsion is filtered just before use to remove solid particles.

The requirements for the masks used in this process are stringent. Many circuits use lines as small as .0002 in. (or 5 microns) which must be defined on the mask and then reproduced on the wafer. Also, each mask must register with up to ten other masks during the wafer manufacture as noted by the sequence in Fig. 1-7. To meet these requirements tolerances as small as 1 micron are necessary, both for the pattern geometries and the pattern-to-pattern spacing.

Equally stringent are the requirements placed on the exposure system. This tool must provide the capability of aligning a mask to a silicon wafer quickly and accurately without damaging the emulsion mechanically and uniformly expose the emulsion to ultraviolet light. When satisfactory alignment is achieved, the wafer is forced against the mask to guarantee intimate contact between the wafer and the mask. This prevents image distortion and parallax. The exposure light must provide uniform intensity over its field of use and be reasonably stable from day to day so that the exposure time is constant for each part of each wafer.

**Epitaxial Growth.** An epitaxial layer is a silicon layer that is grown onto a wafer and is a continuation of the existing crystal lattice. However, the impurity concentration or type is different from that of the substrate. Initially, expitaxial wafers were used for transistors with an $n$ layer of about 1 ohm-cm resistivity on a very low resistivity $n$-type substrate. This material combined the advantage of a high junction breakdown due to the lightly doped epitaxial layer with a low device voltage drop due to the low resistivity of the substrate which composes the major portion of the wafer. In SIC manufacture an $n$-type layer is deposited on a $p$-type substrate to provide part of the circuit isolation.

The epitaxial process starts with wafers being cleaned and then loaded into the epitaxial reactor. Initially, a hydrogen ambient is introduced, and the wafers are heated to 1250°C. Hydrogen chloride gas is then introduced to etch about 0.2 micron off the silicon surface so that the epitaxial layer will have the cleanest possible surface on which to grow. After the etching cycle a mixture of silicon tetrachloride is introduced with a trace of impurity. The silicon tetrachloride decomposes in the hydrogen ambient to form silicon, and the impurity dopes the silicon to the proper level as it forms on the substrate. A layer thickness of 7–9 microns is usual, and the impurity level ranges from $5.5 \times 10^{16}$ to $9 \times 10^{17}$ atoms/cc, depending on the device requirements. Silicon has about $10^{23}$ atoms/cc, so the impurity level is 1 impurity atom for each $10^6$ silicon atoms.

The epitaxial process must be carefully controlled to minimize defects in the new silicon layer because the electrical and physical properties of the silicon are different in the area of a defect, and circuit operation may be impaired. Lattice defects in the epitaxial layer can occur if the substrate has defects in its lattice, and these defects can be caused either by imperfections in the initial ingot or by mechanical damage during lapping and polishing. The hydrogen chloride etch in the reactor is used specifically to remove the mechanically damaged layer. Projections that extend well above the epitaxial surface can also occur if an appropriate nucleation site, such as a foreign particle, is present. Particles may fall on the wafer surface, either prior to placement in the reactor or during deposition if the reactor is not kept clean. Projections tend to scratch the photolithographic masks and to cause local image distortion by preventing intimate contact between the mask and the wafer. All defects in the epitaxial layer tend to reduce wafer yield and must be minimized.

**Solid State Diffusion.** As indicated previously, the circuit components of an SIC are fabricated by generating $n$ or $p$ type silicon in different regions of the chip. Planar solid state diffusion is used to accomplish this, i.e., impurities are selected that dissolve and diffuse into the exposed silicon at elevated temperatures but will not penetrate an oxide film. The usual $p$-type impurity is boron, while phosphorus, antimony, and arsenic are used for $n$-type doping.

In circuit fabrication after the window is etched in the oxide the silicon wafer is placed in a furnace at an elevated temperature (usually from 750°C to 1250°C), and the desired impurity is introduced into the furnace. The impurity forms a doped oxide from which it dissolves in the surface of the silicon at a concentration that depends on the solid solubility of the particular species and the temperature. The wafer may be subsequently heat treated to redistribute the impurity by allowing diffusion deeper into the silicon.

In device manufacture successive diffusions are generally of the opposite type ($n$ into $p$ or $p$ into $n$) so that the depth of diffusion is considered to be the depth at which the $p$-$n$ junction is formed. There is, however, an impurity gradient on both sides of the junction.

The individual circuit components are formed from properly doped silicon regions in which the type and the concentration of impurity are carefully controlled, both at the surface of the silicon and in the epitaxial layer. The active components of many SIC designs are formed in an epitaxial layer that is 10 microns or .0004 in. thick.

### component fabrication

The fabrication of components with their isolation and junction seal requires 7 photomasking steps, 16 furnace operations, an epitaxial deposition, numerous cleanings, and several process evaluations. The flow chart (Fig. 1-7) clearly shows the cyclical nature of the process. In turn, Fig. 1-8 and Table 1-1 show the results of the diffusion processes.

The emitter diffusion forms the transistor emitters and also is used to underpass connectors. The base diffusion forms the transistor base and the resistors. The transistor saturation voltage is greatly reduced by the use of the buried layer and deep collector, both of which are high concentration $n$-diffusions in the $n$-collector region. Together with the substrate, the isolation diffusion forms a $p$-shell about the $n$-regions in which the active components are formed and separates them electrically.

**Junction Seal.** To reduce the device sensitivity to its operating ambient Bell System SIC's are provided with a junction seal. The junction seal, which is a combination of a layer of silicon nitride and a noble metal contact system, prevents the diffusion of impurities from the ambient through the silicon oxide to the silicon surface where they can electrically degrade chip performance. Silicon nitride is impervious to sodium ions and other known contaminating agents. To be effective the layer must be continuous and free of pinholes and cracks. Stringent tests are placed on the systems for depositing the silicon nitride to ensure the quality of the junction seal.

**Component Interconnections.** Within a chip the individual circuit components are connected by thin metal layers. For Bell System SIC's, gold is the prime current conductor but is underlaid with platinum and titanium to provide adherence to the silicon nitride layer. Aluminum contacts are often used by manufacturers with and without vacuum sealing, depending upon reliability objectives. The gold platinum contact together with silicon nitride is believed to be the most reliable surface seal now known.

### testing and inspection

After the metallization is completed, the circuits are tested to guarantee that they meet the electrical requirements. In addition to the basic electrical tests, mechanical and reliability tests may be applied to the chip or package. These tests are designed to guarantee that the chip can be handled in packaging and that the finished unit will meet the required life characteristics. In general, this type of test is done on a sample basis since many of these tests are destructive. For beam lead devices, typical tests are silicon size, bondability, bendability, and anchor strength. Temperature and power aging are the usual stresses applied for reliability testing.

## thin film technology

Thin film devices in their simplest form provide the interconnecting conductor paths between silicon integrated circuits and become thereby an essential complement of the silicon circuit technology. Another type is the tantalum thin film circuit. Passive circuits of high precision resistors and capacitors over a wide range of values with high stability, prescribed temperature coefficients, and the possibility of in situ operational adjustments are the result of this development. Most versatile but also most complex is, finally, the combination of tantalum and silicon integrated circuit technologies which makes high precision active networks and systems possible in the form of hybrid integrated circuits (HIC).

The basic building blocks of thin film circuits are interconnections, crossovers, resistors, and capacitors. Inductors are rarely used except in microwave circuits.

### interconnections

Basic to all thin film circuits are the interconnection patterns (Fig. 1-9a). In thin film technology the conductor material is applied to a substrate made of suitable insulating material, such as glass, ceramic, sapphire, etc. The most widely used conductor material is gold because it has a high conductivity and a high resistance to corrosion. Since gold does not adhere well to glass, ceramic, or tantalum, an intermediate bonding layer must be placed between the substrate and the gold conductor. Titanium, nichrome, chromium, etc., are used as such adhesive layers. Conductor pattern manufacturing yields

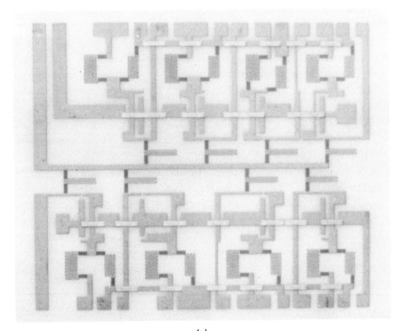

(a)

(b)

Figure 1-9 (a) Conductor pattern with crossovers; (b) close-up of beam crossovers; (c) close-up of stitch bonding; (d) resistor and capacitor pattern; (e) cross section of tantalum integrated circuit.

(c)

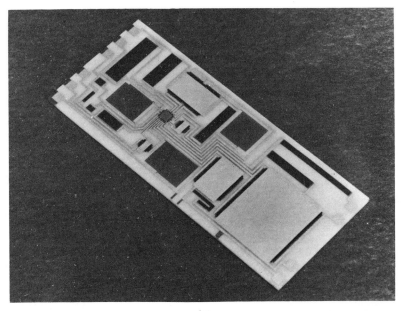

(d)

*Figure 1.9 continued*

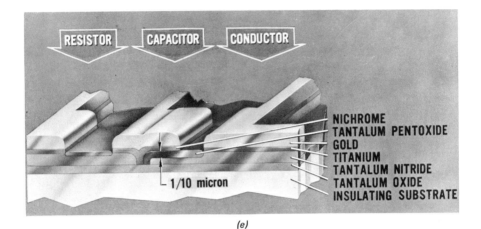

(e)

*Figure 1.9 continued*

depend on line width and length. At line widths below 5 mils the elimination of conductor and substrate irregularities becomes increasingly important. The usual causes of faults in the conductors are defects in the ceramic substrate and faulty pattern generation due to physical and chemical contamination.

### crossovers

The structure which permits one thin film conductor to cross another is called a crossover (Fig. 1-9a). Crossovers are an integral part of the interconnection patterns of thin film circuits and are required in all more complex designs. In the Bell System, beam type (Fig. 1-9b) and stitch bonded (Fig. 1-9c) crossovers are in manufacture. Plated gold is used for the beam type and gold wire is used for the stitch bonded crossovers.

The preparation sequence for the beam type crossover calls for a variety of techniques, ranging from photolithography to high quality plating of copper and gold. Yield requirements are in the order of 99.95 to 99.99% for individual crossovers so as to obtain circuit yields of about 70 to 80% for circuits containing up to 4,000 crossovers. The control of contamination is important to make such high yields possible. Typically, the plating baths for copper and gold must be constantly filtered, and the surfaces to be plated must be very carefully cleaned so that no particulate matter can attach itself to the surface and cause plating defects. The stitch bonded crossover is normally used when there are only a few crosspoints required, but it requires a good bonding technique and clean bonding surfaces to ensure good adhesion.

### resistors

Numerous materials have been used for thin film resistors, but tantalum and tantalum nitride deposited on suitable substrates by sputtering techniques have proved to be the most suitable for Bell System use (Figs. 1-9d and e). Tantalum nitride is more stable than tantalum upon exposure to air and is, therefore, the preferred resistor material. Important properties of the resistor films, such as the temperature coefficient of resistance and the change of resistance itself, can be modified by doping the tantalum or tantalum nitride with oxygen or aluminum. Adjustment of the resistance values to a high degree of precision is possible by anodization trimming. Resistance values are commonly within 1% of the specified rating, and for certain applications the tolerances are held within .05% at end of life. Like conductors, the yield of resistor meander patterns depends on the width and length.

## capacitors

A thin film capacitor consists essentially of a sputtered tantalum film which is partially oxidized to form a dielectric layer. (Fig. 1-9e). In the simplest version the second (or counterelectrode) is then applied by evaporation of Cr-Au, Ti-Au, or NiCr-Au.

The yield of this structure depends on the integrity and stability of the oxide layer. The yield will also be dependent on the anodizability of the tantalum layer which, among other factors, is critically dependent on the purity of the sputtering gases. It will also depend on the presence of particulate contamination on the tantalum surface which prevents oxidation and thus allows conducting paths between the counter electrode and tantalum through the oxide layer.

To reduce the sensitivity of the capacitors to all these possible causes for failure a semiconductive layer is frequently applied. The semiconductive layer increases the yield via its self-healing properties. For example, if $MnO_2$ is used an electrical breakdown generates enough heat to cause a modification of the $MnO_2$ in that region to an insulating layer. The increase in yield obtained with this structure is accompanied by a slight increase in the dissipation factor.

## general processing—thin film

The preparation of thin film devices and circuits requires processing steps involving a wide variety of techniques, such as cleaning, evaporation, sputtering, etching, anodization, bonding, and encapsulation. The operations arranged in the form of a processing sequence are shown in Table 1-2. An important consideration for the establishment of processing sequences is the compatibility of succeeding processing steps. Use of proper procedures during each step is essential for the quality and yield of the final product. Reduction or elimination of contamination is part of this proper processing whenever contamination can have a significant influence on the device's properties.

**Substrate Preparation.** The substrate material most frequently used for thin film devices is 99.5% alumina ceramic. Beryllia substrates have high thermal conductivity but present health hazards because of the toxicity of the beryllia. Glazing of ceramic is gaining increasing acceptance; the thin (0.2 mil thick)

TABLE 1-2.  Thin Film Processing Sequence Circuit Components

| Progress Steps | Conductors | Conductors and Crossovers | Resistors | Capacitors |
|---|---|---|---|---|
| Clean substrate | X | X | X | X |
| Sputter Ta | | | X | X |
| Evaporate TiPdAu | X | X | X | X |
| Photolithography, conductor pattern | X | X | X | X |
| Photolithography, R's or C's | | | X | X |
| Anodize | | | X | X |
| Evaporate Ti-Cu | | X | | |
| Plate Cu | | X | | |
| Electroless Plate Ni | | X | | |
| Photolithography, pillars | | X | | |
| Etching, pillars | | X | | |
| Photolithography, beams | | X | | |
| Plating, beams, Au | | X | | |
| Removal, spacing layer etching | | X | | |

glaze provides a smooth surface which is essential for good capacitor yields and does not substantially reduce the thermal dissipation properties of the ceramic.

Before any substrate can be used it must be adequately cleaned to remove all foreign materials which could interfere with the direct contact of the metallization to the substrate surface. Two types of contamination are to be considered: particulate matter which adheres to the surface and causes pinholes in the metallization, and large area organic contamination which impairs the overall adhesion and purity of the metallization.

**Metal Deposition.**  The major metal deposition techniques are plating, evaporation, and sputtering.  Each has specific applications where its particular features are best put to use.  Plating is used whenever thicker metal layers over large areas are required.  Defect-free plating of surfaces is difficult, and the structure of plated layers is also very sensitive to bath composition and contamination.  Evaporation is the method used to deposit thin layers of metals and certain insulators. It is more difficult to deposit refractory metals by this method even though E-gun evaporation has advanced the state of the art.  Some metals are difficult to evaporate because of splattering, e.g., the ejection of small particles of evaporant onto the surface.  Sputtering is used for the deposition of refractory metals and of insulators.  In general, the adhesion of sputtered layers is excellent, a disadvantage being the slower deposition rate.

**Sputtering.**  If a surface is bombarded with high energy particles, such as positive ions, atoms are ejected which can deposit on nearby solids.  This process is called sputtering.  The ions are formed when a high electric field is applied to a gas under low pressure, creating a glow discharge in the deposition chamber.

A major application of sputtering in thin film technology is the deposition of thin tantalum layers for resistors and capacitors. Argon is here chosen as the sputtering gas because it is inert to the material to be deposited, it has a high sputtering yield, and it is available in high purity at reasonable cost. Doping with a small amount of nitrogen increases the sheet resistance and stability of the sputtered films and is, therefore, used for the preparation of resistor films.

Because of the high reactivity of the glow discharge any impurities will be readily incorporated in the sputtered films and change their properties. It is therefore necessary to clean the walls and fixtures of the sputtering chamber by a bake under high vacuum ($10^{-7}$ Torr) and by a presputter period during which the substrate is shielded by a shutter. The backstreaming of oil vapors from diffusion pumps must be prevented by a liquid nitrogen trap. Turbomolecular pumps eliminate the need for such trapping. The vacuum system must be leak-free so as to prevent any undesired, involuntary doping. The walls of the sputtering chamber and the fixtures must be periodically cleaned to avoid flaking of sputtered material which is only loosely adhering. The substrate must be clean to ensure good film adherence and to avoid pinholes.

**Evaporation.** Evaporation takes place when the thermal energy of a solid or liquid is high enough to break bonds between its atoms or molecules so that some of them can leave the surface. The complete deposition process, consisting of evaporation, transit, and condensation of the material, must take place in a well evacuated chamber so as to eliminate any reaction of the evaporant with residual gases. The walls of the chamber and the fixturing must be clean; contamination from back-streaming pump oil must be prevented by appropriate trapping; and the purity of the evaporant must be high enough to yield films having the required properties.

Adhesion of the films is closely related to substrate cleanliness. Outgassing of the substrate before deposition may be necessary in some cases to ensure good adhesion between substrate and film.

**Plating.** Metal deposition by electroplating (the use of external source of electricity) or electroless plating (chemical reactions provide the electrical potential) is frequently used to build the conducting patterns of thin film circuits. The properties of the plated films depend on the composition of the plating solution, the plating conditions, the purity of the solution, and the geometry of the plating bath. To obtain reproducible films of desired properties and quality all these parameters must be carefully controlled.

The structure, surface, and hardness of a plated layer can be strongly influenced by the addition of small amounts of chemicals. In addition, the concentration of the metal ions must remain within the prescribed limits.

Larger volumes of the plating solution are therefore advantageous to prevent premature depletion of the bath. The current density during the plating process is critical and must also remain within the experimentally determined limits. The structure of the plated layer may otherwise change, elements of the plating bath may be incorporated in the film, or nonuniform surfaces (nodules) may appear.

**Photolithography.** Photolithography is used in the preparation of thin film circuits for the delineation of conductor patterns on substrates of up to 3¾ × 4½ in. in size and repeated up to ten times during the preparation of the most complex circuits. A single pattern generation step consists of the application of photoresist by spinning or spraying on the substrate surface, baking of the resist, exposing through a mask, developing of the photoresist, drying, etching of the exposed metal areas, and finally removing (stripping) the remaining photoresist.

For the present line widths of thin film circuits of 2–5 mils, the less distinct line definition caused by thick photoresist (3–4 microns) is comparatively negligible. In such thick photoresist layers pinholes are not a major problem. It is important, however, that the photoresist be applied in layers thick enough to cover all peaks on rough (compared to a polished silicon wafer) ceramic substrates.

The photolithographic processing is most demanding with respect to the control of the processing parameters and to the cleanliness of the operation. It is best performed in a clean room or in clean hoods with filters ensuring the elimination of most particulate contaminants. The substrates must be carefully cleaned before application of the photoresist to ensure freedom from defects and good adhesion. The photoresist must be filtered and, for critical applications, centrifuged. Masks must be kept in clean containers and must be recleaned before each exposure.

**Anodization.** Anodization is a process used in tantalum film technology to adjust the value of resistors and to form the insulating layer of tantalum capacitors. The anodization process is quite similar to electroplating but takes place at the anode. Oxygen ions from the solution react with the tantalum to form a film of anodic oxide which has high resistance. With increasing oxide thickness the current decreases during the anodization process; the limit for the oxidation process is reached when breakdown occurs in the oxide layer.

The anodization process permits in situ, high precision adjustment of resistor values (trimming) when necessary. The formation of capacitor insulating films by anodization is quite sensitive to defects of the substrate or the tantalum film which may impair the formation of the oxide. Capacitors are, therefore, normally made only on glass or glazed ceramic substrates.

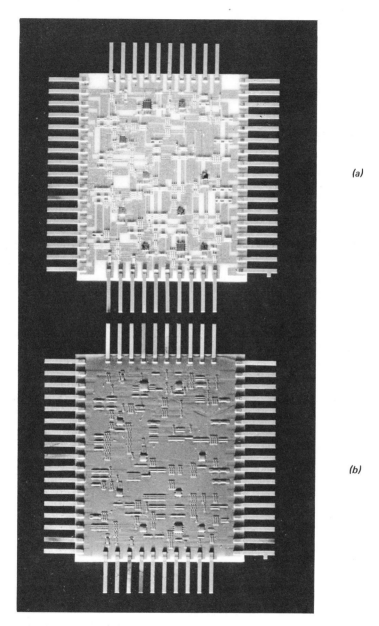

(a)

(b)

*Figure 1-10  Hybrid integrated circuit (a) without encapsulation; (b) with encapsulation.*

## hybrid integrated circuit assembly

Thin film circuit designs range from simple interconnection patterns to combinations of resistors, capacitors, and crossovers on single substrates which then are used as wiring boards for SIC's. Figure 1-10 shows such a circuit.

Processing sequences for complex circuits require up to 50 major processing steps which must be mutually compatible. Also, the compatibility of the thin film circuit with the assembly operations must be considered since after preparation of the thin film parts of a circuit, the finishing operations—bonding of SIC's and lead frames, encapsulation, and the inspection and testing—must not alter the circuit characteristics.

**Bonding.** Attachment of SIC's to thin film circuit is a critical but well established process. The gold beam leads of the SIC's are thermocompression bonded to the corresponding conductor termination areas. The success of the operation depends on the appropriate adjustment of the bonding pressure, of thermode and substrate temperatures, and of the cleanliness of leads and bonding pads. Attachment of leads for outside connections is also done by thermocompressive bonding. The leads are furnished attached to frames which hold them in position. The frames are cut off after the bonding.

The quality of the bond depends not only on the correct adjustment of the bonding parameters but also on the quality of the film conductor. The type of bonding failures can be used to identify weak parts of the system if destructive pull testing is performed. Examples are given in Table 1-3.

TABLE 1-3.   Bond Evaluations by Pull Testing Analysis

| *Failure Mode* | *Analysis* |
|---|---|
| Ceramic pull-out. | Good adhesion between the lead, bottom conductor, and ceramic; indication of ceramic weakness. |
| Detachment of bottom conductor from ceramic | An indication of poor evaporation or sputtering conditions, or of poor cleaning of the substrate. |
| Failure of bond between lead and bottom conductor. | An indication of poor adjustment of bonding parameters or poorly cleaned leads and/or bottom conductors. |
| Failure of gold plating on leads | Poor gold plating will cause this failure mode. |
| Breaking of lead | The desired failure mode, indicating that all other parts are well under control. |

**Encapsulation.** The protection of the circuits against environmental influences and mechanical damage can be provided by hermetically sealed packages or plastic encapsulation. Because plastics are permeable to vapors some applications require hermetically sealed enclosures. However, plastic encapsulation is cheaper and has gained wide acceptance during recent years.

The encapsulant must meet a series of requirements, some of them contradictory. It must not be rigid after curing so that the stresses between the HIC and the plastic are minimized. It must, on the other hand, protect the circuit during handling against mechanical damage. The encapsulant must not contain any chemicals which would damage the circuit. It must penetrate all spaces below SIC's and crossovers so as not to leave any voids in which water vapors can condense. Finally, it must adhere well to the circuit and its adhesion must not deteriorate as a function of time. The last condition requires again a careful cleaning of the circuit, such as vapor degreasing, rinsing in acetone, and immersion in boiling $H_2O_2$. Filling of voids is best obtained by vacuum impregnation whereby the circuit is covered by a solution of the encapsulant in low viscosity form and then placed in a vacuum.

## inspection and testing

Packaged HIC's are then tested for electrical and mechanical properties. Electrical tests are made to guarantee the circuits will operate with the required current, voltage, and leakage levels. Packaged units may again be subjected to either temperature or power stress to ensure reliability. Mechanical tests such as lead pull strength and solderability measure the ability of the device to be assembled with other units into the electronic system. These tests are necessary to ensure the proper function of the final system, both initially and during its useful life.

## future trends

As the microelectronics industry matures, an increasing emphasis will be placed on device cost. This in time will require improved processing for successful competition in the industry. For example, the trend in SIC's is towards smaller individual components and larger chips as a means of reducing over all circuit cost and improving circuit speed. Smaller components tend to reduce circuit costs because the wafer process cost is essentially independent of component size. However, as the component size decreases, the tolerances tighten to almost infinitesimal values. With the present designs geometrical tolerances of 1 micron are usual although lower values are necessary in especially critical cases. New designs to meet circuit and cost objectives will require an even better effort.

Also, as components become smaller or more densely packed, there is a greater probability that the chip will include a defective area. Consider a typical 1.5 in. diameter silicon wafer with 100 defects that will cause device failure if they fall in an active area. If we assume a .06 in. square chip with 50% of the silicon active, there are about 400 potential chips and 50 failures anticipated from the 100 defects or 12.5% shrinkage. If the chip size is increased to .100 in. square, there are only 145 potential chips; and the 50 expected failures now represent a 34% shrinkage. Finally, if the active area is increased to 80% by using a newer technology, 80 defective circuits are now expected; and the shrinkage rises to 55%.

To meet the needs of the future the environment and processes must allow the production of better material, cleaner surfaces, and mechanically more perfect wafers. Towards this end a vast engineering effort is underway to perfect the present processes and to devise new methods that eliminate the shortcomings of the current methods. Among these are ion implantation of impurities, projection printing of mask patterns, and the generation of patterns without masks by electron beams. These processes and many others are expected to contribute to better and cheaper integrated circuits.

## summary

This chapter has shown that the need for control of contaminants is dependent on the sensitivity of the affected product. As product tolerances become more precise, the significance of contaminating influences from the manufacturing environment increases. This is especially true in the production of microelectronic devices because of their small size, exacting tolerances, purity of raw materials, and dependence on sophisticated processes. The descriptions of these characteristics in this chapter illustrate the importance of contamination control to one precision industry, and they provide a point of reference to relate the given definitions of contaminants and their means of transport to other precision industries.

## bibliography

1. A. S. Grove, *Physics and Technology of Semiconductor Devices*, John Wiley and Sons Inc., 1967.
2. Motorola Staff, R. M. Warner, Ed., *Integrated Circuits, Design Principles and Fabrication*, McGraw-Hill, 1965.
3. J. T. Wallmark and H. Johnson, *Field Effect Transistors*, Prentice-Hall Inc., 1966.
4. M. P. Lepselter, "Beam Lead Technology," *The Bell System Technical Journal*, Vol. XLV, No. 2, February 1966.
5. S. S. Hause and R. Whitner, "Manufacturing Beam-Leaded Sealed-Junction Monolithic Integrated Circuits," *The Western Electric Engineer*, Vol. XI, No. 4, Dec. 1967.
6. R. W. Berry, P. M. Hall, and M. T. Harris, *Thin Film Technology*, Van Nostrand, Princeton, 1968.
7. K. L. Chopra, *Thin Film Phenomena*, McGraw-Hill, N.Y., 1968.
8. L. I. Maissel and R. Glang, Eds., *Handbook of Thin Film Technology*, McGraw-Hill, N.Y., 1970.
9. G. Hass and R. E. Thun, Eds., *Physics of Thin Films*, Vol. 1–4, Academic Press, N.Y., 1967.

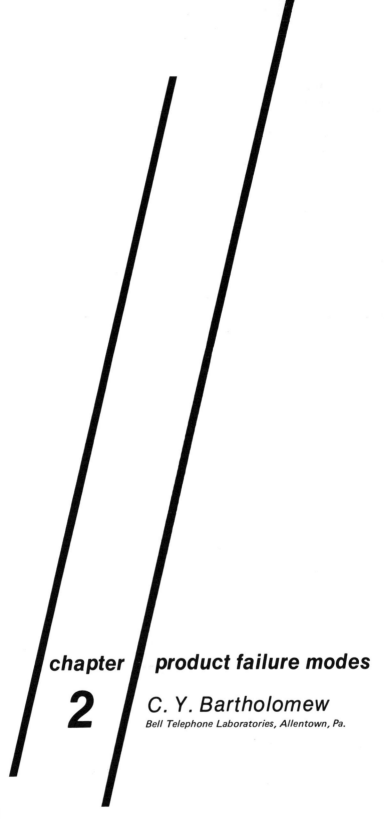

chapter

**2**

**product failure modes**

*C. Y. Bartholomew*
*Bell Telephone Laboratories, Allentown, Pa.*

The term contamination is a relative term, i.e., it must have a sufficient adverse effect on products to warrant the expense of elimination or neutralization. This suggests that a justification based on the mode of product failure be a part of a proposal to control contaminants. If contaminants affect the basic feasibility of a process or product this would be an obvious justification. However, the more common situation is the desire for significant improvement in product yield or reliability as a result of controlling specific contaminants. Proving the latter case is often very difficult, for the reason that yield and reliability are functions of many other variables besides contaminants. However, difficulty does not mean impossibility. With a knowledge of a device's operating characteristics and careful experimentation, it is often possible to establish the influence of various contaminants on product performance. The documentation of an electronic product's failure to operate properly under controlled exposure to specific contaminants is the best possible justification for the expense of environmental control. That such documentation is somewhat limited is testimony to the difficulty of failure mode analysis.

Chapter 1 described the basic manufacturing processes of semiconductor and thin film technologies which are inherently capable of producing reliable products at high yields and in large numbers if contamination problems are kept under control. The reader was told how microelectronic circuits are made and the effects of contaminants on the manufacturing processes, but no information was given on what the final product does and the harmful effects of contaminants on operating performance. The functions of resistors, capacitors, and interconnection circuits are well known, and the shorting or open-circuit effects of contaminants are self-explanatory. However, the electrical functions of semiconductor devices are less obvious, making product operating failures caused by contaminants more difficult to understand.

The purpose of this chapter is to clarify the relationships between semiconductor structure, its electrical operating characteristics, and the major failure modes caused by contaminants. The failure modes of semiconductor devices to be discussed are: (1) the charge in the passivation layer ($SiO_2$); (2) the separation of ionic impurities on the passivation layer surface in wet ambients; (3) the migration of ionic impurities in the passivation layer; (4) metal precipitates in the semiconductor junction; and (5) the effects of small particles. Where it is appropriate, the failure modes of thin film devices will be mentioned. Please realize that many other product failure modes exist; unfortunately, the documentation is fragmentary. For this reason additional references are included at the end of the chapter for those interested in learning more about this subject.

## *semiconductor structure*

The operation of semiconductor devices depends on the use of very pure materials, which contain precise amounts of intentionally added impurities.

Unwanted impurities in the amount of even a few parts per million can completely disrupt the operation of a device. Let us look, then, to what constitutes good operating characteristics of semiconductor devices and then examine the sensitivities to low impurity concentrations.

### operation characteristics

Figure 2-1 includes illustrations of the basic structure of the three most prominent semiconductor devices: diodes, bipolar transistors, and surface field-effect transistors. The diode is essentially a junction of $p$-type silicon and $n$-type silicon. Such a structure, when biased with the $p$-type connected

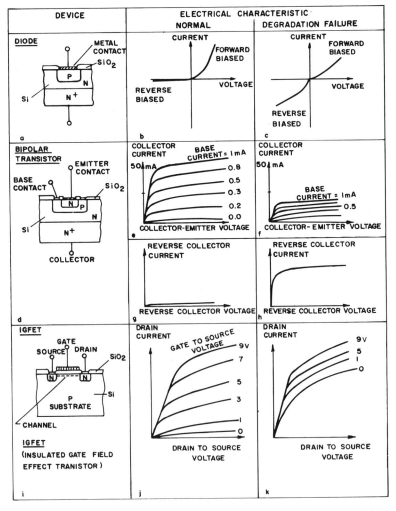

*Figure 2-1  Basic microelectronic devices.*

to positive voltage and $n$-type connected to negative voltage, conducts current easily, but, when the connections are reversed, the structure conducts very poorly. The device is thus a rectifier. A typical current-voltage characteristic curve is given in Fig. 2-1b. A failed device may have a characteristic, as shown in Fig. 2-1c, in which the current under reverse connection is increased, and the forward current is reduced.

Figure 2-1d shows an $npn$ bipolar transistor structure. In normal operation, the collector-base junction is reversed biased, while the emitter-base junction is forward-biased. The emitter, which is $n$-type silicon, injects electrons into the base, and these pass through the base, with some loss, to the collector. The electrons which are lost in the base, and thus do not reach the collector, constitute the base current. The base current is typically 1/10 to 1/100 of the collector current, so that a change in base current can be thought of as producing a 10 to 100 fold greater change in collector current. This ability to use a small current to control a large current is called the gain of the transistor. Some typical collector characteristics of a transistor are shown in Figs. 2-1e and 2-1f, where it is seen that, when gain degrades, the curves of constant base current drop to lower collector current values, and also become more bunched together. Transistor leakage degradation is illustrated in Figs. 2-1g and 2-1h, where it is seen that the reverse collector current increases markedly. This introduces an extra unwanted current which can alter the gain and can cause unbalance in the rest of the circuit.

An insulated gate field-effect transistor (IGFET) structure, which is a surface-effect device, is shown in Fig. 2-1i. In this type of transistor two diffused regions, the source and the drain, are separated by a region called the channel. The conductivity of the channel can be controlled by the voltage applied to the gate, which is an electrode separated from the channel by a thin layered insulator. Typical electrical characteristics for normal and degraded units are shown in Figs. 2-1j and 2-1k. On the degraded device, the curve for zero gate voltage is higher and the gate control of drain current is seen to be reduced. Junction leakage degradation can also occur with such devices.

**Resistivity.**  To get an idea of how important small concentrations of impurities can be, consider Fig. 2-2, which shows how the resistivity of silicon depends on doping impurity concentration.[7] The two curves shown are for $n$-type and $p$-type silicon. A typical resistivity for a silicon transistor starting wafer is from 0.5 to 10 ohm-cm; this is seen to correspond to from 10 ppb to 1 ppm, and this is the desired impurity level in the collector. In the base of the transistor, the impurity concentration may be about 100 ppm and in the emitter the impurity concentration may be up to $10^4$ ppm, or about 1%. Therefore, a change of impurity concentration of a few parts per million in certain regions of the device can produce a drastic change in resistivity in that region. This will, in general, produce a significant change in device characteristics, such as junction breakdown voltage and gain.

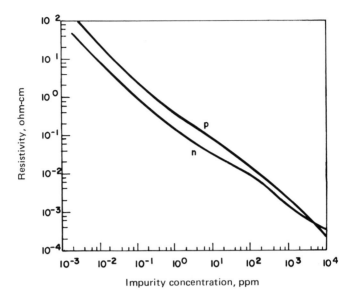

*Figure 2-2 Resistivity vs. impurity concentration for silicon. (Reprinted with permission from S. M. Sze and J. C. Irvin, Solid State Electronics, Vol. 11, pp. 599–602, 1968, Pergamon Press.)*

**Carrier Lifetime.** For another example of the effects of small amount of impurities on the characteristics of silicon, consider the effect of gold on minority carrier lifetime. The lifetime of holes in $n$-type silicon, for example, is the average time that an excess free hole will remain free before it recombines with an electron. Figure 2-3 shows the lifetime of holes in $n$-type silicon as a function of gold concentration in parts per million.[8] As a gold concentration increases from 1 ppb to 1 ppm, the minority carrier lifetime decreases from $5 \times 10^{-7}$ sec to $5 \times 10^{-10}$ sec or from 0.5 microsecond to 0.5 nanosecond. Thus, small changes in gold concentration have a significant effect on minority carrier lifetime. An increase in lifetime produces a corresponding increase in device switching time, which would be detrimental for such applications as computers where circuitry speed is important.

**Surface Charge.** Another example of the effect of a small amount of impurity on silicon devices is the effect of surface charge in or on the oxide layer of a planar silicon device. The surface charge, which is an excess of electric charge, either positive or negative, localized in the surface layer, induces a layer of charge of opposite type in the underlying silicon. A sufficiently positive oxide surface charge can induce enough negative charge in $p$-type silicon to invert it to $n$-type in a thin layer adjacent to the oxide-silicon interface. Figure 2-4 shows the relationship[9,10] between silicon resistivity and the surface charge required to invert it. It is seen that a net surface

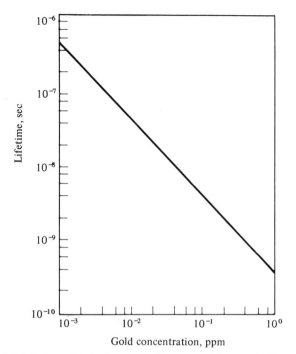

Figure 2-3 Lifetime of holes in n-type silicon vs. gold concentration. (After A. S. Grove, *Physics and Technology of Semiconductor Devices*, John Wiley & Sons, 1967.)

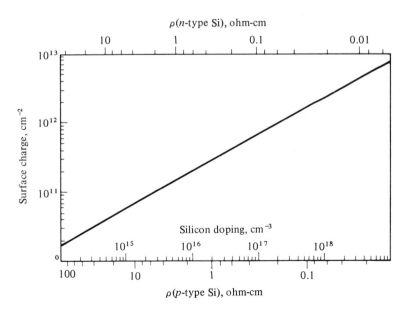

Figure 2-4 Surface charge required for inversion of silicon. (After Fitzgerald and Grove, "Mechanisms of channel current formation in p-n junctions," in: Goldberg and Vaccarro (eds.), Physics of Failure in Electronics, Vol. 4, Rome Air Development Center, 1966.)

charge of $10^{12}$ to $10^{13}$ electronic charges/cm$^2$ is sufficient to invert even highly doped silicon. Since a monolayer is about $10^{15}$ atoms/cm$^2$, $10^{-3}$ to $10^{-2}$ monolayer of impurities, if charged, is sufficient to produce gross degradation of silicon planar devices.

**Impurity Migration.** Studies on MOS diodes[*] have shown that there can be mobile contaminants in the $SiO_2$ which can traverse 1000-angstrom thick oxides in a time of about 10 minutes,[11] with an electric potential difference of 10 volts across the oxide. Raising the device temperature shortens the time required for transit. The most common mobile impurity has been identified as sodium, although the other alkalis behave in a similar way. Migration over long distances would take correspondingly longer times.

Often such migration results in significant degradation of device characteristics, such as increase in junction leakage and reduction of gain, and, in such cases, it can be seen that the presence of the contamination leads to shorter service life. For example, the presence of $10^{12}$ atoms/cm$^2$ on the outer surface of the oxide might not cause an MOS device to fail initial tests, but as soon as an appreciable fraction of these atoms migrate to the silicon-silicon dioxide interface, the device would probably fail.

**Junction Effects.** Silicon junction devices suffer from a form of junction degradation when metal precipitates form in the junction region.[12-14] This produces leakage currents with a nonsaturating (soft) characteristic as contrasted with a saturating characteristic of a channel. The two types are shown in Fig. 2-5 along with the characteristic for a good device. Excess leakage currents can upset the electrical balance of a circuit and cause failure to operate. When the emitter of a transistor has a soft characteristic, the gain is also degraded, particularly at low values of collector current.[2] Degraded gain may result in loss of the signal in an amplifier or failure of a switching transistor to switch.

An example of degradation of transistors due to metal precipitates is copper,[15-18] which is a component of some device packages. Copper precipitates preferentially in highly doped regions of the device when the device is operated under power at elevated chip temperatures. Presence of hydrogen in the environment accentuates this process.[16] The precipitate size may be about 1 cubic micron, which is about $9 \times 10^{-12}$ g or about $7 \times 10^{10}$ atoms of copper.

**Effect of Particles.** Small particles present during fabrication of microelectronic devices can result in pinholes and other defects of a size comparable to the original particle. In Fig. 2-6a, two small particles, 0.3 micron in diameter, are shown embedded in the photoresist layer of a silicon planar device just prior to pattern formation. In (b) the particles have led to exposure of oxide

[*]A metal-oxide-semiconductor diode is a surface space-charge device similar in concept to an IGFET.

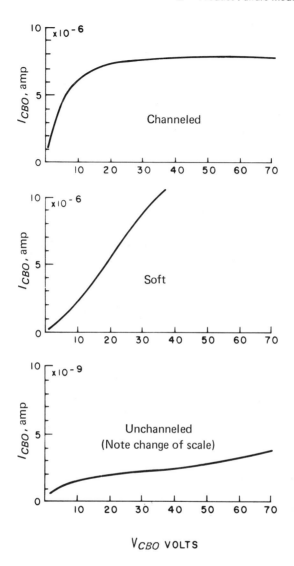

Figure 2-5 $I_{CBO}$ vs. collector-based voltage for channeled, soft, and unchanneled devices.

when the normal photoresist pattern is formed. Etching of the oxide in the normal pattern thus exposes the silicon $p$-$n$ junction as shown in (c). The exposed junction, then, is generally subject to excessive electrical leakage.

High contact resistance and low current carrying capacity for conductor patterns of both semiconductors and thin film devices can result because of reduced conducting area caused by the physical presence of particles during metallization. These same two defects can result also from chemical corrosion caused by particles or residues from some chemical process. Larger particles can actually short-out the electrical paths of microelectronic devices. Fig-

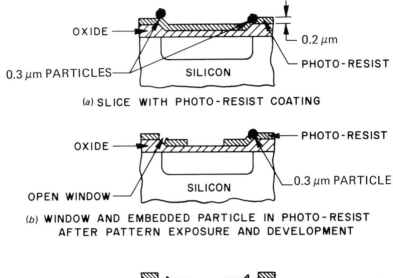

(a) SLICE WITH PHOTO-RESIST COATING

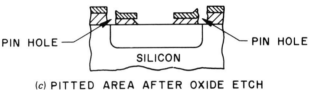

(b) WINDOW AND EMBEDDED PARTICLE IN PHOTO-RESIST
AFTER PATTERN EXPOSURE AND DEVELOPMENT

(c) PITTED AREA AFTER OXIDE ETCH

*Figure 2-6 Particle contamination effects in silicon circuit pattern generation process.*

ure 2-7 illustrates such a case in a microscopic view of a tantalum thin film resistor. This is a gross example since the resistor is built with 5 mil lines and 5 mil spacings between lines.

## specific failure analyses

The previous section has shown the range of sensitivity of device resistivity, lifetime, and other performance factors to minute impurities. Let us now turn to specific analyses of product failures for the documentation on product-contaminant relationship either from material experiments or from manufacturing experience. These examples are representative synopses of the type of theoretical and experimental analyses necessary to prove what actually happens to microelectronic devices when exposed to various contaminants.

### water

Water vapor can lead to degradation of semiconductor devices by the Atalla effect.[19-24] Here, ions on the surface of the protective oxide covering the *p-n* junction are separated by the junction fringing field under reverse bias, under high humidity conditions. Negative ions move to the *n*-side and posi-

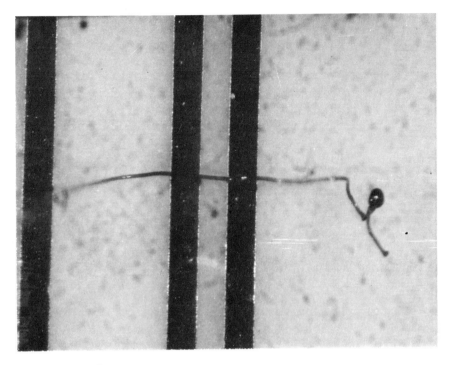

*Figure 2-7  Photograph of particle on thin film resistor.*

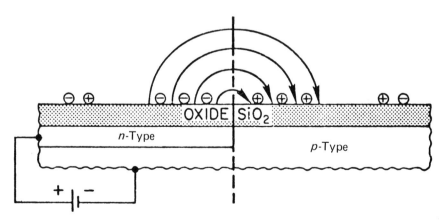

*Figure 2-8 Separation of charges on silicon dioxide surface in wet atmosphere by fringing field of reverse-bias p-n junction. (After Schroen, "Accumulation and decay of mobile surface charges on insulating layers and relationships to reliability of silicon devices," in: Goldberg and Vaccaro (eds.),* Physics of Failure in Electronics, *Vol. 4, Rome Air Development Center, 1966.)*

tive ions move to the $p$-side as shown in Fig. 2-8. If the surface concentration of the ions reaches the range of $10^{12}$ to $10^{13}$ ions/cm$^2$, the underlying silicon can be inverted and excess junction leakage currents can occur. This effect can readily occur when the device is encapsulated in a metal can so that any water generated in the package is trapped in the enclosure. Water can form in metal packages if the inert encapsulant gas contains traces of oxygen. The oxygen can react with hydrogen, which may be present in or on the metal, to produce water vapor. The dew point may be sufficiently raised such that the protective oxide surfaces adsorb a sufficient number of water molecules to permit residual contaminating surface ions to become mobile. In effect, the presence of the water molecules acts as an electrolyte to permit the mobility of the contaminating ions. Some early transistor designs contained porous glass or BaO pellets to adsorb water vapor and thereby reduce the moisture level in the transistor package.

### alkali metals in SiO$_2$[25-36]

Degradation in MOS devices and planar transistors can result from ionized sodium and other alkali metal ions in the oxide layer which covers the silicon surface. Sodium ions with a net positive charge can cause depletion or inversion of $p$-type silicon adjacent to the silicon-silicondioxide interface. Normal thermal oxides on silicon have sodium contamination levels of $10^{16}$ to $10^{18}$ atoms/cm$^3$ (0.5–50 ppm), with higher concentrations at the air-oxide and silicon-oxide interface. The best oxides produced to date have $10^{15}$-$10^{16}$ atoms/cm$^3$ of Na (50–500 ppb). A typical profile is shown in Fig. 2-9. Oxides with $10^{17}$-$10^{18}$ atoms/cm$^3$ (5–50 ppm) are not uncommon.

Not all the sodium in SiO$_2$ is electrically charged and has electrical effect on the silicon. In fact, most of the sodium introduced during oxide growth is inactive both as to net charge and mobility.[27,34] It has been reported that only the sodium in the thin layer of oxide next to the oxide-silicon interface has an effect on the silicon.[31]

**Sodium Contamination from the Atmosphere.** Samples stored under ordinary atmospheric conditions quickly become contaminated with sodium-bearing compounds as shown by the following experiment. Oxidized silicon wafers stored for 2 days in ordinary laboratory environment collected up to $10^{15}$ atoms/cm$^2$. In a filtered laminar flow clean bench the level was $6 \times 10^{12}$ atoms/cm$^2$. After 20 days storage sodium levels ranged from $8 \times 10^{12}$ to $5 \times 10^{15}$ atoms/cm$^2$. After 120 days levels ranged from $2 \times 10^{14}$ to $1.5 \times 10^{16}$ atoms/cm$^2$. Potassium was also detected. These levels could well lead to inversion layer problems if the wafers were not adequately cleaned before utilization in devices.

The source of the sodium from the environment was not determined. However, it was surmised to have come from environmental particles of sufficient size to settle onto the wafers in opposition to the horizontal laminar

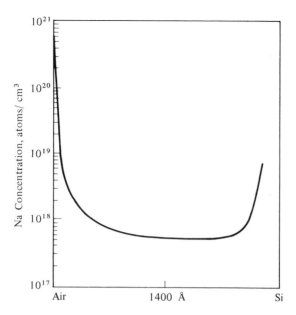

*Figure 2-9  Residual sodium in grown SiO$_2$ on silicon.  (After Fowkes and Burgess, Surface Science, Vol. 13, 1969, North-Holland Publishing Co., Amsterdam.)*

air flow.  Such particles typically contain sodium or have sodium adhering to their surfaces.

### factory environment correlation

Earlier, particles were shown to be a contributary cause to semiconductor failure.  An actual correlation of environmental conditions to failure rates of silicon transistors was achieved some years ago when the planar processing techniques were first being applied.  Initially, the manufacture of an *npn* planar transistor was started on a pilot basis in a normal factory ambient. Under these conditions the dust count levels were commonly $10^6$ particles/ cu. ft for particles $\geqslant 0.5$ micron.  Later, a manufacturing line was set up with a relatively dust-free atmosphere at critical work positions ($\approx 2,000-10,000$ particles/cu. ft).  The results of accelerated life tests performed on product from the two areas are shown in Fig. 2-10.  Accelerated life tests are run with about twice the rated power input, which results in a chip temperature in excess of the rated value.  By use of experimentally determined temperature dependence of failure rate factors, the failure rate under rated power can be predicted using the failure rate measured under accelerated testing conditions. The predicted failure rates at a junction temperature of 150°C (the maximum rated condition) would be 0.03% per thousand hours for the pilot line product, and 0.001% per thousand hours for the product made in the low-dust content ambient.  The initial test yield was also higher in the low-dust ambient product.

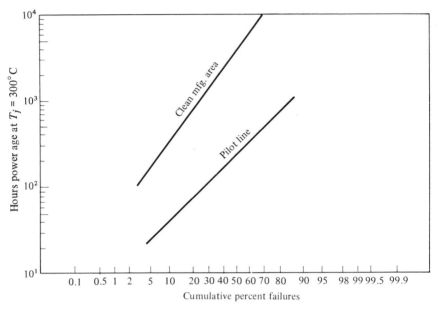

*Figure 2-10 Accelerated power age life test results for npn planar transistors made in two different locations.*

### product design correlation [9,37,38]

There are many cases where the effects of contaminants can be avoided by modification of the product design. A definite correlation between product design, electrical performance, and contaminant sensitivity is clearly demonstrated in the case of excess leakage currents due to surface inversions of silicon planar transistors.

Devices with geometry shown in Fig. 2-11 are subject to excess collector currents, due to inversion regions reaching from the base-collector junction to the edge of the chip. This effect, called a channel, has the collector-voltage characteristic shown in Fig. 2-1h. Current can flow from the base region along the inversion layer and enter the underlying silicon at the edge of the chip. The chip edge is a position of high defect density where breakdown from the inversion layer to the collector region occurs easily. Such inversion layers can be rendered harmless by the addition of a highly doped diffused guard ring or "channel stopper," as shown in Fig. 2-11. These guard rings work on both *npn* and *pnp* transistors.

The guard ring works as follows. The inversion layers are caused by net electric charge in the oxide layer used over the device surface for passivation, as part of the planar process. The oxide charge induces a charge of opposite sign in the silicon at the Si–SiO$_2$ interface. The degree of inversion produced

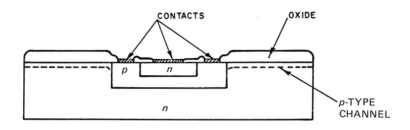

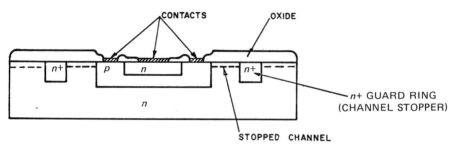

*Figure 2-11  Guard ring design for silicon planar transistors.  Top: Location of channel.  Bottom: Stopping of channel by guard ring.*

by a given amount of oxide charge decreases as the underlying silicon is made more highly doped.  Therefore, a highly doped guard ring will not be inverted, and the channel will be interrupted.

TABLE 2-1.   Relationship of Major Failure Modes of Microcircuits
to Contamination

| Microcircuit Subclassification | Failure Mode | Contaminant-related Cause Factor |
|---|---|---|
| Bipolar transistor | High junction leakage | 1. Oxide or interface charge.<br>2. Metal precipitates in junction.<br>3. Diffusion defect because of oxide pinholes. |
| | Low gain | 1. Oxide or interface charge.<br>2. Metal precipitates in junction.<br>3. Diffusion defect because of oxide pinholes.<br>4. High surface recombination velocity due to impurities at Si-SiO₂ interface. |

**TABLE 2.1 (continued)**

| Microcircuit Subclassification | Failure Mode | Contaminant-related Cause Factor |
|---|---|---|
| | High saturation voltage | 1. High resistance metal interconnect bond because of reduced area of contact. |
| Bipolar diode | High junction leakage | 1. Oxide or interface charge.<br>2. Metal precipitates in junction.<br>3. Diffusion defect because of oxide pinholes. |
| | High forward resistance | 1. High resistance metal interconnect bond because of reduced area of contact. |
| Schottky barrier diode | Improper I-V characteristics | 1. Foreign contamination. |
| IGFET (Insulated Gate Field-Effect Transistor) | Improper threshold voltage | 1. Oxide or interface charge. |
| | Shift in threshold voltage | 1. Shift in oxide or interface charge. |
| | Low gain | 1. Low channel mobility due to interface contamination.<br>2. Improper channel area due to particulate contamination during fabrication. |
| Interconnections | 1. Short Circuit | 1. Foreign conducting particle.<br>2. Oxide pinhole<br>3. Corrosion |
| | 2. Open circuit | 1. Poor bond due to foreign contamination.<br>2. Reduced cross-sectional interconnect area due to particulate contamination during fabrication.<br>3. Corrosion |
| Tantalum thin films | 1. Resistance increase | 1. Oxidation of tantalum |
| | 2. Resistance decrease | 1. Aklali metal contamination. |
| | 3. Short circuit in tantalum capacitor | 1. Contaminants in capacitor dielectric. |

## summary

Contaminants, both as surface deposits and as material impurities, can drastically reduce the yield and reliability of microelectronic circuits even when present only at relatively low levels. The effect of these contaminants is seen in degradation of the electrical characteristics, such as open circuit, short circuit, low gain, or other shifts of device parameters, and high leakage. Table 2-1 summarizes the relationships of major failure modes of semiconductors and thin film devices in a general manner. In order to realize the inherent uniformity, low cost, and reliability of microelectronics, understanding of contaminant effects and stringent contaminant control is necessary.

## references

1.  A. B. Phillips, *Transistor Engineering and Introduction to Integrated Semiconductor Circuits*, McGraw-Hill, New York, 1962.
2.  A. S. Grove, *Physics and Technology of Semiconductor Devices*, John Wiley and Sons, Inc., New York, 1967.
3.  S. M. Sze, *Physics of Semiconductor Devices*, Wiley-Interscience, New York, 1969.
4.  R. M. Warner, Jr. and J. N. Fordemwalt, *Integrated Circuits, Design Principles, and Fabrication*, McGraw-Hill, New York, 1965.
5.  Staff of Integrated Circuit Engineering Corporation, *Integrated Circuit Engineering, Basic Technology*, Fourth Edition, Boston Technical Publishers, Inc., Cambridge, Mass., 1966.
6.  R. M. Berry, P. M. Hall, and M. T. Harris, *Thin Film Technology*, Van Nostrand Company, Princeton, N.J., 1968.
7.  Adapted from S. M. Sze and J. C. Irvin, "Resistivity, Mobility, and Impurity Levels in GaAs, Ge, and Si at 300°K," *Solid State Electronics*, *11*, 599–602 (1968).
8.  Adapted from reference 2, p. 142, Fig. 5.16.
9.  D. J. Fitzgerald and A. S. Grove, "Mechanisms of Channel Current Formation in Silicon *P-N* Junctions," in *Physics of Failure in Electronics*, Vol. 4, 1966, Edited by M. E. Goldberg and J. Vaccaro, Published by Rome Air Development Center. Available from The Clearing House for Federal Scientific and Technical Information (CFSTI), Springfield, Va.
10. E. Kooi, *The Surface Properties of Oxidized Silicon*, Springer-Verlag, New York, 1967, p. 20.
11. Reference 2, pp. 337–340.
12. A. Goetzberger and W. Shockley, "Metal Precipitates in Silicon *P-N* Junctions," *J. Appl. Phys.*, *31*, 1821–1824 (1960).
13. S. M. Henning and L. E. Miller, "Emitter Softening in Diffused Silicon Transistors," in *Physics of Failure in Electronics*, Vol. 3, 1965, Edited by M. F. Goldberg and J. Vaccaro. Published by Rome Air Development Center. Available from The Clearing House for Federal Scientific and Technical Information, Department of Commerce.
14. E. J. Mets, "Poisoning and Gettering Effects in Silicon Junctions," *J. Electrochem. Soc.*, *112*, 420–425 (1965).
15. A. A. Bergh and G. H. Schneer, "The Effect of Ionic Contaminants on Silicon Transistor Stability," IEEE Transactions on Reliability R-18, 34–38 (1969).
16. A. A. Bergh, "Some Failure Modes of Double Diffused Silicon Mesa Transistors" Proceedings of Third Symposium of Physics of Failure In Electronics, 421–432 (1964).

17. J. E. Lawrence, "Electrical Properties of Copper Segregates in Silicon *P-N* Junctions," *J. Electrochem. Soc.*, *112*, 796–800 (1965).
18. F. Barson, P. A. Totta, and J. Overmeyer, "Copper Junction Poisoning as a Reliability Risk," Fall 1969 Meeting of Electrochem. Soc., Detroit, Mich., Abstract 196.
19. M. M. Atalla, A. R. Bray, and R. Lindner, "Stability of Thermally Oxidized Silicon Junctions in Wet Atmospheres," *Suppl. Proc. Inst. Elec. Engrs.* (London), Pt. B, *106*, 1130–1137 (1959).
20. E. D. Metz, "Silicon Transistor Failure Mechanisms Caused by Surface Charge Separation," in *Physics of Failure in Electronics*, Vol. 2, Edited by M. F. Goldberg and J. Vaccaro.
21. W. Shockley, H. J. Queisser, and W. W. Hooper, "Charges on Oxidized Silicon Surfaces," *Phys. Rev. Letts.*, *11*, 489–90 (1963).
22. W. Shockley, W. W. Hooper, H. J. Queisser, and W. Schroen, "Mobile Electric Charges on Insulating Oxides with Application to Oxide Covered Silicon *p-n* Junctions," *Surface Science*, *2*, 277–287 (1964).
23. W. Schroen, "Accumulation and Decay of Mobile Surface Charges on Insulating Layers and Relationship to Reliability of Silicon Devices," in *Physics of Failure in Electronics*, Vol. 4 (1966), Edited by M. F. Goldberg and J. Vaccaro, Published by Rome Air Development Center. Available from the Clearinghouse for Federal Scientific and Technical Information (CFSTI), Springfield, Va.
24. P. Ho, K. Lehovec, and L. Fedotowski, "Charge Motion on Silicon Oxide Surfaces," *Surface Science*, *6*, 440–460 (1967).
25. P. V. Gray, "The Silicon-Silicon Dioxide System," *Proc. IEEE*, *57*, 1543–1551 (1969).
26. R. P. Donovan, "The Oxide-Silicon Interface," in *Physics of Failure in Electronics*, Vol. 5 (1967), Edited by T. S. Shilliday and J. Vaccaro, Published by Rome Air Development Center.
27. A. G. Revesz and K. H. Zaininger, "The Si-SiO$_2$ Solid-Solid Interface System," *RCA Review*, *29*, 22–76 (1968).
28. T. M. Buck, F. G. Allen, J. V. Dalton, and J. D. Struthers, "Studies of Sodium in SiO$_2$ Films by Neuton Activation and Radio-Tracer Techniques," *J. Electrochem. Soc.*, *114*, 862–866 (1967).
29. F. M. Fowkes and T. E. Burgess, "Electric Fields at the Surface and Interface of SiO$_2$ Films on Silicon," *Surface Science*, *13*, 184–195 (1969).
30. H. G. Carlson, G. A. Brown, C. R. Fuller, and J. Osborne, "The Effect of Phosphorus Diffusion in Thermal Oxides on the Elevated Temperature Stability of MOS Structures," in *Physics of Failure in Electronics*, Vol. 4 (1966), Edited by M. F. Goldberg and J. Vaccaro, Published by Rome Air Development Center. Available from the Clearing House for Federal Scientific and Technical Information (CFSTI), Springfield, Va.
31. E. Yon, W. H. Ko, and A. B. Kuper, "Sodium Distribution in Thermal Oxide on Silicon by Radiochemical MOS Analysis," *IEEE Transactions on Electron Devices*, *ED-13*, 276–280 (1966).
32. A. B. Kuper, C. J. Slabinski, and E. Yon, "Positive and Negative Ion Motion in Thermal Oxide on Silicon by Radiochemical and MOS Analysis," in *Physics of Failure in Electronics*, Vol. 5 (1967), Edited by T. S. Shilliday and J. Vaccaro. Published by Rome Air Development Center.
33. H. G. Carlson, C. R. Fuller, D. E. Meyer, J. R. Osborne, V. Harrap and G. A. Brown, "Clean Metal Oxide Semiconductor Systems," in *Physics of Failure in Electronics*, Vol. 5 (1967), Edited by T. S. Shilliday and J. Vaccaro. Published by Rome Air Development Center.
34. S. R. Hofstein, "Stabilization of MOS Devices," *Solid State Electronics*, *10*, 657–670 (1967).
35. W. W. Smith, Jr., and A. B. Kuper, "Phosphosilicate Glass Passivation Against

Sodium Impurity in Thermal Oxide on Silicon," Sixth Annual Reliability Physics Symposium Proceedings, IEEE, New York, pp. 40–45 (1968).
36. E. H. Snow, A. S. Grove, B. E. Deal, and C. T. Sah, "Ion Transport Phenomena in Insulating Films," *J. Appl. Phys.*, *36*, 1664–1673 (1965).
37. P. Coppen, "Causes, Effects, and a Cure for Channeling in Silicon Planar Transistors," *SCP and Solid State Technology*, July 1965, pp. 20–22.
38. B. Kurz, "Degradation Phenomena of Planar Si Devices Due to Surface and Bulk Effects," in Sixth Annual Reliability Physics Symposium Proceedings, IEEE, New York, pp. 47–65 (1968).

## *bibliography*

1. L. E. Miller, "Basic Mechanisms of Failure in Diffused Silicon and Germanium Transistors," Supplement to Proceedings of 6th Annual New York Conference on Electronic Reliability, May 21, 1965.
2. H. Kressel, "A Review of the Effect of Imperfections on the Electrical Breakdown of *P-N* Junctions," *RCA Review*, June, 1967, pp. 175–207.
3. M. L. Embree and K. K. Ferridun, "Detecting Metal Particles in Semiconductor Device Housings," *Bell Laboratories Record*, October, 1967, pp. 295–298.
4. M. R. P. Young and D. A. Peterman, "Reliability Engineering," *Microelectronics and Reliability*, *7*, 91–103 (1968).
5. D. L. Cannon and O. D. Trapp, "The Effect of Cleanliness on Integrated Circuit Reliability," Sixth Annual Reliability Physics Symposium Proceedings, IEEE, New York, pp. 68–79 (1968).
6. D. A. Hope, "Contamination Problems in the Semiconductor Industry," *Environmental Engineering*, November, 1968, pp. 9–12.

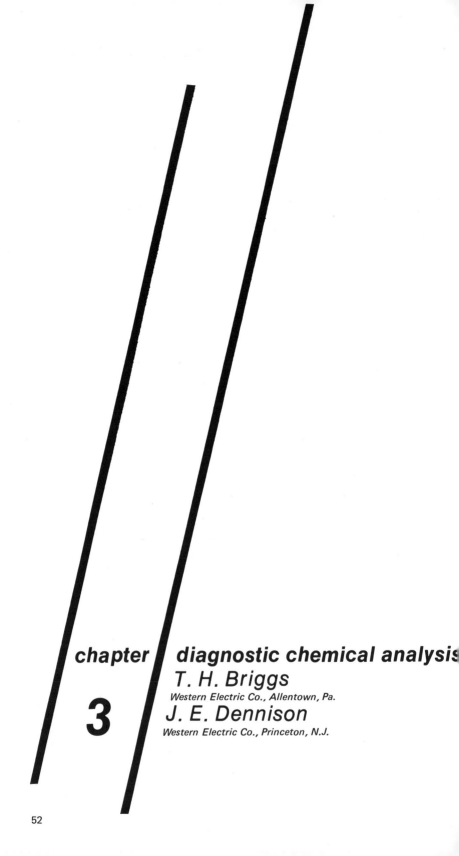

chapter

3

*diagnostic chemical analysis*

*T. H. Briggs*
*Western Electric Co., Allentown, Pa.*

*J. E. Dennison*
*Western Electric Co., Princeton, N.J.*

The identification and quantitative measurement of contaminants are the first steps in planning their elimination or control. One must know if the impurities present have an effect on the product being manufactured and what the sources of impurities are. Through analytical chemistry one can determine the identity and quantity of the impurities, in what form they exist, and how they are distributed. Therefore, an understanding of this subject is essential to anyone interested in microelectronics manufacture and factory environmental control. Particularly in microelectronics, materials investigations are centered about trace contaminant levels. With only limited amounts of sample available for analyses, the investigation usually is handled with sensitive, sophisticated, and costly instruments. This chapter is intended to discuss diagnostic analytical problems in terms of the basic needs of microelectronics, including basic approaches used to solve contamination problems.

## *diagnostic methods*

Two different approaches are used to solve contamination problems: specific investigations of products to identify contaminants and general investigations of known contaminants to relate them to product problems. In the first case one begins with finding differences between "good" and "bad" samples and relating these results to product problems through theory and past experience. An example of this involved transistors with low amplification factors. Comparison of silicon material used for them showed that 0.05 ppm gold was present in the "bad" ones but not in "good" ones. The source of gold in the process was located in one of the transistor material processing furnaces. Removal of the contaminated furnace part, which contained 50 ppm gold, eliminated the problem.

In the second case one attempts to establish correlations between suspected contaminants and product defects. For example, a situation arose in which an organic coating was polymerizing, thus inhibiting proper thin film etching. Ozone, which is known to cause polymerization, was monitored throughout several days and was found to be correlated with product defects. Reduction of ozone levels to 15 ppb eliminated the problem.

### *device purity requirements*

Tolerable impurity levels in and on devices and the capability of instruments to measure levels are highly variable. Some cases do not tax an instrument's sensitivity, but in others the most sensitive analytical techniques may not be capable of detecting harmful impurities. In particular, some impurities in silicon are difficult to detect. Other factors which are usually considered in selecting an instrumental method for a problem include the number of samples to be analyzed, whether only a few known elements are to be measured and the need to use nondestructive tests.

An example of device sensitivity to an extremely small quantity of contamination is the presence of an alkali metal ion on a silicon device.[1] As little as $10^{-11}$ g/cm$^2$ sodium or potassium on a transistor can produce detrimental reverse current leakages. (With the sensitive parts of a transistor being as small as $10^{-6}$ cm$^2$, in a critical place $10^{-19}$ g of an alkali can cause device deterioration.) To date, this quantity on a small transistor or integrated circuit is not detectable analytically. Sodium and potassium are both common elements in contaminating sources such as human perspiration, air pollution particles, and water residues.

Not all problems demand such limits of sensitivity for their solution. One should be aware, however, that a report showing a given element as "not detected" means just that—but a given element may still be present at levels below the detection limit of the instrument used. Since these limits vary widely between instruments and elements, a selection of methods may depend heavily on what contaminants are sought and their suspected concentrations. There are no fixed rules which can be applied to the selection of a technique for analysis. Each product problem is different, and thus each must be considered in terms of the nature of the problem and the characteristics of the instrument. Such items as analytical sensitivity, instrument "blind spots," and usable kinds of samples will vary with the combinations of elements and compounds present and with the manner of sampling needed for the instrument. Some guidelines exist and should be used to select analytical techniques according to the problem presented. Before discussing this, some definitions of nomenclature are needed.

### nomenclature

In the majority of situations dealing the contaminants, it is necessary to state the amount present. The terms used should be meaningful to a given problem. Impurities are most commonly expressed as a concentration of the material in question, as a percent, or so many parts per million (ppm), or occasionally in parts per billion (ppb). This method of expression is satisfactory for homogeneously distributed impurities in the bulk of the host material. It is necessary to use this type of expression when checking raw materials against specified impurity limits. Concentrations expressed in ppm imply that impurities are uniformly distributed in bulk materials. This is usually not the case at the microscopic level, and in microelectronics such variations can become extremely important.

Being able to measure and express amounts of surface contaminants is a matter of some concern in diagnostic chemistry for microelectronics. The term "parts per million" requires that a volume of material must be taken into account, and that the impurity being measured is related to that volume. Its use misrepresents surface concentrations. In Fig. 3-1a, impurities on a surface can be present as uniform films, discrete particles, or something in between, such as droplets of a liquid which have partially spread out over the surface. A better way of expressing all of these surface contaminants is to use weight per unit area, such as micrograms/square centimeter. Other informa-

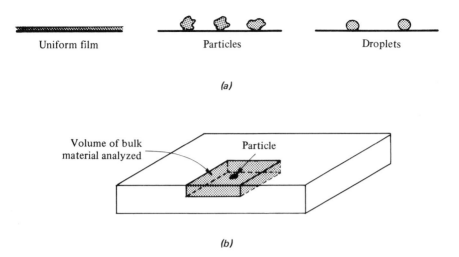

(a)

(b)

*Figure 3-1  Surfaces and contaminants.  (a) Surface impurities; (b) small particle on surface of bulk material.  Surface impurities are not a part of the bulk material, although they may appear to be when a surface is analyzed.  If there is a change in the volume of substrate being considered, the same particle will appear as a different impurity concentration relative to the bulk material sampled.*

tion about the form of the contaminant can be included as another dimension, e.g., expressing particle size and population, or layer thickness.

A particle can be considered a point source contaminant to a surface. Point source contaminants can have serious effects in microelectronics since these devices do not depend on an overall average concentration of a contaminant for their performance. A particle, even though microscopic, can influence a small part of an active integrated circuit region as though 100% concentration of its constituents were present in that area, which in fact they are. From an analytical point of view, if a 1 $\mu m^3$ particle is analyzed on a substrate by an instrument which can only resolve a volume 100 $\mu m$ X 100 $\mu m$ area, 10 $\mu m$ deep, that particle will appear to be only 10 ppm of the substrate material (Fig. 3-1b).  By using an instrument which can work with 1 $\mu m^3$ volumes, the particle will now appear to be 100% of its constituents. Spatial resolution is an important consideration in analyzing contaminants.

## analytical instruments and equipment

Analytical instruments are based on the measurement of some physical or chemical property: absorption or emission of electromagnetic radiation, adsorption and decay of unstable nuclei, the mass of an ion or molecule, solubility in or adsorption on a substrate, boiling points, heats of fusion, vaporization, specific heats, chemical reactivity, and so on.  Chapter 4 will deal with the instruments of microscopy; Chapter 5 specializes in particle measurement; this chapter will discuss electromagnetic radiation instruments plus other methods having major application in microelectronics technology.

## electromagnetic radiation instruments[2]

Of the physical phenomena used in analytical chemical instruments, those involving electromagnetic radiation form an important group. The types of radiation utilized extend from ordinary visible light toward both the long and short wavelengths. These instruments reveal the degree of emission or absorption of energy radiated as a function of wavelength (Fig. 3-2). The wavelength of the radiation absorbed or emitted is a characteristic of the material being studied, while the degree of absorption or emission is characteristic of the quantity of element or compound in the sample. From this it follows that each chemical element and compound will have its own characteristic response in terms of energy, and thus spectral wavelength.

### material behavior characteristics

Every response, either energy emission or absorption, is based on the change of state between two energy levels. The energy of any given state can be formally expressed by Eq. (3-1) as a sum of energy terms where each term assumes certain discrete quantitized values:[2]

$$E_{total} = E_{electronic} + E_{translation} + E_{rotation} + E_{vibration} \qquad (3\text{-}1)$$

For radiation to be absorbed by a material the energy of the incident radiation must be such as to cause the system to pass from a low energy state to a higher state. The emission of radiant energy is the reverse process; namely, the system passes from a state of high energy to lower energy states. This transfer is accompanied by the emission of a photon whose energy is equal to the difference between these two discrete or quantized states. The wavelength of the absorbed or emitted photon is given by:

$$E_2 - E_1 = h\nu = hc/\lambda \qquad (3\text{-}2)$$

where $\nu$ is the radiation frequency, $h$ is Planck's constant, $c$ is the velocity of light, $\lambda$ is the radiation wavelength, and $E_1$ and $E_2$ are the initial and final energy states.

Transitions between electronic energy states fall in different energy or wavelength ranges than those transitions between rotational or vibrational states. Consider the electronic transition between a orbital near the nucleus and a free electron; this is in the X-ray region of the spectrum. Transitions involving K and L shells (Fig. 3-3) require more energy than those between M and L shells. Radiation in the ultraviolet and visible range is associated with electron transitions within the valance shell where electrons are held more loosely than in the K shell. In the infrared analysis range, light absorption and emission are the result of changes occurring between vibrational energy states of molecules, while in the far infrared region changes occur

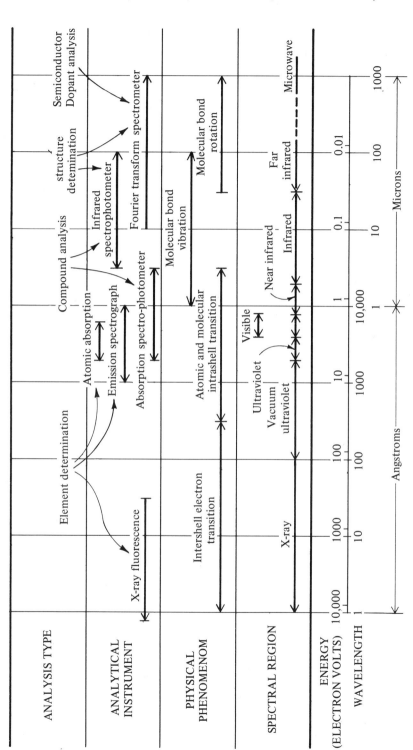

Figure 3-2 Scale of wavelength, energy, and spectroscopic phenomena.

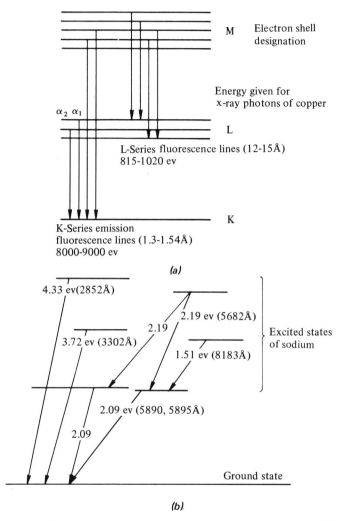

**Figure 3-3** (a) Intershell transitions on copper producing x-rays, and (b) intrashell transitions producing visible radiation for sodium. Compare energies of lines shown here with intershell energies of copper. Energy requirements here are similar in magnitude with possible transitions within a shell in the x-ray diagram. Spectrographic display of lines is shown in Fig. 3-5b.

between rotational energy states. To use each of these wavelength ranges for analytical purposes requires different types of spectrographs, spectrophotometers, or spectrometers.

### fundamentals of spectroscopy

Spectroscopic methods of analysis depend on emission, absorption, or fluorescence as the basic mechanisms to produce characteristic radiation spectra (Fig. 3-4). Emission occurs when a sample is excited by an external

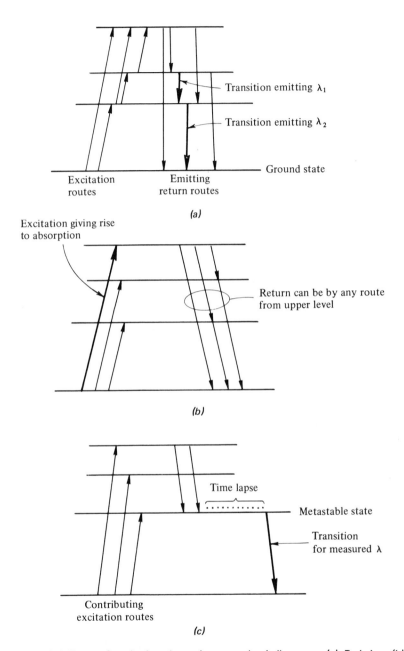

Figure 3-4  Types of excitation shown in energy level diagrams.  (a) Emission; (b) absorption; (c) fluorescence.

source of energy.  After excitation, the sample itself emits photon characteristics of elements or compounds present.  The external energy sources are
generally thermal or electrical.  Samples are excited by electrical discharges in
optical emission spectroscopy but are thermally excited in flame photometry.

Absorption spectra are produced when radiation from a primary source is passed through a sample, and the amount of absorption is measured as a function of wavelength. Of the types of spectroscopy listed in Fig. 3-2, this principle applies to infrared, visible, ultraviolet, and atomic absorption spectroscopy instruments, and it can be used with X-ray spectroscopy.

Fluorescence spectroscopy is based on a sample absorbing radiation from a primary source and re-emitting that energy at a longer wavelength (or lower energy) as the excited sample returns to lower energy levels. In this process the measured radiation corresponds to the quantity emitted between two specific levels. The atom temporarily dwells in an excited "metastable" state before returning to some lower level. X-ray fluorescence spectroscopy, which is very common for inorganic analysis, typifies this principle. Ultraviolet fluorescence analysis which also falls in this category is applied to trace analysis of some organic compounds.

### emission spectrograph[3]

The emission spectrograph is used for detecting and measuring small amounts of metallic elements in a variety of sample types.

**Principle.** The elements may be present as metals or as compounds. A commonly used technique is to introduce a sample into a d-c arc discharge by using a graphite electrode with a cavity drilled into its end to contain the sample. Other types of discharges exist, and techniques for liquid as well as solid analysis are well known. For the case shown the opposite electrode is a pointed graphite electrode (Fig. 3-5). A d-c arc of 5–10 amp is used to vaporize and excite the sample. During excitation, valence electrons of metallic elements absorb energy and move into higher energy levels. When the electrons fall back into lower levels, light is emitted at discrete wavelengths which are characteristic of each element. The light emitted in the d-c arc is focused on the entrance of the spectrograph. The light passes onto a diffraction grating where it is dispersed into its wavelengths, and projected as a series of lines at a focal plane. A photographic plate, or a bank of photomultipliers positioned for each line of interest, is located at the focal plane. Interpretation is carried out with the aid of concentration standards for comparison. Wavelength (line position) is used to identify elements, and line intensity is used to determine concentration.

**Sensitivity.** The detection sensitivity is quite variable (Table 3-1), with a few elements measured directly to about $10^{-9}$ gram, or 0.1 ppm in a $10^{-2}$ gram sample (approximately a normal size sample). Most elements have detection limits of about $10^{-8}$ gram. Sensitivity for a given element depends heavily on the major sample components, or matrix, and has to be determined for each sample type. Each element has lines highly variable in sensitivity. Interferences occur when lines of two elements appear at the same wavelength. This can require finding another, perhaps less suitable line for both elements.

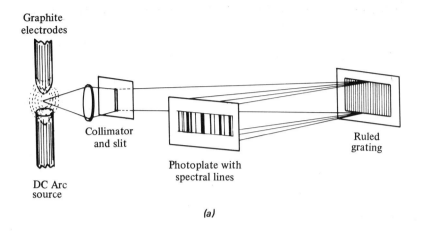

Graphite electrodes

Collimator and slit

DC Arc source

Photoplate with spectral lines

Ruled grating

(a)

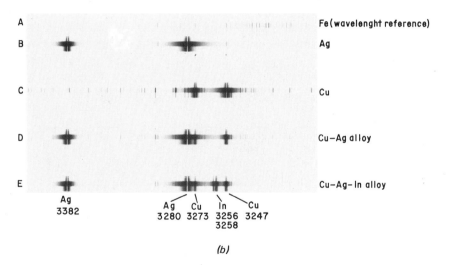

A    Fe(wavelenght reference)

B    Ag

C    Cu

D    Cu–Ag alloy

E    Cu–Ag–In alloy

Ag
3382

Ag   Cu   In    Cu
3280 3273 3256 3247
          3258

(b)

Figure 3-5 (a) Diagram of emission spectrograph. Each line is an image of the slit, displaced along the photoplate according to wavelength. Grating and photoplate are enclosed in light-tight cabinet. (b) Portion of emission spectrograph photoplate from 3200–3450 angstroms (negative image). Each band is one sample. Band A = iron; B = silver; C = copper; D = silver-copper alloy; E = silver-copper-indium alloy. White streak in center of each major line is self-absorption of emitted radiation by cooler atoms.

**Limitations.** Emission spectroscopy has some "blind spots." Generally speaking, nonmetals (halides, sulfur, oxygen, nitrogen) are not detected. Selenium and tellurium are so insensitive that they are not useful. Certain

TABLE 3-1.  Comparison of Typical Instrument Sensitivities for Selected Elements (in grams)

| | Emission Spectrograph[1] | X-ray Fluorescence[2] | | Spark Source Mass Spectrograph[3] | Neutron Activation Analysis[4] | Atomic Absorption Spectrophotometer | |
|---|---|---|---|---|---|---|---|
| | | Coprex | Micro Spot | | | Burner[5] | Graphite Crucible[6] |
| Na | $3 \times 10^{-7}$ | X | X | $4 \times 10^{-12}$ | $6 \times 10^{-8}$ | $5 \times 10^{-9}$ | NA |
| K | $3 \times 10^{-5}$ | $4 \times 10^{-7}$ | $5 \times 10^{-10}$ | $1 \times 10^{-11}$ | $7 \times 10^{-8}$ | $5 \times 10^{-9}$ | $4 \times 10^{-11}$ |
| Cu | $3 \times 10^{-8}$ | $7 \times 10^{-7}$ | $1 \times 10^{-7}$ | $1 \times 10^{-10}$ | $3 \times 10^{-8}$ | $5 \times 10^{-9}$ | $6 \times 10^{-13}$ |
| Au | $3 \times 10^{-7}$ | $4 \times 10^{-6}$ | $3 \times 10^{-9}$ | $3 \times 10^{-10}$ | $3 \times 10^{-10}$ | $1 \times 10^{-7}$ | $1 \times 10^{-12}$ |
| Zn | $3 \times 10^{-8}$ | $4 \times 10^{-7}$ | $5 \times 10^{-10}$ | $1 \times 10^{-10}$ | $1 \times 10^{-6}$ | $5 \times 10^{-7}$ | $3 \times 10^{-14}$ |
| B | $3 \times 10^{-7}$ | X | X | $6 \times 10^{-12}$ | NA | $6 \times 10^{-6}$ | $2 \times 10^{-10}$ |
| P | $3 \times 10^{-6}$ | X | X | $2 \times 10^{-11}$ | $4 \times 10^{-8}$ | $1 \times 10^{-6}$ | NA |
| S | X | $3 \times 10^{-7}$ | $5 \times 10^{-10}$ | $6 \times 10^{-10}$ | $1 \times 10^{-6}$ | X | X |
| Cl | X | $8 \times 10^{-7}$ | $1 \times 10^{-8}$ | $4 \times 10^{-11}$ | $4 \times 10^{-8}$ | X | X |

X—Not sensitive enough for analytical use.
NA—Data not available.
[1] Unpublished data, G. A. Ziegenfuss, Western Electric, Allentown, Pa.
[2] Unpublished data, J. E. Kessler and S. M. Vincent, Bell Telephone Laboratories, Murray Hill, N.J.
[3] Unpublished data, D. F. Lesher, Western Electric, Allentown, Pa.
[4] W. S. Lyon, *Guide to Activation Analysis*, Van Nostrand, 1964, Princeton, N.J.
[5] Normal AAS Solution Technique, Perkin-Elmer Handbook.
[6] B. V. L'vov, "The potentialities of the graphite crucible method in atomic absorption spectroscopy," *Spectrochemics Acta*, V 24B, pp. 53–70.

metals have extremely complex spectra (e.g., rare earths, tungsten, tantalum) which interfere with many elements, and samples of these materials sometimes require special techniques when trace impurities are sought.

**Sample Type.**  The most favorable sample form is a powder that is loaded directly into a cup electrode.  Liquids can be evaporated to leave residues. Special techniques for preconcentrating impurities have been used many times and have been reported in numerous sources.[3-6]  Sample size is commonly 10–20 mg, but up to 100 mg of some materials have been handled directly. Smaller sizes do not present serious problems.

**Applications.**  The emission spectrograph generates both qualitative and quantitative results.  It is a rapid, excellent general survey technique for detecting and obtaining a rough idea of metallic composition without any prior knowledge of the sample.  Routine quantitative analyses can be performed rapidly for many elements at one time in one sample.

### atomic absorption spectrophotometer[7]

This technique is used for measuring small amounts of metallic elements in solutions.  It is best applied to the measurement of a group of predetermined elements rather than as a general survey instrument to see what is present. It is capable of performing routine analyses rapidly, simply, and with good precision.

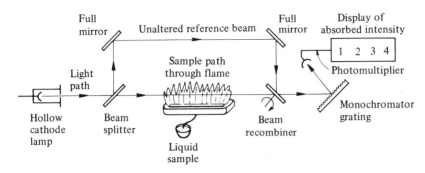

*Figure 3-6 Optical path of typical atomic absorption spectrophotometer. Double beam system is shown. System display is adjusted to show zero before sample is introduced into flame. Absorption in sample path unbalances signal and causes a number to be displayed.*

**Principle.** A lamp emitting the atomic spectrum of the element of interest is used as a source of radiation. The light is sent through a monochromator which sorts out one discrete wavelength emitted by the lamp and permits a photomultiplier tube detector to measure the intensity of just that one line. A long flame is placed in the optical path (Fig. 3-6). A solution of the sample or a standard is aspirated into the flame which vaporizes the liquid and reduces its dissolved components into atomic species. The energy of the flame excites a few atoms to emit light, but most (over 90%) remain in the "ground state" (Fig. 3-4) where they can absorb light emitted by the selected lamp source. One of these lines is monitored for measurement. These atoms reduce the intensity of the wavelength being measured from the lamp source; this attenuation is related to the concentration of the element in the original solution. Quantitative results are obtained by comparing unknowns with standards.

**Sensitivity.** Detection limits depend on the element (Table 3-1) and vary from approximately 0.05 ppm for sodium to several hundred ppm for such elements as tantalum and tungsten. Sensitivity is not strongly dependent on matrix or solvent but is controlled by flame fuel mixture and position with respect to optical path. Interferences are generally small, a great benefit in routine analysis.

**Speed and Precision.** Atomic absorption methods can be very rapid, in excess of 60 samples/hour for one element, and the methods are adaptable to automation. Multiple elemental analyses require re-running samples for each element. Precision can be better than 1% of the amount present. These two factors have made atomic absorption valuable where large numbers of samples must be checked for a few elements.

**Limitations.** The same elements, i.e., the nonmetals, which cannot be handled directly by emission spectroscopy also cannot be directly checked by atomic absorption. Solutions with variable viscosity or high concentrations of salts also can cause erratic results. If an element to be checked is at a high concentration, its absorption of the light beam will be large enough that the equipment will not respond linearly to additional amounts; high concentrations generally can be handled by diluting the sample to bring it within range of standards. The method is generally limited to solutions and materials which can be dissolved, but methods for directly handling some solid samples are being developed.

### flame emission spectrometer[7]

Flame emission spectrometry is used primarily for solutions, particularly for alkali metals, which are very easily excited to emit their most sensitive lines.

**Principle.** In the flame emission spectrometer a flame system similar to that used in atomic absorption is used as an emitting source. The light from the flame is dispersed in a monochromator so that a wavelength or "line," emitted by the element of interest, falls on the detector. The emitted rather than absorbed intensity is measured and related to composition. Flame emission spectrometry is highly sensitive to a few elements, notably alkalies. These are common environmental contaminants to which semiconductors are extremely sensitive.

**Sensitivity and Applications.** Sensitivities of 0.01 ppm of Na and K in solution are readily obtained directly on solutions. It has been reported that modified instruments have detected less than 0.001 ppm Na.[7] Other elements are detectable but at less sensitive levels. Speed and precision are similar to that for atomic absorption spectrometry. The technique is simple, rapid, and used for routine measurements of Na and K. As with atomic absorption, it is not satisfactory for general survey studies.

### x-ray fluorescence spectrometer[8]

X-ray fluorescence spectrometry offers a reasonably sensitive method for detecting elements with atomic numbers greater than magnesium and with rapidly decreasing sensitivity can measure elements as light as fluorine. The method is not as sensitive as emission spectroscopy for most metallic elements, but it can detect nonmetals as easily as metals.

**Principle.** X-ray fluorescence spectrometry is carried out by bombarding a sample with X-rays from an X-ray tube. This primary beam causes all the constituent elements in the sample to omit their characteristic X-ray fluorescence spectrum (Fig. 3-7). The fluorescence is due to intershell electron transitions. The fluorescent, or secondary, X-rays emitted from the sample are diffracted by an oriented crystal whose regular crystal planes act as a

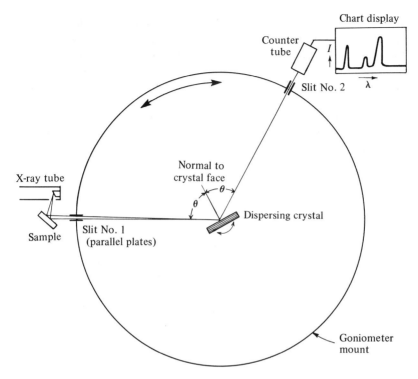

*Figure 3-7 Typical x-ray fluorescence spectrometer. X-ray tube causes sample to fluoresce, giving off radiation characteristic of sample. This radiation passes through slit No. 1, is diffracted by crystal, and each wavelength is detected by the counter tube. Display can be a chart, or a pulse accumulating circuit. Angles θ must meet diffraction equation nλ = 2d sin θ, where λ is x-ray wavelength, d is interplane crystal spacing of dispersing crystal. Thus crystal must rotate and counter tube must move along goniometer circumference.*

diffraction grating. The X-rays are detected as pulses by a Geiger counter, a scintillation counter, or a similar device. By counting pulses due to photons of a given wavelength for a fixed period of time it is possible to compare quantitatively amounts of an element in samples and standards.

**Sensitivity and Applications.** X-ray fluorescence spectrometry is capable of high precision when at least moderate quantities of sought elements are present (Table 3-1). Bulk samples can be accurately analyzed for impurities above 100 ppm and samples containing contaminant amounts above several micrograms can be checked with accuracy of ±5% of the amount present. X-ray spectroscopy is generally nondestructive, permitting subsequent analyses by other instrumentation.

Although the method can be used for qualitative analysis by scanning through a full spectrum, such scanning is relatively time-consuming (about 1 hour for a normal scan) and generally less sensitive than emission spectroscopy. Because of this there is little advantage over emission spectroscopy in checking for the presence of many elements in one sample by X-ray fluorescence. X-ray fluorescence spectrometry is most frequently applied to quantitative analysis or comparison of particular element concentrations in similar samples.

In recent years there have been important advances in the use of energy dispersive detectors. The result is a reduction in the time needed for general qualitative analysis. The resolution is not as good as in wavelength dispersive instruments, but with improvements in resolution the technique may become competitive with emission spectroscopy in many cases. The energy dispersive techniques can be used for low atomic number elements plus elements not easily detected by emission spectroscopy.

### ultraviolet-visible absorption spectrophotometer[2]

Absorption spectrometry was among the first techniques to become instrumented and is useful in both qualitative and quantitative analyses. The essential components of spectrophotometric systems are similar in concept: a radiant energy source; associated optics and slits; dispersing devices; detectors; and data acquisition systems. The difference in materials and construction of the systems is dictated by the wavelength span of the electromagnetic spectrum being measured. The UV-visible equipment functions in the visible band (780–380 nm) and the ultraviolet region (380–200 nm).

**Principle.** When molecules or ions interact with radiant energy in the UV or visible region, the absorption results in the displacement of electrons from one electron orbit to an orbit of higher energy. At shorter wavelengths, low UV or vacuum UV, the energy may be sufficient to remove the electron completely or to break molecular bonds. The absorption spectrum is a function of the whole energy structure of the molecule, and not the property of a single bond or energy state.

The absorption spectrum is specific for a given molecular structure. Where the differences in electronic energy levels are small, the absorption takes place at longer wavelengths, often in the visible region. In Table 3-2 there are several examples of absorption bands in the range 185–800 nm for representative chromophores which can be used for identification.

A diagram of a typical UV-visible spectrophotometer is shown in Fig. 3-8a. Light from a source is collimated, and passed through a scanning monochromator which is driven continuously to change wavelength. The light is again collimated, and passed through the sample. The light is detected by a photosensitive cell, the output of which is plotted on graph paper as a function of wavelength.

**Sensitivity and Applications.** The detection limit for UV-visible spectrophotometry is of the order of $10^{-6}$ moles/liter, which translates to about 0.5 $\mu$g/l. Samples are most often solutions in any of a number of suit-

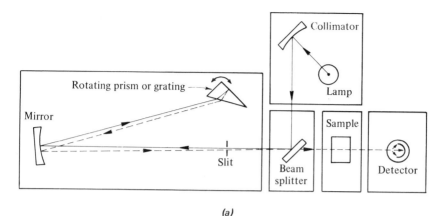

(a)

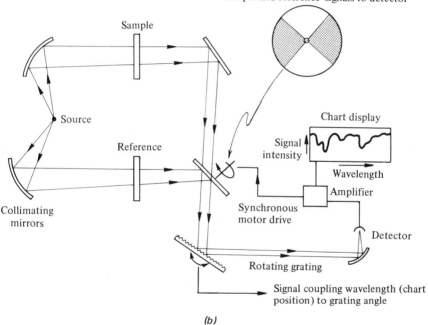

(b)

*Figure 3-8  (a) UV-visible spectrophotometer.  (b) IR spectrophotometer.. Emitted radiation from source is collimated through sample and reference cells. Light passing through them is alternately compared for signal strength, dispersed by grating, and sensed by detector, which sends signals to chart display.*

able solvents, water being the most common. The volume of the solution is usually 1–5 ml. Analysis of transparent solids for impurities is also well known. Spectrophotometric methods for measuring most elements in either atomic or molecular form are now available. A spectrophotometric method is

**TABLE 3-2.  Electronic Absorption Bands for Representative Chromophores**

Wavelength in Nanometers
Extinction Values in cm$^{-1}$

| Chromophore | System | $\lambda_{max}$ | $\epsilon_{max}$ | $\lambda_{max}$ | $\epsilon_{max}$ | $\lambda_{max}$ | $\epsilon_{max}$ |
|---|---|---|---|---|---|---|---|
| Ether | −O− | 185 | 1000 | | | | |
| Thioether | −S− | 194 | 4600 | 215 | 1600 | | |
| Amine | −NH$_2$ | 195 | 2800 | | | | |
| Bromide | −Br | 208 | 300 | | | | |
| Iodide | −I | 260 | 400 | | | | |
| Nitrile | −C≡N | 160 | − | | | | |
| Acetylide | −C≡C− | 175–180 | 6000 | | | | |
| Ethylene | −C=C− | 190 | 8000 | | | | |
| Ketone | >C=O | 195 | 1000 | 270–285 | 18–30 | | |
| Esters | −COOR | 205 | 50 | | | | |
| Aldehyde | −CHO | 210 | strong | 280–300 | 11–18 | | |
| Nitrite | −ONO | 220–230 | 1000–2000 | 300–400 | 10 | | |
| Azo | −N=N− | 285–400 | 3–25 | | | | |
| Nitrate | −ONO$_2$ | 270 (shoulder) | 12 | | | | |
| | −(C=C)$_2$− (acyclic) | 210–230 | 21,000 | | | | |
| | −(C=C)$_3$− | 260 | 35,000 | | | | |
| | −(C=C)$_4$− | 300 | 52,000 | | | | |
| Benzene | | 184 | 46,700 | 202 | 6,900 | 255 | 170 |
| Naphthalene | | 220 | 112,000 | 275 | 5,600 | 312 | 175 |
| Quinoline | | 227 | 37,000 | 270 | 3,600 | 314 | 2,750 |

From "Instrumental Methods of Analysis, 4th Ed., W. W. Willard, L. L. Merritt, Jr., and John A. Dean, Van Nostrand Reinhold, New York, 1965.

one of the few techniques which will determine the oxidation state of elements and organic functional groups. It can be accurate and precise and is adaptable to automation. However, spectrophotometric techniques are often slow and are not suitable for survey analysis. A typical use of UV-visible spectrophotometry is in monitoring certain broad classes of organic compounds, which can be detected to about 1 ppm.

### infrared spectrophotometer[9]

Most commercial infrared spectrophotometers function in the 2.5-25 $\mu$m wavelength range, the "fingerprint region" of organic compounds. Like the UV-visible spectrophotometer, IR equipment is composed of an appropriate radiation source, dispersion devices, optical equipment, detectors, and a data acquisition system.

**Principle.** Radiant thermal energy is generated by a glow bar or Nernst glower, which by suitable optics, slits, and a monochromator, is passed through the sample onto a thermopile or bolometer detector. Figure 3-8b illustrates schematically a typical arrangement of a spectrophotometer. As in UV-visible spectrophotometry the wavelength range is scanned for qualitative work, and intensity measurements at specific wavelengths are used for quantitative work.

Only $E_{ROT}$ and $E_{VIB}$ (Eq. (3-1)) are of importance in the IR region because translational energy has little effect on the molecular structure and the quanta of IR energies are insufficient to cause changes in electronic energy states. All molecules connected by chemical bonds vibrate much like balls connected by a spring, with each vibrational mode having several possible energy states. Molecules can absorb IR energy and be raised to a higher energy state in the process. At ambient temperatures the ground state population is high, but higher states also have significant populations. In addition, there may also be simultaneously changes in rotational energy states accompanying changes in vibration states. High resolution IR shows these changes in rotational states as fine structure in the vibrational absorption band, and these rotational energy state transitions in liquids or gases at high pressure cause broadening of the pure vibrational spectra, much the same as vibrational energy state transitions broaden the electronic spectra.

**Qualitative Analysis.** The primary value of IR spectrophotometry is in the identification of unknown compounds. Experimental spectra are compared with spectra obtained for pure compounds. There are numerous spectra published;[10-13] in many cases identifications can be made directly by simply observing the experimental spectra. Figure 3-9 shows the absorption bands of numerous groups along with the intensity and bandwidths.

**Sensitivity.** IR spectrophotometry is not noted for high sensitivity; at least 1 $\mu$g samples are usually required. An example of high sensitivity is the detection of a deposited organic film by IR where about 50 ng/cm$^2$ or about 1-5 monolayers are required. Detection limits for gas samples are less than 1 ppm with long path gas cells for strongly absorbing components such as CO or $NO_2$.

### fourier transform spectrometer

Fourier transform spectroscopy is based on the use of an interferometer and a computer to transform data from a fast scan, nondispersive interferogram to the usual analog IR signal. Since no slits are employed, the fundamental sensitivity is much higher than usual IR spectroscopy. It is being applied to infrared spectroscopy from 0.6 nm to 1000 nm, a range wider by far than any other instrument.

**Principle.** Figure 3-10 shows a schematic of a Fourier transform spectrometer. Light from the IR source passes into the interferometer where the beam splitter passes half of the light through a movable mirror while half is reflected to a fixed mirror. Upon striking the beam splitter the second time, each beam is again split so that half returns toward the source and half continues through the optical system. However, the beam at this point consists of overlapping interferograms of all incident wavelengths. Starting with both mirrors equidistant and with the movable mirror moving toward the beam splitter, monochromatic photons will give an interferogram consisting of alternating constructive and destructive interference bands. Polychromatic

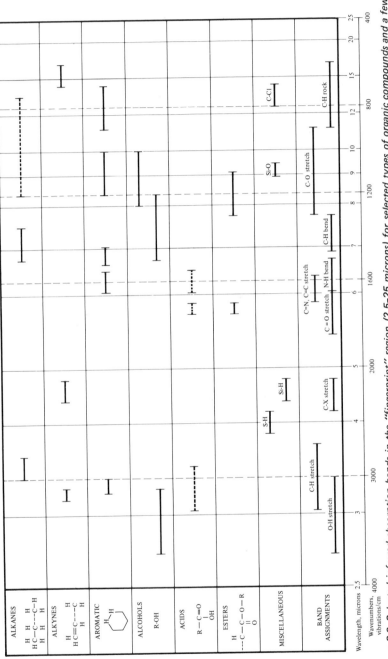

Figure 3-9 Selected infrared absorption bands in the "fingerprint" region (2.5-25 microns) for selected types of organic compounds and a few inorganic materials. Bands which may not appear due to some additional structure are shown as dashed lines. Regions characteristic of a particular type of band are indicated in "Assignments." (Courtesy Journal of the Optical Society of America, Vol. 40, p. 397, 1950.)

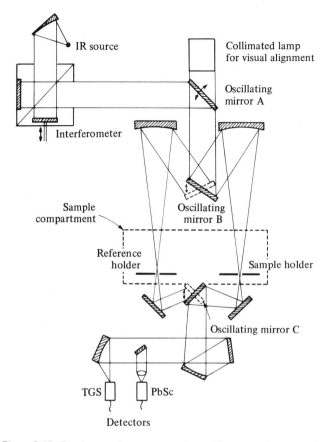

*Figure 3-10 Fourier transform spectrometer. (Courtesy Digilab, Inc.)*

radiation gives an intensity as an integral function of the equation over all wavelengths.

The light beam is focused alternately on the sample and the reference cell and then passes to the detector. The polychromatic radiation intensity data can be converted into a meaningful absorption spectra through the efforts of a small computer (manual reduction of the data would be tedious).

**Sensitivity.** Fourier transform spectroscopy employs no slits. This improves by 100 fold the amount of useful light available for spectral measurements. In addition, all the light falls on the detector, not just the wavelength selected by the dispersing device in conventional IR. The result is at least 1000 fold increase in sensitivity.

The resolution is usually less than conventional high resolution IR spectrometers, 0.3 cm$^{-1}$ versus 0.1 cm$^{-1}$. Scan speeds of less than one second are easily obtained. Typically full range spectra are scanned in 1-2 minutes, whereas 20 minutes are needed for conventional IR scans, or much longer for high resolution scans. Changes in the beam splitters which can be accomplished in 20–30 minutes, are required for far IR measurements.

**Applications.** There are numerous applications, most of them new, because of the fast scan speed, wide range, and high sensitivity. On-the-fly spectra of gas chromatogram effluents have been obtained. In this case the scan must be completed in 1–2 seconds, and the total weight of the sample is about one microgram. Sensitivity at this level is impossible with conventional spectrometers. The unit can also be used for in situ identification and quantitative measurement of many atmospheric impurities. IR spectra of contaminating films can be measured to less than a monolayer thick on microelectronic components because of the high sensitivity. Sensitive measurements in the far infrared, beyond 25 nm, can be done with a simple change in optics.

## other major instrumentation

The following instruments are treated separately because they do not depend on electromagnetic radiation for their basis of operation. No unifying principle can be used to cover their performance, and each principle must be discussed separately.

### mass spectrometry

In mass spectrometry materials are differentiated according to mass. All mass spectrometers act on gas phase ions but vary in the means of forming the ions, types of samples, and results obtained. These differences are great enough to warrant separate sections on gas mass spectrometry, as opposed to spark source, or solids, mass spectrometry.

### gas mass spectrometer[14]

The gas mass spectrometer is used to identify gases and components of gas mixtures and other materials which can be conveniently vaporized.

**Principle.** The instrument itself (Fig. 3-11a) is evacuated to better than $10^{-6}$ torr. The molecules of a gaseous sample are introduced into an evacuated sample inlet chamber where the pressure can rise as high as $10^{-4}$ torr as the sample is allowed to expand into it. The sample vapor is then allowed to leak through a calibrated aperture into the ionizing section, where it is bombarded by a beam of electrons.

The molecules of the sample are ionized by the electron beam and fragmented in a repeatable way. The fragmentation pattern is characteristic of the molecule irradiated by the electron beam.

Ions are accelerated out of the ionizing compartment by a potential gradient and then deflected by a magnetic or an electric field according to their mass to charge ratio. The mass of a molecule, fragment, or atom is expressed in atomic mass units (amu). For instance, carbon $12 = 12.000$ amu, and the charge is given as the number of electrons removed by ionization ($e = 1, 2, 3, \ldots$).

By varying the accelerating electric potential or the strength of the deflecting magnetic field, ions of different mass/charge ratios are brought to a detector, which is usually an electron multiplier sensitive to currents as low as $10^{-14}$ amp. If the mass/charge ratio selection is scanned, the output of the detector can give a plot (Fig. 3-11b) of the ion masses detected from mass = 1 up to the limit of the mass spectrometer. For permanent gases this limit need be only mass = 100, while for larger molecules an instrument capable of detecting masses up to 400–600 is highly desirable.

**Sensitivity.** The mass spectrometer is generally sensitive to about $10^{-10}$ torr partial pressure or about 1 ppm of the molecule or fragment being detected, depending on the ion species. For organic molecules this will depend partly on how the molecules of the species break up during ionization. The pattern of fragmentation is determined largely by the energy of the ionizing electrons and bonding energies (or resistance to breaking) of each bond within an atom. For these reasons, an experienced spectroscopist is needed to interpret the mass spectra.

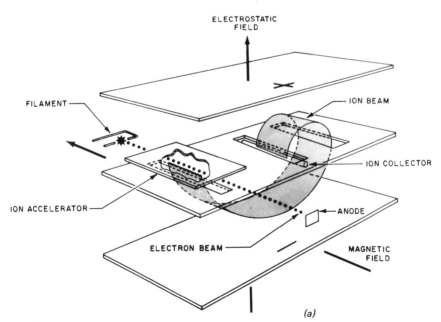

*Figure 3-11 (a) Gas mass spectrometer with high sensitivity. Gas sample is injected into compartment containing filament, anode, and ion accelerator. Ions pass through slit in plate, following a path determined by ion mass and crossed electrostatic and magnetic fields. Varying the electrostatic field changes which mass arrives at ion collector. Data is displayed on a chart. (Courtesy E. I. DuPont.) (b) Chart from gas mass spectrometer showing permanent gases in mass 2–45 mass range of a nitrogen filled furnace.*

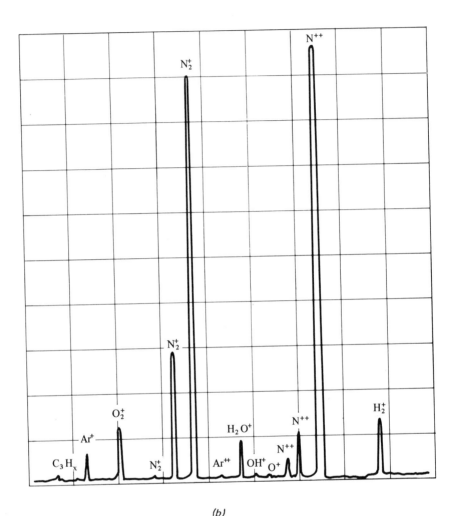

(b)

Figure 3-11 continued

The mass spectrometer is useful both as a qualitative and quantitative tool. In much the same way as an IR spectrum the fragmentation pattern yields a "fingerprint" of the molecule and helps to identify it as well as yield information about its structure and bonding energies. Samples must be gases or materials which can be volatilized. Often, multicomponent mixtures can be analyzed readily.

Instrument background usually consists of water vapor, oxygen, nitrogen, carbon dioxide, and all common atmospheric constituents; and this tends to impair sensitivity for these materials. Samples high in oxygen ($> 10\%$) disrupt the detector by oxidizing it, and water vapor is preferentially adsorbed on the inside walls of the mass spectrometer, where it creates a long-lasting background pressure within the system.

**Application Example.** The gaseous mass spectrometer has been used to advantage in analyzing the character of the gas environment inside metal encapsulated transistors. Metal encapsulations have small volumes (often well below 1 cm$^3$) and thus have only a small amount of gas available for analysis. A device is sampled by placing it in an evacuated fixture with a puncturing needle, opening the device under vacuum, then letting the gas from the sealed unit expand into the sample chamber. At various times device difficulties have been correlated with the presence of excess hydrogen, hydrocarbons, and water vapor. There is sometimes indirect evidence of other contamination difficulties such as organic residues or insufficient rinsing of plated parts. Evidence of gaseous contaminants must be reviewed in conjunction with other data to determine if the gas atmosphere is itself contaminated, or has become polluted indirectly by other sources within the transistor.

### spark source mass spectrograph [15]

The spark source mass spectrograph is used for the analysis of solids for trace impurities, both within the bulk of the sample and on its surface. It is frequently used as a general survey tool for traces of impurities.

**Principle.** A sample, usually consisting of two pieces of the material to be analyzed, is placed in a sample chamber such that a small gap is left between the two pieces (Fig. 3-12a). A high vacuum is required within the entire instrument. A high voltage spark discharge is passed through the gap, eroding the sample and producing ions. The ions are accelerated through a slit and into the electric sector. Those with the proper energy exit from the electric sector through a slit. Those with too much or too little energy collide with the walls and are lost. The ions passing from the electric sector enter a magnetic field which bends them into circular trajectories with radii increasing with the mass/charge ratio of the species. The ions strike a photoplate in the vacuum system and leave a record of the atomic masses detected, from the lightest (lithium) to the heaviest (uranium). Figure 3-12b shows a typical photoplate. Alternatively, an electrometer can be used to monitor a single mass and thus provide a direct and instant electrical reading.

**Sensitivity and Applications.** The mass spectrograph is one of the most sensitive instruments available. Nominally, its sensitivity is about 1 ppb; in terms of weight of material, it is capable of detecting less than $10^{-11}$ gram of an element. Virtually all elements are detectable within a factor of 3 of the same sensitivity (Table 3-1). The spark source technique has been especially suited for checking high purity materials such as semiconductor silicon for very low level contaminants, for detecting surface contaminants on substrates as a result of processing, and for measuring impurities in thin films.

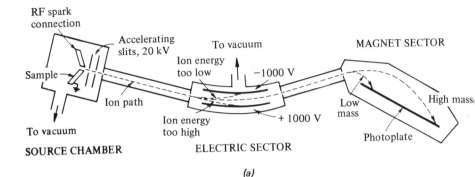

(a)

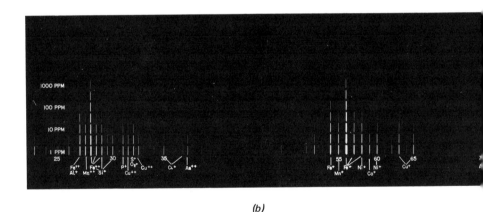

(b)

Figure 3-12 (a) Spark source mass spectrograph. RF spark at sample liberates ions which are drawn through slit system. Ions with proper energy pass through electric sector into magnetic field (perpendicular to plane of paper). Ions with different masses arrive at the photoplate, ranging from low to high as shown. A mass range of 7-250 amu is commonly used. (b) Photoplate section of series of exposures, mass 25-75, of a steel sample. Exposures increase from top to bottom. Bottom exposure has sensitivity of about 1 ppm. Note strongest line group centered on mass 56 (major iron isotope). Group is repeated by doubly charged ions at mass 28. Also note arsenic at mass 75 and doubly charged ions at mass 37.5.

**Limitations.** The spark source mass spectrograph cannot handle more than a few samples in a day if low detection limits are required. Materials must be conductive and in a self-supporting form, or handled in ways to make them so. For example, nonconductive powders must be compressed into rigid pellets with a conductive binder such as graphite or silver. Such treatment reduces sensitivity and introduces a high probability of contamination. Liquids cannot be handled, and moist or volatile materials will degrade the vacuum so that operation is impossible. Organic materials and those substances which evolve gases in the spark discharge also degrade the vacuum. Organic materials also produce a multiplicity of lines due to extremely complex molecule fragmentation that occurs in a spark discharge. The organic fragments are recorded on the photoplate and can create difficulties in interpretation.

### *gas chromatograph*[16]

The gas chromatograph combines separations through dynamic sorption techniques with detectors of physical characteristics to provide an extremely versatile analytical tool.

**Principle.** The basic principles apply to all types of column chromatography: gas-liquid chromatography, gas-solid chromatography, gel permeation chromatography, ion exchange chromatography, etc. Figure 3-13 shows a diagram of the instrument and a schematic of the chromatographic process. A sample mixture, in this case a binary mixture, is injected by some means into a column which consists of a tube filled with packing material (usually a solid support coated with a liquid). The column's solid support and liquid coating constitute the stationary phase while the carrier gas (a nonreactive gas such as helium, argon, or nitrogen) is the mobile phase. The mixture is carried down the column by the mobile phase; the least volatile substance, the substance having the greatest solubility in the liquid coating or the substance having the greatest heat of adsorption on solid surfaces, will pass through the loaded column more slowly and, ideally, will be separated from the other components. As a consequence, the components will emerge individually into suitable detectors.

The time required to elute a component at a given temperature is called the retention time, and the volume of carrier gas used is the retention volume. When these are measured relative to a standard material, the relative values obtained become independent of the equipment used. Such data are published,[17] and any laboratory can use the data to identify compounds. However, many compounds may have the same retention volume, in which case the identification of the compound with a nonselective detector would fail. A mass spectrometer is often used as a selective detector for this purpose.

Many nonselective detectors are used in gas chromatography. Each may be optimum for a particular case but may not be suitable for another. A thermal conductivity detector responds to all components and requires essen-

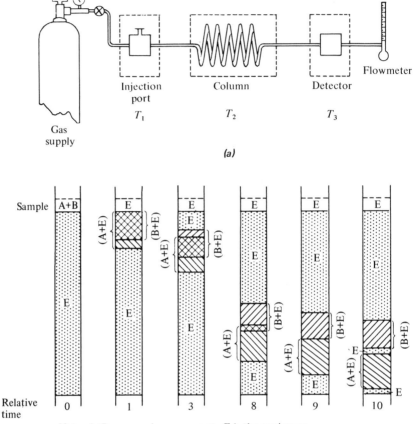

Note: A+B are sample components; E is the carrier gas.

*(b)*

*Figure 3-13 Gas chromatograph. (a) Equipment diagram; (b) process diagram. (From A. I. Keulemans: "Gas Chromatography," 2nd ed., Van Nostrand Reinhold, New York, 1959.)*

tially no maintenance. The sensitivity is adequate for many applications. A hydrogen flame ionization detector is more sensitive and has a modest selectivity for hydrocarbons and related compounds. Electron capture detectors are extremely sensitive to some compounds (e.g., sulfur hexafluoride) and are selective to compounds which capture low energy electrons. Other important detectors are the ultrasonic detector, photo-ionization detector, and the helium ionization detector. The latter is very sensitive and nonselective but has a limited linear range. In environmental characterization work all of these detectors find use.

**Sensitivity.** Discussions on sensitivity are difficult because it depends on resolution, the nature of the substance being analyzed, and the instrumentation. The detection limit in grams for some common atmospheric contam-

**TABLE 3-3.  Common Gas Chromatograph Detection Ranges for Selected Materials**

(Detection Limits[1] of Gas Chromatograph Detectors to Atmospheric Contaminants)

| Compound | Thermal Conductivity | Hydrogen Flame Ionization | Electron Capture | Flame Photometric |
|---|---|---|---|---|
| Phosphorus | $10^{-7}$ | ND[2] | ///// | $10^{-11}$ |
| SO$_2$ & Sulfur | | | $10^{-7(3)}$ | $10^{-11}$ |
| NO$_2$ | | | | ND |
| NO | | | | |
| Volatile Acids | | | | |
| Freons | | $10^{-7}$–$10^{-8}$ | $10^{-6}$–$10^{-14(4)}$ | |
| Chlorinated hydrocarbons | | $10^{-7}$–$10^{-8}$ | $10^{-8}$–$10^{13(4)}$ | |
| Xylene | | $10^{-11}$ | $10^{-5(5)}$ | |
| Toluene | | $10^{-11}$ | | |
| Hydrocarbons | | $10^{-11}$ | | |
| Acetone (Ketones) | | $10^{-10}$ | $10^{-6(5)}$ | |
| Esters | | $10^{-10}$ | | |
| Alcohols | | $10^{-10}$ | | |

1. Detection limits are given in grams.
2. ND—Not detected even as a major component of a sample.
3. The evidence for the sensitivity of the electron capture detector for inorganic gases is conflicting but is within an order of magnitude of $10^{-7}$ g.
4. The electron capture detector is extremely sensitive to compounds which capture electrons easily such as $CFCl_3$ and $CCl_4$, and less sensitive to many other chlorinated hydrocarbons.
5. The sensitivities for hydrocarbons and oxygen-containing materials varies little between different compounds.

inants are given in Table 3-3.  It is apparent that different detectors offer widely varying detection limits for the same compound, and a single detector may have widely varying limits for different compounds.  For example, the detection limit under ideal conditions is about $10^{-14}$ gram of $CFCl_3$ (a Freon) for the electron capture detector, while the same detector does not sense hydrocarbons.  For thermal conductivity detectors the detection limit is uniformly about $10^{-7}$ gram.

**Applications.**  In environmental control, gas chromatography has numerous applications.  Air samples containing hydrocarbons, chlorinated hydrocarbons, and other organics can be analyzed.  A chromatogram of a mixture of chlorinated hydrocarbons is shown in Fig. 3-14.  The sample contains nine identified components.  In this case the gas stream is split between the hydrogen flame ionization and thermal conductivity detectors.  The responses do not overlap since there is a 12 second offset between the two traces.

**Limitations.**  In gas chromatographic analysis the sample must be volatile at column temperatures operating without decomposition (the vapor pressure must be at least 1 mm at $400°C$) and generally must not react with the

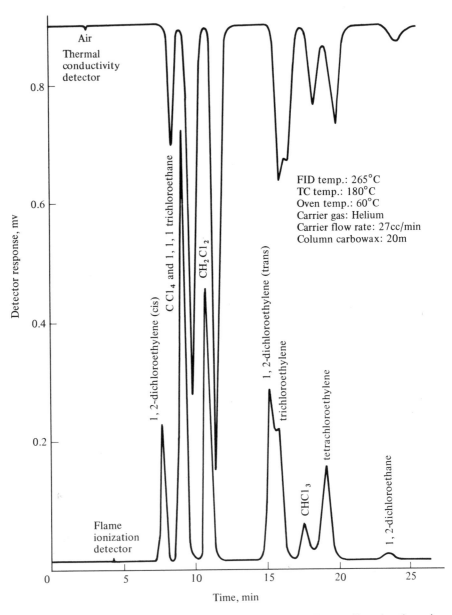

*Figure 3-14 Chromatogram of chlorinated hydrocarbon mixture. Note that thermal conductivity is offset from that of flame ionization because of detection sequence of the particular instrument.*

materials of construction. The inability to identify the component without excessive effort is another major limitation which is partially overcome by using a mass spectrometer directly combined to a gas chromatograph.

### combination of analytical instruments

Diagnostic chemical analysis of contamination problems in microelectronics often requires more than one analytical technique in order to obtain two different types of answers about the same sample.

**The GC-MS.**  In one notable case, the gas chromatograph and mass spectrometer have been combined into one instrument, with the mass spectrometer acting as a detector for the gas chromatograph.  The combination provides a far more powerful means of separating and identifying organic contaminants than the two units could separately, even though they might be adjacent to each other in a laboratory.  In the combination instrument a sample is injected into the entrance of the gas chromatograph, and each separated component yields a mass spectrum as it leaves the GC column and passes through the mass spectrometer.

There are two fundamental problems to this procedure.  Fast scan instruments are necessary because the entire mass spectrum scan must be completed before a second peak enters the ion source.  Sample dilution, however, is a greater problem.  For an average sample on packed columns, each component is diluted with carrier gas as it is carried along the column by as much as two orders of magnitude.  Many instruments either must split the stream, taking only a small sample, or use sample enrichment systems.  Actually, sample enrichment up to 50 times is desirable, if not necessary, to gain a corresponding enhancement of detection sensitivity.

### comparison of analytical instruments

Tables 3-4 and 3-5 are included as a summary of diagnostic instrument capabilities, limitations, and approximate cost.

## sample collection and preparation

Sample collection and preparation are critical stages in diagnostic chemical analysis.  Care must be taken to ensure that the sample is not contaminated during its collection.  A common practice is the use of dummy samples.  Two identical samples are selected: one is reserved as a control, and the other one is treated as required.

On many occasions one suspects that impurities may be below detection limits for available instruments when straight-forward techniques are used.  Several possibilities exist for improving sensitivity by preconcentrating the impurities of interest.  These include solvent leaching or extracting a particular group of compounds, filtering off precipitated materials, evaporating solutions to lower volumes or dryness, removing impurities on a suitable adsorptive type of filter such as an ion exchange column or activated charcoal, and fractional distillation.

Collecting samples can introduce contaminants and make results suspect.  A common problem is handling samples with fingers.  Frequently, the first

**TABLE 3-4.  Instrumental Principles**

| Instrument | Measured Characteristic | Excitation Source | Dispersion Means | Detection Method | Cause of Energy Interaction |
|---|---|---|---|---|---|
| Emission Spectrograph | Light at discrete wavelengths, emitted by excited atoms UV-visible range | Thermal energy of d-c arc or a-c spark | Diffraction grating (ruled lines on glass, commonly 500–1000 lines/mm | Photographic film or photomultiplier | Transition atom valence electrons between energy levels within outer electron shell. |
| X-Ray Fluorescence Spectrometer | X-rays at discrete wavelengths, emitted by excited atoms in sample 0.5–10Å | Primary beam of high voltage X-ray tube | Single crystal (lattice spacings of crystal act as grating) | Geiger counters, scintillation counters, gas-filled counters | Transition of atomic inner electrons from one shell to another. |
| Atomic Absorption Spectrophotometer | Absorption of light by sample at discrete wavelengths produced | Sample aspirated into flame in optical path | Diffraction grating commonly | Photomultiplier | Transition of atom valence electrons from ground state to an energy level. |
| Infra-red Absorption Spectrophotometer | Absorption of light by sample in relatively broad bands in 2–50 $\mu$m range | Radiation from hot solid being passed through sample | Diffraction grating or prism | Thermopile, Golay cell | Match of rotational & vibrational resonance frequency and radiation frequency. |

| Instrument | | | | | |
|---|---|---|---|---|---|
| UV-visible Absorption Spectrophotometer | Absorption of light be sample in relatively broad bands in 1850–8000Å range | Radiation from lamp being passed through sample | Diffraction grating or prism | Photomultiplier | Transitions of electron between electron energy levels of molecules. |
| X-ray Diffraction Spectrometer | Monochromatic X-rays diffracted by structure of sample | X-rays filtered to give one discrete wavelength | Crystal lattice spacings of sample acting as diffraction grating | Photographic film or X-ray scintillator detector tubes | Diffraction effect caused by sample crystal structure |
| Spark Source Mass Spectrograph | Mass/charge ratio of atomic ions | RF spark vaporization and ionization of sample, plus acceleration of ions | Magnetic field preceded by electric energy sorter | Photoplates or electrometer | Magnetic field bends ion trajectories according to mass/charge ratio |
| Gaseous Mass Spectrometer | Mass/charge ratio of ionized molecules and molecular fragments | Electron beam bombarding gas or vaporized sample | Magnetic field | Electrometer | Same |
| Gas Chromatograph | Time required for compound to be transported through column | Column temperature and helium flow through column and thermal vaporization of sample | Different rates of transporting different compounds through column. | Thermocouple, flame ionization detectors & others | Coating of inside of column retards various compounds selectively according to compound/coating chemical affinity. |

**TABLE 3-5.   Instrumental Use and Costs**

| Instrument | Approximate Equipment Cost Range (in $1000 units) | Information Obtained | Sensitivity | Sample Size & Type |
|---|---|---|---|---|
| Optical Microscopy | 10 | Surface, cross-section structure, film thickness measurements, contaminant identity. | Resolution to 3000Å | Highly variable. Usually solids, powders, films, cross-sections, etc. |
| Infra-red Absorption Spectrophotometer | 20 | Identity & concentration of compounds, usually organics | 1000 ppm, some surface films to 100Å. | 0.1–1 ml liquid, 10 mg solid. Liquid, solid, gas, surface film. |
| Ultraviolet-Visible Absorption Spectrophotometer | 20 | Concentration of compounds. | 1 ppm or $5 \times 10^{-10}$ gram | 2–25 ml liquid and gas. Usually liquids and gases, some solids. |
| Gas Chromatograph | 10 | Separation, concentration but not identification. | $10^{-7}$ to $10^{-14}$ gram | 1–5 mg, and smaller with loss of sensitivity. Volatile liquids, gases. |
| Gaseous Mass Spectrograph | 50 | Identity & concentration | To 1 ppm. | Volume as required 0.1–10 ml Gases only. |
| Differential Thermal Analyzer | 10 | Detection of phase and crystalline structure changes with respect to temperature. | Not applicable | 1–10 mg solids |
| Emission Spectrograph | 30 | Identity & concentration of metallic elements. | 0.1–10 ppm or 1.0 ng; varies greatly for different elements. | Usually 10 mg; may be smaller. Any solid, also dried residue from liquid. |

| Method | | Information obtained | Sensitivity | Sample requirements |
|---|---|---|---|---|
| X-Ray Fluorescence Spectrometer | 50 | Identity & concentration of elements, both metals and nonmetals. | 10–100 ppm or 10 µg. Light elements have poor sensitivity | Highly variable. Any solid, also some liquids. |
| Atomic Absorption Spectrophotometer | 10 | Identity & concentration of metallic elements. | 0.05–10 ppm in at least 5 ml of solution. Varies greatly with element. | More than 2 ml of solution required. Liquids. Solids must be dissolved. |
| Spark Source Mass Spectrograph | 100 | Identity & concentration of metallic elements. | 1–10 ppb. More sensitivity to low atomic wt. elements | Conductive solids. Destructive, very slow, poor precision |
| Electron Microprobe Analyzer | 100 | Identity, distribution and concentration of elements. | 100 ppm or $10^{-15}$ gram in a few cubic microns. Light elements poor. | Up to 20 mm. Solid, after complete device or metallographic section. |
| Neutron Activation Analysis | 500 | Identity & concentration of elements. | To 1 ppb, or $10^{-10}$ gram. Highly variable | About 1 gram. Also smaller, but with less sensitivity. Solid or liquid. |
| X-Ray Diffraction Spectrometer | 20 | Crystal & compound form of elements detected by other means. | Highly variable with application. | From ½ mm up or 1000Å film up. Crystalline solid. Not applicable to liquids, amorphous solids. |
| Electron Diffraction Camera | 20 | Same as X-ray Diffraction | Penetrates very thin 100Å surface films. More sensitive than X-Ray diffraction | Same as X-Ray diffraction. |
| Electron Microscopy Transmission | 50 | Surface texture, morphology structure with extreme detail. | Does not apply. Resolution to 5Å | From 5 mm down. Solids, powders, replicas of surfaces. |

evidence of a problem is submitted by individuals not familiar with the sensitivity capabilities of modern laboratory instruments, and "people dirt" can be a real problem when even moderately low levels of K, Ca, Na, Cl, and S and organics are suspected. Investigations have indicated that these elements are also present in normal air. They have been detected as a result of leaving samples exposed, rather than carefully stored, even for short periods of time.

Storing samples can introduce contaminants through interaction between container and sample. Liquids stored in glass can leach out significant amounts of sodium and possibly other materials. Some elements present at low levels in aqueous solutions have been known to be adsorbed on the walls of containers, so that the concentration left in solution is significantly lowered after several days. Some types of plastic containers have plasticizers which will contaminate stored liquids, and plastic tubing can also introduce plasticizer into liquids flowing through them.

Containers and sample handling materials should be selected according to the type of sample, problem, and instruments to be used. If both organic and inorganic materials are to be checked, it may be necessary to use two types of samples and handling systems. Emission and X-ray fluorescence spectroscopy are not sensitive to organic materials, while infrared spectrometry and gas chromatography are generally insensitive to inorganic impurities.

Frequently, analytical methods are carried out using a substrate as a carrier or collector for a sample. Common substrates are glass slides, small sheets of high purity metal such as aluminum, specially processed graphite rods and sheets, and in the microelectronics industry, silicon wafers.

Sample cleaning before analysis, while sometimes necessary, can be detrimental. When surface contaminants are sought, cleaning may remove ones which are significant. Cleaning has often been shown to leave residues, usually one of the major cleaning ingredients. For example, hydrofluoric acid leaves a trace of fluoride residue on a silicon surface during an etching process. Thorough rinsing will not remove the residue; it must be removed by a high temperature treatment. In preparing samples for analysis it is necessary to consider not only what one needs to remove, but what else (impurities of interest, for example) may also be preferentially removed, and what may be left as an interfering residue.

Methods of holding samples during processing must also be considered. The holder should not chemically interfere with analysis. Several instances of this have occurred during the use of some types of plastics and adhesives. Zinc oxide is sometimes used as a filler, and its detection can and has caused false alarms in searching for contamination in samples held by adhesives of this type.

In brief, the main goal of obtaining samples is to collect material which accurately represents the contaminants present without adding contaminants to them in amounts sufficient to measure by the analytical method selected. The manner of obtaining samples depends on what impurities are to be

checked and on the instrument to be used. The considerations entering into selecting a method of evaluation can be summarized as follows:
1. Type of contaminant expected:
   a. Organic or inorganic;
   b. Surface or bulk;
   c. A few specific ones vs. a general survey.
2. Type of sample available:
   a. Liquid, gas, metal part, powder, residual film, etc.;
   b. Direct investigation of product or environmental media;
   c. Need for dummy samples;
   d. Need for impurity enrichment;
   e. Collection and storage limitations.
3. Type of sample preparation:
   a. Direct examination vs. extensive sample preparation;
   b. Need for sample cleaning;
   c. Mounting media for samples.
4. Sensitivity requirements vs. available analytical techniques.

## case histories

It has been implied that no one instrument is likely to give a complete diagnosis of a product or environmental contamination problem. The analyst must select the instruments with the appropriate impurity sensitivities for a particular sample. Just how this is done is best illustrated in a series of case histories.

### faulty thin film

Hazy sputtered deposits of tantalum thin films were being produced by an "in-line" sputtering system in which glass or ceramic substrates are fed continuously into a vacuum system, and tantalum metal is deposited on them. In this process cleaned substrates are placed in holders and sent through the sputtering operation without breaking the vacuum. A specially designed holder and rail assembly is used to send the substrates through the system. Since it operates under vacuum, no lubrication is permitted.

The formation of haze was noted only on the material produced in the "in-line" system and not on substrates coated in the slower batch-type equipment. The haze was also associated with poor capacitors which were made by anodizing the tantalum in a pattern and then coating it with another metal as the opposite capacitor plate. Other clues were noted: rubbing the hazy films left a gray deposit on a tissue while a clean film did not generate any visible dirt; rubbing hazy films left pinholes in the films, and rubbed hazy films made better capacitors than ones without rubbing.

Up to this point no analytical work had been needed. Examination by an

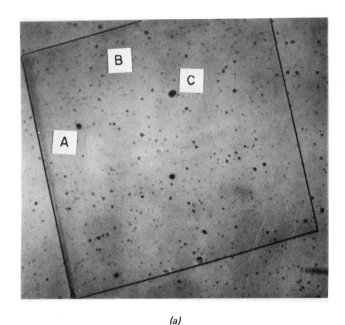

(a)

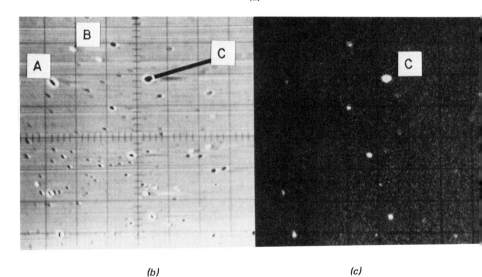

(b)                                        (c)

*Figure 3-15 Photomicrographs of particles imbedded in thin film. (a) Optical pre-sentation at 200X showing particles imbedded in thin tantalum film. Outlined area is region of electron microprobe examination; (b) EMA sample current image at 200X; (c) EMA x-ray image with bright spots showing location of iron-rich particles. Note that particle C is iron rich, which is not the case for particles A and B. However, A and B cannot be distinguished from iron particles in the optical or sample current images.*

optical microscope revealed the presence of many small particles apparently embedded in hazy films (Fig. 3-15). A first approach was to compare the compositions of hazy and clear films. The spark source mass spectrograph was selected as the method of analysis since it could give the answer with the limited amounts of thin film samples. Rough comparisons showed that iron was one of the principal differences. Electron microprobe examination correlated the presence of iron with the particles. A second approach was tried to see if the presence of iron could be associated with any process step. Pure graphite points were rubbed across the samples for each major step from purchased raw materials to the finished thin film deposit. Rubbings were used to duplicate the cloth rubbings mentioned earlier. This showed that the "in-line" sputtering system films yielded iron with other impurities, suggesting a tool steel. About this same time the in-line system was opened for repairs. Black deposits were found in most of the chambers, but mainly in the sputtering chamber. They were largely iron as well as magnetic. Since the chambers were made of nonmagnetic stainless steel this source was eliminated from consideration. It was suggested that the iron particles came from galling action between the boats and the track. Such action was found to have occurred. Buffing of the films and frequent cleaning of the chambers of the in-line machine were changes which minimized the problem, while designs for subsequent machines have taken the galling problem into account.

In this case the defect cause was visible as unwanted particles, which quickly alerted production personnel to the existence of a problem. Data other than analytical results aided in interpreting the spectrographic information. Information from other than analytical sources was used to hypothesize on the mechanisms accounting for the problem. The combination of process and analytical data can lead to valid corrective action and a better understanding of the product or process.

### faulty epitaxial silicon

Not all contaminants are visible, and they may not produce detectable defects until near the end of a manufacturing process. Another incident began with finding one contaminant introduced accidentally at an early process step, and since it did not cause difficulties for some time, it went unnoticed until a great deal of product was manufactured. In the process of tracking down this one contaminant other contaminants and sources previously unsuspected were found.

Electrical difficulties were noted in some integrated circuits at an intermediate process step. Analysis of the surfaces of the defective devices showed that the silicon of the defective ones was contaminated with germanium at levels from 150-50,000 ppm. This naturally led to a further search for the source of the germanium, and many analyses were performed by spark source mass spectroscopy, emission spectroscopy, and electron microprobe analyzer in locating the source of germanium.

The source of germanium was found in the epitaxial process. Two of the reactors used for epitaxial deposition contained silicon tetrachloride contaminated with germanium, and these were generating germanium-bearing epitaxial silicon deposits. A further search was made to determine how germanium could have been introduced. The purchased silicon tetrachloride was analyzed by emission and spark source spectroscopy and found to be free of germanium. Thus, the contamination of the reactors was probably caused by an inadvertent operating error in handling the silicon tetrachloride. Weekly checks for the presence of germanium were made for a six month period to ensure that the error was not repeated.

While this surveillance precluded any further errors in chemical handling, a germanium contamination case from another source was discovered during this check. Germanium was found on both sides of one of the weekly samples. Wafers from other lots processed before and after the initial sample showed sporadic occurrences, and electron microprobe pictures showed that germanium was scattered in streaks, and did not show up on every batch of wafers.

It was found that some silicon ingots and germanium bars had been sliced into wafers on the same saws. One would have thought that further processing, which involves removal of 20–25 microns of material from each side, would have taken off any contaminant transferred from the saw to the wafer. Apparently, this is not always the case. Evidence from many other sources suggests that damage is caused to silicon wafers by cutting, polishing, and even holding with tweezers, producing dislocations. These dislocations become the areas where rapid diffusion occurs, and it is speculated that migration of germanium and possibly smaller amounts of other materials may have occurred as a result of mechanical surface damage. The solution to this germanium problem was direct: stop cutting both materials on the same saws.

### sample handling problem

The analysis of the purchased silicon tetrachloride just mentioned will illustrate how spurious results can be obtained in the handling of the sample. Sample preparation dictated that the silicon tetrachloride be combined with water to form silica powder and hydrochloric acid. The reaction was carried out in a platinum dish, and the dried silica powder was analyzed. By emission spectroscopy a trace of platinum was found from the dish, but spark source mass spectroscopy tests found platinum, iridium alloyed with the platinum, and some nickel from the preceding sample run in the dish. By cross-checking with other instrumentation the platinum, iridium, and nickel were identified as sample preparation additions rather than as impurities in the silicon tetrachloride.

## new frontiers

Both microelectronics and instrumentation used in diagnostic chemistry are dynamic fields. The ever more critical needs of microelectronics has placed great pressures on instrument developers to search for methods which provide better sensitivity, spatial resolution, surface discrimination, and evaluation of chemical bonds. All of these are needed along with further improvements in operating convenience, reliability and speed, accuracy, and freedom from interferences due to similar responses from different materials. While tremendous advances have been made, the tools of analytical chemistry are not fully adequate for the more critical problems facing microelectronics. This situation is not likely to be fully rectified.

Recent technological advances, such as generating images for forming integrated circuits by using finely focused electron beams, have produced pattern tolerances smaller than the best resolution of optical microscopes and electron microprobes. The need still exists to determine chemical composition to spatial resolutions of the order of the device pattern resolution. Thus, some means of analysis at ever smaller sizes is still desirable. As understanding of impurities develops, it becomes increasingly important to measure lower levels of impurities. Surface chemistry is recognized as playing an increasingly important role in determining characteristics of newer devices. The demands of microelectronics are placing emphasis on greater purity, spatial resolution, and surface chemistry.

New instruments are being developed to meet needs of these types. Ion microprobe analyzers are able to achieve part per billion sensitivities in areas of several hundred microns, which are the sizes of conventional patterns of integrated circuits. Low energy electron diffraction is used to study changes at the surface of a sample, including those of chemical bonding at the surface. Auger electron spectroscopy has offered promise of measuring surface impurities and studying chemical bonds at material surfaces. Each of these new instruments offers an extension of technologies to provide a better understanding of microelectronic technologies.

Improvements from other technologies will undoubtedly find applications in analytical instrumentation. The laser, for example, has already found some analytical applications in emission spectroscopy and in Raman spectroscopy. Similar developments in the areas of electronic detection, low energy detection for infrared, and high vacuum improvements will help to extend analytical instrument technologies.

All of these instruments still require human judgment to obtain the most value from them. It is becoming increasingly important for the analyst to be able to understand both the problem giving rise to the samples and what interpretation should be placed on the results. In this manner the greatest benefit can be derived from the instruments, and experiments can be designed to make best use of characteristics of both the instrument and the product or process.

## references

1. J. E. Barry, H. M. Donega, and T. E. Burgess, "Flame Emission Analyses for Sodium in Silicon Oxide Films and on Silicon Surfaces," *J. Electrochem. Soc.*, Vol. 116, No. 2, February (1969).

2. H. H. Willard, *Instrumental Methods of Analysis*, 4th ed., Van Nostrand Reinhold, New York, 1965.
3. N. H. Nachtrieb, *Principles and Practice of Spectrochemical Analysis*, p. 265 ff, McGraw-Hill, N.Y., 1950.
4. G. L. Clark, *Encyclopedia of Spectroscopy*, p. 331ff, Reinhold, New York, 1960.
5. ASTM, "Index to the Literature on Spectrochemical Analysis 1920–1955," STP 41A-D, Issued 1959 ASTM, Philadelphia.
6. C. E. Harvey, *Spectrochemical Procedures*, p. 359, Applied Research Laboratories, Glendale, California, 1950.
7. John A. Dean and Theodore C. Rains, *Flame Emission and Atomic Absorption Spectroscopy*, Dekker, New York, 1969.
8. Eugene P. Bertin, *Principles and Practice of X-Ray Spectrochemical Analysis*, Plenum, New York, 1970.
9. A. D. Cross, *Introduction to Practical Infrared Spectroscopy*, Butterworths, London, 1960.
10. ASTM, Wyandotte Index STP 131. 1962, ASTM, Philadelphia.
11. ASTM, Wyandotte Index STP 131A, 1963, ASTM, Philadelphia.
12. M. B. B. Thomas, ed., "An Index of Published Infra-red Spectra," Vol. 1, 2, British Information Service, New York, 1960.
13. "Catalog of Infra-red Spectrograms," Sadtler Research Laboratories, Philadelphia.
14. F. A. White, *Mass Spectrometry in Science and Technology*, Wiley, New York, 1968.
15. A. J. Ahern, ed., *Mass Spectrometric Analysis of Solids*, Elsevier, Amsterdam, 1966.
16. S. Dal Nogare, R. S. Juvet, *Gas-Liquid Chromatography*, Interscience, New York, 1962.
17. ASTM "Gas Chromatographic Compilation" AMD 25A and 25A-Supp. 1, ASTM, Philadelphia, 1971.

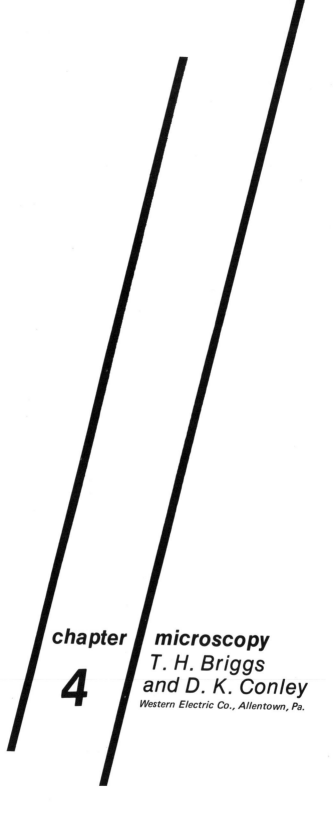

**chapter** **microscopy**
*T. H. Briggs*
**4** *and D. K. Conley*
Western Electric Co., Allentown, Pa.

This chapter treats several types of microscopy and their application in identifying contaminants on microelectronic products. The ability to visualize what is on a surface at micron size levels makes microscopy a powerful diagnostic technique for understanding the interrelationships of contaminants and device structure. Microscopic instruments are treated in two groups: those producing an image simultaneously over the entire focal plane, and those using a point focus which is moved over a defined area to produce an image. The optical light microscope and the transmission electron microscope comprise the simultaneous imaging group and will be discussed first.

## optical light microscope

### optics of lighting

Optical microscopes are rapid and comprehensive tools for directly observing defects and foreign particles. Particles can often be identified directly by microscopic observation, and possible cause and effect relationships can be determined between particles and defects. Reflected and transmitted lighting systems are both used (Fig. 4-1). A microscope has a series of lenses: condenser, objective, and eyepiece. In reflected light the objective lens can serve as both condenser and objective. Several types of illumination available for either reflected or transmitted light are: bright field, dark field, polarized light, phase contrast, and interference contrast.

**Bright Field.** Bright field illumination is the type of lighting shown in Fig. 4-1. Light is focused by the condenser (or objective in reflected light), and if it is unmodified by the specimen, it will produce a bright featureless image. It is the brightest, most generally used, and simplest to interpret of all the lighting techniques used. Most investigations begin with low-power examinations, and this involves lenses with low numerical aperture. Since these lenses have a relatively narrow cone of light acceptance, the image obtained is similar to one using a narrow spotlight beam. Surfaces must be nearly normal to the optic axis to return light and appear bright. High power objectives, on the other hand, have a large cone of acceptance. One consequence of this is that a purely topographical feature, such as a dimple or a hillock in or on a reflecting surface, can appear dark in low power. This can lead to a false conclusion that a stain is present.

**Dark Field Lighting.** In dark field illumination, light is directed to the specimen at angles outside of the lens acceptance cone (Fig. 4-2). As a result, light reflected from a flat surface normal to the optical axis will appear totally dark, since light coming from an angle outside of the acceptance cone will not be reflected into it.

Surface texture and particles will, however, reflect or scatter light into the cone, and will appear bright against a dark background. This method of illumination is a powerful tool in searching for fine particles and irregularities

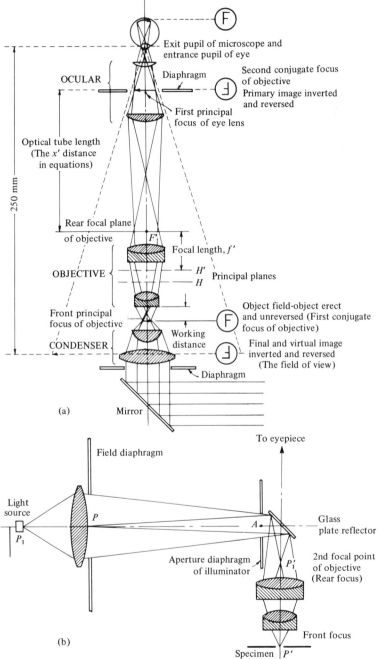

Figure 4-1 (a) Diagram of light path through microscope using transmitted light; (b) diagram showing modification of microscope objective for both condenser and magnifying lens. Positions of field and aperture diaphragms are shown. Ocular portion as shown in (a) is unchanged. (Courtesy C. P. Shillaber, Photomicrography in Theory and Practice, John Wiley & Sons, 1944.)

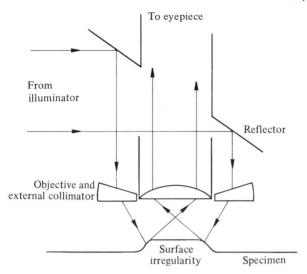

*Figure 4-2 Darkfield illumination. Light is brought in from outside the cone of acceptance of the objective and reflected by irregularities into the objective.*

which need to be emphasized. Since particles and textures are revealed not only by reflection but also by scattered light, particles and surface irregularities smaller than the resolution limit of an objective can be detected. This is an important distinction in capability between bright and dark field lighting. Surface irregularities, such as a scratch or edge of a film, less than 1000Å deep and particles of about 1000Å size can be detected in this manner as compared to approximately 3000Å as the lower limit of visibility in bright field.

**Polarized Light.** Polarized light interacts with materials to yield valuable information about crystalline form, bulk color of material, and the appearance of a substance when surface reflections are eliminated. Ordinary light is considered to be composed of waves with randomly oriented planes of vibration. After such a light beam passes through a polarizing system, the emerging light waves vibrate in only one plane.

One material property which is readily revealed in polarized light is refractive index as related to orientation of a transparent crystalline material. In a single crystal molecules are oriented periodically in the same direction. Light is retarded on passing through a substance by the amount of interaction it has with the electronic bonding field density along its direction of passage; the stronger the fields, the greater is the retardation of light, or the less the velocity of light within the substance. A detailed discussion of this is given in Jenkins and White.[1] A material is said to be "optically active" if the degree of retardation is different with changing direction of light passage through the material. Refractive index is also a measure of the change of light velocity passing from one substance to another.

By placing polarizing filters before and after the specimen with polarizing directions crossed at a right angle, light which is undisturbed by the specimen is blocked, causing these regions to appear dark. Light which interacts with the specimen will have its plane of polarization rotated and appear bright against a dark background. When an optically active specimen is rotated 360° between crossed polarizing filters, the specimen will pass alternately from dark to light twice.

Light can also be polarized on reflection from nonmetallic surfaces. This permits "glare," or specular reflection, to be eliminated and the color of a surface to be revealed. This can be important in determining whether a surface is that of a metallic material or a nonmetal.

**Phase Contrast.** Phase contrast is a modified form of bright field optics in which small height differences in a specimen surface produce a phase shift between the light reflected from different parts of the specimen. In this modification, an annular diaphragm is positioned in the front focus of the

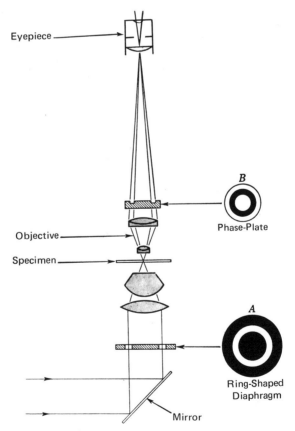

*Figure 4-3 Phase contrast system for optical microscope using transmitted light. Phase plate B blocks unretarded light, permitting light with phase retardation. (Courtesy D. F. Lawson, The Technique of Photomicrography, Newner Technical Books, London, 1960.)*

condenser, and a phase plate (negative of the annular diaphragm) is positioned at the rear focus of the objective (Fig. 4-3). The annular diaphragm is focused on the phase plate. Light which is unchanged in phase will be partially blocked, while phase changes will cause constructive or destructive interference, depending on the direction of wave front shift.

Phase contrast is useful for detecting very thin surface layers and shallow surface anomalies down to about 200Å. Phase contrast has an upper limit of discrimination of about 2500Å, or ½ wavelength for visible light. Larger height differences again lose phase contrast, but become reflections discernible through more common lighting techniques.

**Interference Contrast.** Interference contrast optics (the Nomarski system[2]) employs an image splitter and polarized light. The image is split, and each half is polarized oppositely (Fig. 4-4). A given difference in path length be-

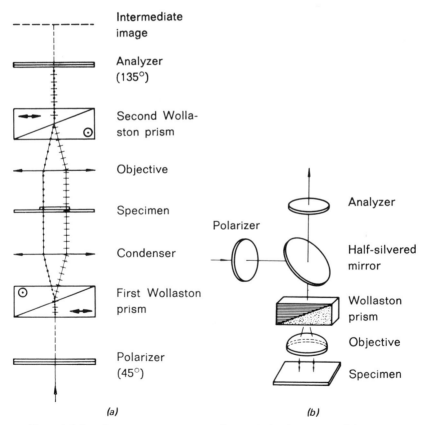

Intermediate image

Analyzer (135°)

Second Wollaston prism

Objective

Specimen

Condenser

First Wollaston prism

Polarizer (45°)

(a)

Polarizer

Analyzer

Half-silvered mirror

Wollaston prism

Objective

Specimen

(b)

Figure 4-4 Interference contrast system for optical microscope. (a) Transmitted light; (b) the same system modified for incident lighting. The interference arises from the path difference introduced by the specimen and sensed as an interference color in the recombined beams. (Courtesy Lang and Walter, Zeiss Information, Issue 70, 15 July 1970.)

tween corresponding parts of the polarized images creates a degree of elliptical polarization, which shows up as color imposed on the magnified view of the specimen. The color changes according to the amount of difference. This technique is the most sensitive way of revealing surface topography for the types of specimens common to microelectronics and environmental control. Interference contrast is used to advantage to determine whether a detail is raised or depressed, and to reveal topographical detail not visible in other methods of lighting.

### optics of materials

In using microscopy for environmental control studies, one is frequently faced with using the microscope for characterizing a stain, residue, particle, or some other form of contaminant. Optical effects in materials are of prime importance in determing the nature of what is being observed. Several effects are particularly important and are discussed below.

**Refractive Index.**[3] A basic technique for identifying light transmitting particles, the refractive index is an easily measured property, and it can be used to compare small particles from known and unknown sources. Refractive index is commonly measured by using immersion oils, i.e., finding an oil of known refractive index which matches that of the sample using transmitted light. Immersion oil sets are available from $n = 1.300$ up to $n = 2.11$ in steps of .002 for most of the range. Measurements to ±.001 can be readily made.

**Color.** Color is an important and useful characteristic for distinguishing various features on a sample. In addition to being an inherent bulk property, color is often produced in thin layers of transparent films as the result of destructive interference of a light beam reflected from the upper and lower surfaces of a film. When the thickness of the film is $N\lambda/2$, with $\lambda$ being the light wavelength, the reflections reinforce each other, while at $(N + \frac{1}{2})\lambda/2$, they are out of phase and cancel each other. These colors change considerably for small thicknesses, and also for small variations in thickness of films under two microns, with the change becoming less evident with increasing thickness. Films over five microns have very weak interference colors. If a film of 5000 Å is viewed, thickness changes of only 100 Å can be detected. Thus, even microscopic spots of very thin contamination can be detected by a color change at this level.

### optics and microscope components

A microscope suitable for metallurgical and general purpose examinations is shown in Fig. 4-5. An instrument of this type illuminates the specimen by using light sources and separate collimating and/or condenser systems, one for transmitted light, the other for reflected light. The specimen is observed through magnifying objectives, which are usually mounted on a rotating

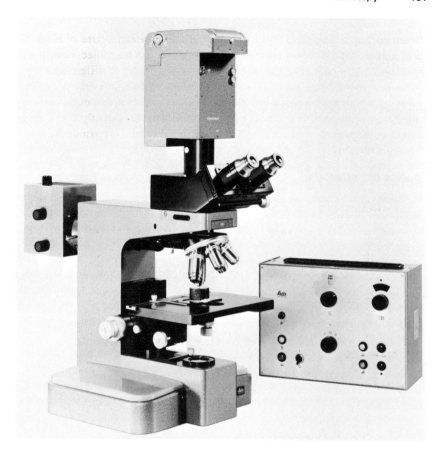

*Figure 4-5 Optical light microscope. (Courtesy E. Leitz, Inc.)*

turret for convenient changing, and magnifying oculars, or eyepieces. Frequently, some form of camera is included for recording images.

Condenser systems for light from the illumination source are designed to
make the light converge into an evenly illuminated, concentrated spot on the
specimen (Fig. 4-1). The vertical illuminator portion of the optical system
produces collimated light and brings it to a focus at the rear focal plane of
the objective. The objective then concentrates the light onto the specimen.
In both condenser systems there are two apertures: the diaphragm, which is
adjusted to control light intensity and resolution; and a field aperture which
is adjusted to just fill the field of view to reduce stray light and improve
image contrast.

**Magnification.** Objectives produce a magnified image of a field of view of
specimen. This image in turn is magnified by an ocular, or eye piece. The
overall magnification of the image is the product of the magnifying power of
the combination: a 20X objective and a 10X ocular will produce a 200X final

image at a given distance from objective to ocular. Objectives are rated by both power and numerical aperture.

**Numerical Aperture.** This optical characteristic determines the theoretical resolution of the objective, its depth of focus, and to some extent the appearance of surface relief features. Numerical aperture (N.A.) is defined as N.A. = $n \sin \theta$, where $\theta$ is the angle between the focused cone of light and its optical axis and $n$ is the refractive index. An aperture in the vertical illuminator is focused at the rear focus of the objective, and this opening determines the actual numerical aperture used up to the limit of the lens. Resolution of a lens as a function of numerical aperture is expressed by the equation $R = \lambda/2$ (N.A.), where $R$ is resolution, $\lambda$ is wavelength of the light used, and N.A. is numerical aperture. Depth of focus is a function of numerical aperture (Fig. 4-6). It can be seen that depth of focus becomes extremely critical at high magnifications. This is a serious limitation of the optical microscope—one which is inherent in its optical theory. Scanning electron microscopy by its superior depth of focus offers one way to circumvent this problem. Oil immersion objectives have been used to improve resolution by increasing the numerical aperture of the lens. This is shown in Fig. 4-6 as a discontinuity at N.A. > 1.00 because of a somewhat improved depth of focus.

By introducing oil ($n = 1.5$) or other suitable liquids between the specimen and an appropriately designed objective, greater resolution can be obtained than is possible in air. Such a technique is inconvenient to use, and the immersion fluid can contaminate the specimen, preventing further analyses.

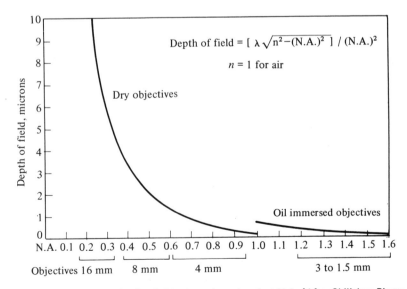

*Figure 4-6 The field depth of objectives plotted against N.A. (After Shillaber,* Photomicrography in Theory and Practice, *John Wiley & Sons, 1944.)*

The resolving power of an objective can be considered qualitatively in terms of the diffraction of light by an object. Light falling on an object is diffracted by that object in such a way that a series of alternately light and

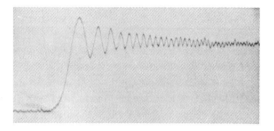

(a)

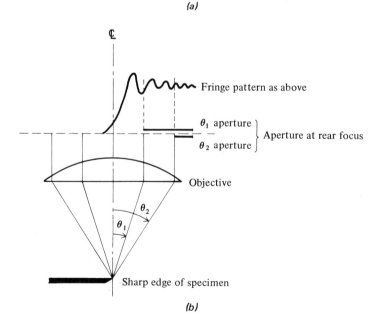

(b)

Figure 4-7 (a) Diffraction of light around a straight edge. Each fringe, when combined to form an image, helps to determine the sharpness of that image. (b) Objective forms diffraction pattern at rear focus. Aperture governs number of fringes used, and thus image sharpens. (Part (a) courtesy Jenkins and White, Fundamentals of Optics, McGraw-Hill, 1950.)

dark bands are produced. By recombining these bands an image of the original object is formed. The diffraction bands are equivalent to terms in an infinite converging series, with each term having progressively less influence on the total. Adding these terms is mathematically similar to recombining the diffraction bands with a lens. The sum defines the position of the object, both mathematically and physically. Mathematically, the more terms that are used, the more precisely the object's position will be defined. Similarly, in the physical model, when more diffraction bands are recombined, the image will be sharper, or have better resolution. An objective lens is used to recombine these diffractions bands, and the angle of acceptance for light (Fig. 4-7) determines the number of diffraction bands which that lens uses to produce an image. It is this angle which is described by numerical aperture.

## transmission electron microscope (TEM)

The optical microscope's limit of resolution has been a hindrance in many fields. In microelectronics it is becoming even more so. Smaller electron devices require knowledge of finer details, some of which require better resolution than the optical microscope can yield. The transmission electron microscope (TEM) was first used heavily in biological applications. Later its use was gradually developed in metallurgical and related fields.

There are two general characteristics which make the TEM useful: high magnification with good resolution, and the ability to do electron diffraction which permits studies of crystalline materials as well as identification of components. Both of these characteristics are important in environmental studies.

### fundamentals

The transmission electron microscope (Fig. 4-8) is capable of achieving magnifications from about 500–100,000X, with resolution limits commonly as low as 5 Å. In addition to good resolution, the microscope also has relatively great depth of focus. As in the optical microscope large depth of focus is dependent on a small numerical aperture, and resolution is inversely dependent on wavelength.

In the transmission electron microscope, a thin sample is placed in the specimen stage (Fig. 4-8), and a beam of electrons is concentrated on it after passing through a condenser lens. Some of the electrons pass through the sample. Others are absorbed, and some are diffracted out of the optical axis of the microscope. The ones which pass through the specimen form an image at the back focal plane of the objective. This image is magnified by an intermediate lens, and finally by a projection lens, either onto a fluorescent screen for immediate viewing or a photographic plate for permanent recording and best resolved detail. The entire image is formed simultaneously, as opposed to the scanning electron microscope where a fine electron beam is scanned across the surface of the sample to form an image by a raster technique. Image contrast in the transmission electron microscope is achieved by:

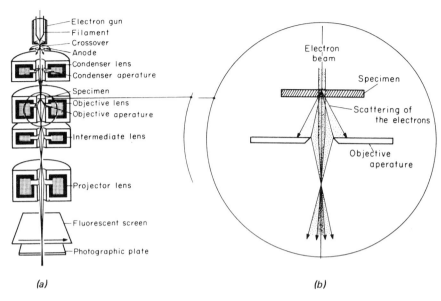

*(a)*                                    *(b)*

*Figure 4-8  (a) Diagram of transmission electron microscope; (b) detail of specimen area, showing scattered electrons.  Not shown are diffracted electrons, which can be used to form their own image by moving objective aperture sideways, blocking beam shown, and passing diffracted electrons. (Courtesy Carl Zeiss, Inc., New York, N.Y.)*

(1)  Absorption of electrons by sample thickness.
(2)  Scattering of electrons by the sample.
(3)  Diffraction out of the beam by crystalline structure.
(4)  Shadowing a replica of a sample with carbon or metal and using the contrast thus produced to reveal texture.

Electron absorption and scattering have been illustrated in Fig. 4-8b; the remaining methods to achieve image contrast deserve additional comment.

### electron diffraction

Electron diffraction occurs when a beam of electrons encounters a crystal oriented at an angle that satisfies the Bragg diffraction condition.[4] In such cases, the specimen produces a diffraction pattern at the back focus of the objective lens.  By proper adjustment of the intermediate and projection lenses the diffraction image is produced on the photoplate or screen.  These diffraction patterns are useful to identify materials, observe extra crystal phases, or establish the existence of chemical modifications. A typical pattern is shown in Fig. 4-9.

Images can be obtained from samples which are tipped at a grazing angle (less than $10°$) to the electron beam.  They are highly distorted, require considerable skill to interpret, and do not have as good resolution as normal TEM images.  Electron diffraction by reflection can also be done, and this is a valuable technique for identifying surface residues since the electrons are

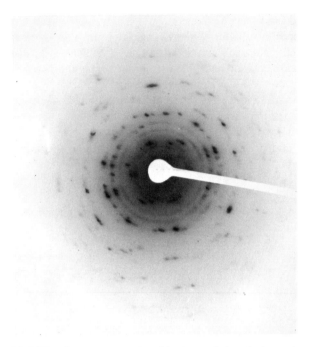

*Figure 4-9  Diffraction pattern produced by transmission electron microscope. Ring pattern produced by tantalum thin film, with spots due to discrete crystals of tantalum carbide (Ta$_2$C and TaC).*

diffracted by only the first 50-100 Å of surface material in this mode of operation.

## specimen thinning and replication

The extremely small penetration of the electron beam into a sample (less than 1000 Å to 5000 Å depending on the material and electron gun voltage) imposes several limitations on the type of sample which can be introduced into the TEM. Sample thickness influences image intensity and resolution; as the beam penetrates the specimen it loses energy by collision processes, becomes diffuse, and resolution is degraded. Also, the emergent electrons have a greater range of energies with increasing thickness; the magnetic lenses used for forming images affect each energy differently, thus degrading resolution further. Generally, resolution is no better than 1/10 thickness of the specimen.

The basic types of preparative techniques are specimen thinning and surface replication. Specimen thinning requires removing material from the sample to a thickness suitable for electron beam transmission. This is usually accomplished through etching a sample to the point of perforation, then examining the sample adjacent to the holes. Thin films are sometimes sufficiently thin to be used directly in the electron microscope, provided that they can be removed from their substrates.

Thinned specimens offer the advantage of being able to examine bulk structure directly for such characteristics as crystalline size, dislocations, crystal orientation, and chemical identity of inclusions and phases. Thinning techniques, however, are destructive and are not applicable to surface topography examination. Sample preparation can be difficult and depends on the availability of an appropriate etchant for a given sample material.

Replication is a somewhat simpler preparation technique in which a solution of a plastic in solvent, such as polyvinyl formal or collodion, is allowed to dry on a surface. The film formed in this manner is shadowed with a thin layer of metal such as platinum to improve contrast effects over what a film would give by itself.

Replication methods offer a way of examining a surface for fine detail, frequently without destroying the specimen for other analytical measurements. The technique is relatively rapid and simple, and it can be used on rough specimens. If particles of a contaminant are present on a surface they can be removed with the film. In these cases identification of the particles can be attempted by electron diffraction. Replication has the advantage of reproducing surface textures which must be examined critically. Depth of focus is superior to that of the optical microscope to such a degree that even moderately rough textures can be viewed without difficulty. Replica techniques limit resolution to about 50 Å, which is somewhat better than the scanning electron microscope.

## focused beam microscopes and analyzers

In the microscopes discussed previously the image has been formed simultaneously. Another principle distinguishes three other instruments—the scanning electron microscope (SEM), the electron microprobe analyzer (EMA), and the ion microanalyzer (IMA). In each of these instruments a finely focused beam of accelerated particles, electrons in the first two cases and gas ions in the last, bombards the specimen. The focused beam is scanned across the specimen in the same way that a television tube has a raster scan. The displayed image from these instruments appears on a television tube whose raster is synchronized with the instrument beam scan. At any given instant only one small spot within the examined area is emitting a signal.

The electron microprobe analyzer (EMA) was the first of these instruments to become commercially available. This was followed by the separate development of the scanning electron microscope. As these two instruments have been refined and new capabilities incorporated into them, it has become increasingly evident that they are overlapping. The EMA was at first designed to produce analytical data on a microscopic basis. The SEM was primarily a magnifying instrument. Both use focused electron beams. By reducing the electron beam diameter of the EMA, and by adding X-ray spectrometers to the SEM, the two instruments have overlapped more and more in capabilities. It is anticipated that in the future single instruments will be produced with the full capabilities of both present instruments. For the purpose of this discussion, however, each will be treated separately.

### scanning electron microscope (SEM)

The SEM is a relatively new member of the microscope family. It is capable of yielding a magnified image from 20X up to 100,000X at present, with resolution as fine as 100 Å. In addition to substantial magnification and resolution advantages over the optical microscope the SEM also has 100–300 times better depth of focus than the light microscope. This coupled with convenient operation has made the SEM a valuable addition to many laboratories.

Features which are readily apparent in the SEM include surface texture and topography, shallow subsurface irregularities, chemical heterogenieties, and regions which vary in conductivity. The SEM can be used in different modes which emphasize one characteristic or another as desired.

**Fundamentals.** The SEM consists of an electron gun source, magnetic focusing lenses, a sample electron detector, and optional X-ray spectrometers for elemental analysis (Fig. 4-10). The entire system except for the X-ray spectrometers is evacuated to better than $10^{-4}$ torr. Since the X-ray spectrometers are similar in operation to those of the electron microprobe, discussion of them will be reserved for that section.

The electron beam is formed by accelerating electrons to 2–30 KV, and focusing the beam to a spot size of 100 Å onto the specimen using a series of magnetic lenses and limiting apertures. The beam is scanned over an area controlled by deflection coils or plates.

The electrons interact with the specimen forming: (1) back-scattered electrons which are scattered from the surface with almost as much energy as the incident beam; (2) secondary electrons emitted from the specimen surface; and (3) conduction or absorbed electrons which enter the bulk of the specimen and are removed by an electrical connection to ground. These interactions permit images to be obtained from the SEM in several manners. The selected detection mode influences the depth from which meaningful signals can be obtained. These will be discussed using terminology proposed by P. R. Thornton.[5]

**Emissive Mode.** Back-scattered electrons of higher energy will yield information about surface topography and some limited data about average atomic number, since scattering efficiency improves with atomic number. Secondary electrons (lower energy electrons emitted by the sample) are less affected by surface topography. These two modes are indicated by the two shaded portions in Fig. 4-10c. By altering the orientation of the specimen with respect to incident electron beam and the detector, images can be interpreted by

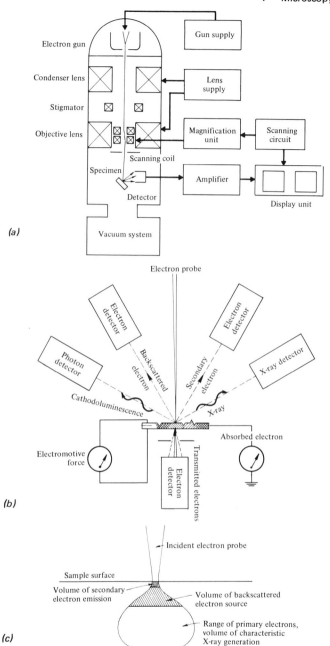

Figure 4-10 (a) Diagram of scanning electron microscope. Primary use of SEM is to obtain information by means of electron-sample interactions. (b) Different modes of signal deflection and display available to scanning electron microscope users. Special properties of each mode determine its usefulness. (c) Relative range of effectiveness of electrons for different operating modes. Deepest penetration is 0.5–1.5 microns, depending on elements present. (Courtesy JEOL News, Japan Electron Optics Laboratory, Medford, Massachusetts, 1969.)

considering the beam and detector orientations as though they were light source and observer directions with respect to an object.

**Conductive Mode.** Electrons which are absorbed by the sample will upset the electrical charge balance of the specimen. If the specimen is even slightly conductive the excess electrons can be bled off through a ground connection. If a specimen has regions which vary in their conductivity they can be revealed by using the current passing to the ground connection to produce the image, as shown in Fig. 4-10b. Highly insulating samples such as glass, ceramic, and biological specimens will build up a surface charge under electron beam bombardment, rather than permit it to be dissipated. If a thin film of carbon or a metallic conductor such as gold is deposited and grounded, the sample will lose its charge through the film, and a satisfactory image can be obtained.

Another effect in the conductive mode occurs specifically for semiconductors. Electrons bombarding the semiconductor create hole-electron pairs which can be bled off and used for forming the image. Such a condition is useful in searching for channels (or reverse current leakages) in transistors and integrated circuits. These channels are often associated with electrical defects created by alkali metal ions. A semiconductor device must be electrically biased to show these regions, either by SEM electron beam bombardment or external power connection. By varying the biasing voltage these electrically "inverted" regions vary in extent, and this can be observed in the SEM.

Conductive mode operation depends on migration of electrons through the specimen and the effect of sample structure and composition at or just slightly below the surface. Even though a beam spot of 100 Å can be achieved the extended interaction of the beam with sample features reaches beyond this point and limits resolution to 1000–3000 Å. Although this resolution is only slightly better than that of a light microscope, it should be noted here that the features observed in the conductive image are not necessarily seen by light optical microscopy. It is thus possible to look for associations between structural and contaminant features in comparison with electrical anomalies revealed by the conductive mode operation.

**Sample Preparation.** In contrast to the TEM, sample preparation for the SEM is relatively simple. The primary consideration is that the sample must be grounded to prevent surface charge accumulation. Samples are usually glued to a mounting stud with conductive silver paste. Specimens with nonconductive portions must have a very thin conductive film of carbon, gold, or some other metal evaporated over the entire surface. Layers that are thin enough to be almost transparent do not interfere with image formation. Elaborate preparations are generally not required. This is a major advantage of the method.

**Limitations.** The SEM offers limited electron diffraction capability at present as compared to the TEM. Elemental analysis is possible in a manner similar to that of the electron microprobe, but with some sacrifice in sensitivity. Since a vacuum must be used, sample size is limited and manipulation of the specimen must be accomplished by remote controls. Moist samples can become dehydrated and contaminate the vacuum system at the same time.

### electron microprobe analyzer (EMA)

The electron microprobe analyzer was designed primarily for producing elemental analyses over microscopic regions. Its magnification range is usually limited to 100–7000X. Operationally the EMA (Fig. 4-11) is similar to the SEM. An electron beam is accelerated, focused on a specimen, and scanned

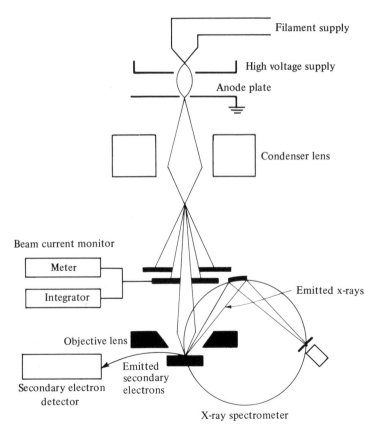

*Figure 4-11  Diagram of electron microprobe analyzer. Production of electron beam is similar to that of scanning electron microscopes. Primary function of electron microprobe is to use emitted X-rays to obtain information about sample. Scanning electron microscope also uses X-rays for image production but not as efficiently. (Courtesy Applied Research Laboratories, Sunland, California.)*

over the area with deflection coils. The electron beam excites the sample to produce X-rays with wavelengths characteristic of the elements present and which are detected by X-ray spectrometers similar to those described in Chap. 3. The X-ray signal from a spectrometer modulates the image so that image brightness indicates the concentration of a selected element when the electron beam is scanned over an area of interest.

**Fundamentals.** The X-rays produced by each element have wavelengths which are dependent on the energy needed to move an electron between electron shells of the atom. In general, the same considerations which apply to X-ray spectroscopy as discussed in Chap. 3 also apply to EMA work. One important exception is that the electron beam penetrates into a specimen less than a beam of X-rays would, and thus surface details of the specimen produce larger effects than would sample X-ray fluorescence by incident X-rays. The depth of penetration of electrons into a given sample is a direct function of the electron beam voltage. This characteristic permits differentiation of materials present as surface concentrations on a bulk substrate. It also means that materials beneath thin deposits such as some electroplates, thin films, and stains can be detected. Depth discrimination is only qualitative unless extensive calibration is carried out and a thorough understanding of the nature of the deposit is available. (For example, submicroscopic porosity or granularity could alter the results by changing the path length which will absorb both electrons and X-rays for the given thickness of material.)

Spatial resolution is dependent on the nature of the sample, the accelerating potential used, and on the use of the electron signal rather than the sample X-rays emitted under bombardment. The electron beam is focused to a 0.3 micron spot size and scattered from this diameter. X-rays, on the other hand, are emitted from the volume over which diffusing electrons can collide and exchange enough energy to produce characteristic X-rays of the element under consideration. The X-ray resolution will depend on the atomic number of the element or the average electron density of the sample.

Direct excitation by electron bombardment occurs in the EMA, as opposed to secondary fluorescence, which is the phenomenon in X-ray fluorescence analysis. In direct bombardment the number of element A atoms excited in the sample region will be proportional to the number of element A atoms present in that region provided that the bombarding electrons have sufficient energy to stimulate a characteristic X-ray of element A.

For a particular set of beam current conditions, to a first approximation, the excited volume times its average mass density will contain the same analyzed mass or number of atoms independent of its composition. This occurs because the retardation of the electrons, which depends on the electron density, and the backscatter of the electrons have opposite effects and tend to cancel one another.

By keeping instrument operating conditions constant, characteristic X-ray intensities are approximately proportional to weight fraction of elements in a sample and to the signal obtained from the pure element. Electron microprobe analysis offers a much more linear relationship between signal intensity and concentration than is the case with X-ray fluorescence or optical emission spectroscopy. Deviations from linearity may be calculated from theoretical considerations, thus improving quantitative accuracy in this technique. Extensive effort has produced several satisfactory computer methods of obtaining good quantitative data from the same comparison method by using absorption and fluorescence corrections.[6]

**Sample Preparation.** The specimen must be somewhat conductive to avoid the accumulation of excessive charge from bombarding electrons, but otherwise preparative techniques can be flexible. The depth of focus of the EMA is similar to the SEM and much greater than that of the optical microscope. This permits device structures to be left in place. In searching for contaminants one must be careful not to remove material of interest nor to add contaminants during sample preparation. Metallographic cross-sectioning can introduce considerable contamination either by smearing one material over another or by trapping foreign material from the polishing steps in cracks in a given device.

**Differences—SEM and EMA.** Although the primary purpose of the SEM is to show the physical structure of a surface and the primary purpose of the EMA is for elemental analyses of a small surface volume, these two instruments are very similar in design (see Figs. 4-10 and 4-11).

The major difference arises in the electron lenses and, thereby, in the operating currents. The first condenser lens is strong and the objective lens is weak in the case of the EMA while these conditions are reversed for the SEM.

In the SEM the electron beam can be focused down to 100 Å with a specimen current from $10^{-12}$ to $10^{-7}$ amp, depending on operating mode. In the case of the EMA the electron beam is focused down to 0.3 micron with a specimen current normally about $10^{-8}$ ampere. It is necessary to operate the EMA at a higher value of specimen current in order to obtain a sufficient X-ray signal. This is the case since at a particular value of sample current there are many more reflected, absorbed, and generated (secondary) electrons available than there are X-rays. This is particularly true at lower accelerating potentials (typically 2-5 kV).

Two other differences in the instruments are that the SEM has a much more versatile stage, and in some models the specimen can be tilted continuously to 90° from the normal incident electron beam. This latter feature is advantageous in that it allows a greater number of electrons to escape from the sample and assists in the ability to effectively "see" or image areas around corners or which are not in a direct line of sight with the electron collector system.

### *ion microanalyzer*

Microelectronic contamination problems frequently demand the utmost in both sensitivity and spatial resolution. When both demands occur simultaneously, neither the spark source mass spectrograph (SSMS) nor the EMA can provide the necessary capability. In these cases the ion microanalyzer offers another two orders of magnitude in sensitivity compared to the SSMS, in an area only slightly larger than that analyzed by the EMA, and a depth considerably shallower than that required by either the SSMS or EMA. The IMA thus offers a "weight detected" sensitivity of four to five orders of magnitude better than either the SSMS or EMA.

**Fundamentals.** The IMA (Fig. 4-12) is a microanalytical instrument which generates a finely focused beam of accelerated monoenergetic ions (usually oxygen or argon) to erode and ionize surface layers of a sample. The ions yielded by the sample are characteristic of its composition in the area being examined. These ions are sorted in a mass spectrometer section according to their mass/charge ratio and are then used in any of several different operating modes:

(1) Image display formation:
An image of the area being considered can be obtained with a resolution of about two microns. Image brightness is a function of the signal intensity of the element isotope of interest. The brightness of regions in the image indicates the distribution of the element being considered within the selected region (see Fig. 4-15).

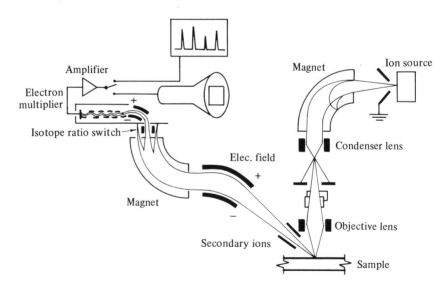

*Figure 4-12 Diagram of ion microprobe mass analyzer. Instruments similar in function to electron microprobe analyzer, but the microprobe mass analyzer uses ion sputtering to remove secondary ions from sample, and a double-focusing mass spectrometer to detect and identify them. (Courtesy Applied Research Laboratories, Sunland, California.)*

(2) Mass analysis:
Ions removed from the sample can be sorted in a double focusing mass spectrometer according to their mass-to-charge ratio. A mass spectrum can be recorded by scanning the mass range and recording detector signal strength.

(3) Depth profile:
By using the detector to monitor one isotope and by eroding the sample, a depth profile of an element's concentration can be obtained by plotting the detector output as a function of erosion depth (also a function of time).

In the manufacture of ion mass analyzers two different design concepts have been developed. In one concept, ions bombard all points in a relatively large area of the sample simultaneously. Means are provided to analyze the ions from a smaller region within the bombarded area. This type of instrument where the image is formed as a whole is analogous to the electron microscope and can be thought of as an ion microscope.

The second concept that has been developed is analogous to the electron microprobe, and it can be thought of as an ion microprobe. This instrument produces its image from the signal derived from a finely focused ion beam which is performing a scanning raster over a selected area in some interval of time.

**Instrument Merit.** The IMA has been compared to the electron probe X-ray microanalyzer with an ion probe being substituted for the electron probe and a double-focusing mass spectrometer substituted for the X-ray spectrometer. Either instrument can be used for a point analysis or to form a scanning image. Some of the advantages of the IMA over the EMA are:

(1) The extremely fine resolution of the depth of this instrument allows one to differentiate between surface impurities and bulk impurities. Surface layers can be removed by ion sputtering and thereby analyzed at the rate of fractions of a monolayer per second to hundreds of monolayers per second. The ability to determine changes of composition within such small increments of depth with high sensitivity is the outstanding feature of this instrument.

(2) Its sensitivity limit normally is in the part per billion region. There is no limitation with regard to atomic number.

(3) The method can function equally well for conductors, semiconductors, or insulators, which is not the case for the SSMS, SEM, or EMA.

(4) Peak-to-background ratios are improved as contrasted to electron probe X-ray microanalysis since ion bombardment produces less of a background signal. Ion bombardment does not produce a phenomenon analogous to the relatively high X-ray continuum background produced by electron bombardment.

# instrument applications

## sealed contact switch

Sealed contact switches are designed to switch and carry signals within an electronic switching system for telephone central offices. They consist of two flexible reeds of magnetic material, plated with gold and silver, and sealed into a glass sleeve so that they overlap without touching (Fig. 4-13). The switch is closed by an external magnetic field. The electrical resistance of the contact path formed on closing the switch is critical to the switch performance, and it degrades gradually with the life of the switch, due to wearing of the contact. At one time switches were found to be defective due to too high a contact resistance. Attempts were made to correlate these defects with dirt particles from air fallout, abrasive particles, residues left from incomplete rinsing of the reeds after plating, and particles which might be introduced from the furnace treatment used to alloy the gold and silver plates. Each of these operations contributed some material, but there were still batches of switches which were free of particles, yet had high contact resistance.

Microscopic examination of these reeds showed that the high resistance ones had what looked like droplets of a glassy material in regions which were not plated (Fig. 4-13). It was realized that the plating "bridged" over these particles and buried them in thickly plated areas such as at the contact tip. Their effect and source were not recognized immediately. The droplets were observed to be most numerous in the grain boundaries of the metal and to form during the decarburization step.

A good correlation between switch performance and the droplets was obtained with the light microscope with the smaller size and number of droplets indicating a better product. The electron microprobe identified manganese and silicon as the principal ingredients of the droplets while the spark source mass spectrograph showed that the droplets also contained a percentage of sodium. Refractive index of the droplets was measured by reflected light using immersion oils and oil immersion lens at high power. This confirmed that the particles had a refractive index of 1.68, the same value as for manganese silicate. It was realized that decarburizing the iron-nickel alloy generated a film of manganese and silicon oxides, since these two elements were added to the iron-nickel alloy as deoxidizers, and decarburization heat treatment is mildly oxidizing. It was also noted that a different furnace did not produce droplets under the "same" conditions. Samples from both furnaces were analyzed with the result that the furnace producing droplets was contaminated with sodium but that manganese and silicon surface enrichment took place in both furnaces. Also, increasing temperature and time increased the size and quantity of droplets formed.

A model was proposed for the formation of the droplets based on the information gathered from all these sources. The decarburization step, designed to reduce the carbon content, also brings manganese and silicon to the

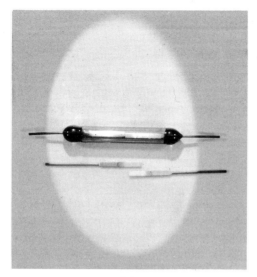

(a)

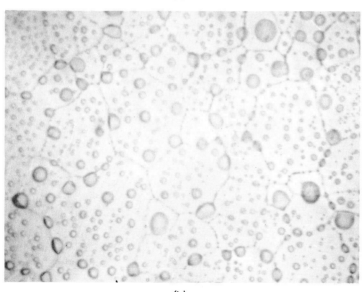

(b)

Figure 4-13 (a) Reed switch and plated reeds prior to assembly, slightly larger than natural size. (b) Glassy beads of manganese-silicon oxides which developed on alloy surface (× 1000).

metal surface. They are also oxidized as is carbon, but not being volatile they remain on the surface. The droplets were once molten and required a "flux" furnished by sodium in the furnace. This forms a glassy material which coalesces into the observed droplets rather than leaving a thin film of oxide on the surface. The droplets developed from ½ to 2 microns thick.

Not every furnace was contaminated with sodium. This accounted for differences observed between furnaces. Standard cleaning operations failed to remove the droplets, and since they were still attached, gold and silver plating covered them. During switch operation the force holding the reeds together is applied by an external magnetic field. Contact between the reeds is made only at a few points which become load-bearing surfaces. If the contact is on gold-silver alloy surfaces, contact resistance remains low. If the droplets share a portion of the contact force, contact resistance will rise. The droplets may be exposed initially, but more commonly they are partially buried and become exposed through contact wear. At this time the switch performance degrades at various points and early failure can occur. This problem was corrected by lowering the decarburization temperature and periodically cleaning the furnace.

### thin film

Tantalum thin film is used for making capacitors by anodizing the tantalum to tantalum oxide of a desired thickness and depositing a layer of conductive metal on the anodized tantalum. Ordinarily, the oxide film is able to withstand a voltage stress of 300 volts. Sometimes, however, a much lower voltage will cause the oxide film to break down and pass current. It was desired to determine why this occurred so that remedial action could be taken.

Replicas of tantalum film surfaces indicated that some rupturing of the oxidized film had occurred. Spark source mass spectrograph analysis of the film did not reveal outstanding amounts of any strange elements although carbon was possibly higher than expected. An experiment was designed with tantalum film deposited on a substrate of oxidized silicon; then the substrate was etched away leaving only a thin tantalum film for examination by a transmitted electron beam. Inclusions were observed at 100,000X in the thinned sections of film, representing material with low voltage break-down (Fig. 4-14). Electron diffraction patterns were made, indicating that the inclusions were tantalum carbide (Fig. 4-9). The formation of tantalum carbide was caused by backstreaming of pump oils into the sputtering chamber and the reaction of these hydrocarbons with highly active tantalum ions in the plasma discharge used for sputtering. This led to corrective steps to minimize the backstreaming of pump oils. This problem illustrates that environmental difficulties can involve well controlled vacuums as well as normal atmospheric conditions.

### milliwatt transistor

Reverse current leakage, or channeling, in a milliwatt transistor (Fig. 4-15) is at times a significant cause for rejection of product. Causes of reverse current have been explored theoretically and experimentally under laboratory conditions. On many occasions devices have been examined analytically to check for the presence of detrimental impurities. Examination at the microscopic level has failed to reveal differences by either spark source mass spectroscopy

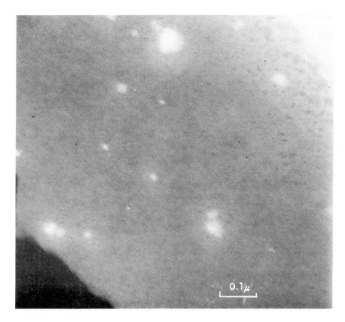

*Figure 4-14  Electron micrograph of tantalum thin film showing .01-.08 µm inclu-sions of tantalum carbide.  Film is tapered in thickness, and both free edge and opaque support are shown.*

or electron microprobe analysis.  Elements suspected of causing channeling are primarily potassium and sodium.  Sodium in particular is known to cause channeling, and its mechanism has been studied thoroughly in controlled experiments.

Directly relating a channel defect with the presence of sodium requires being able to observe which part of a transistor is actually channeled, and then determining if any visible feature is associated with it.  Several channeled units were examined with reverse bias voltage applied, and using the conductive mode.  Under these conditions the channeled regions were clearly visible in the SEM.  No defects or extraneous material could be associated with the channels.

The ability to "see" and locate a channel permitted analytical work to be done on a region known to be defective and to compare it to other areas known not to have reverse current leakage.  This illustrates the use of the SEM as a screening tool for other analytical techniques which are slower in their operation.  In this case the electron microprobe (EMA) was tried first, and neither sodium nor potassium was detected in the channeled area.  This suggested that the ion microanalyzer (IMA) should be tried.  The IMA produced images of the transistor showing that some of the defective regions observed with the SEM were correlated with the presence of potassium and sodium contamination.  However, most of the contamination spots were not associated with reverse current leakage areas. In addition to the spots of high alkali content there was a general low-level concentration of potassium and sodium. Much more alkali was found on defective devices, and in both cases the contamination disappeared before sputtering removed the entire oxide layer.

(a)

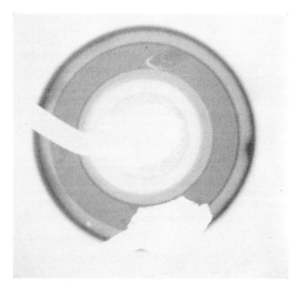

(b)

Figure 4-15

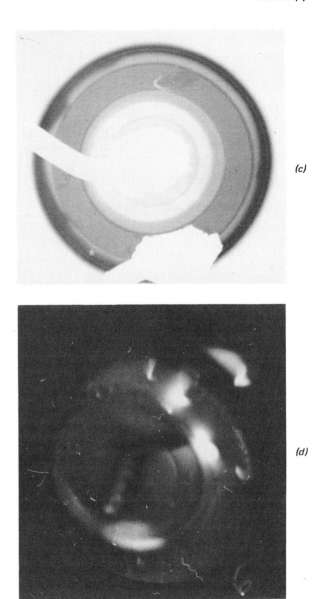

(c)

(d)

*Figure 4-15 Milliwatt transistor with reverse current leakage defect. (a) Scanning electron microscope photograph of active area of device; bright leads are 0.025 mm gold wires connected to ring and dot aluminum thin film metal contacts (X 300, 35° tilt off axis). (b) Similar device, shown in conductive mode; no bias applied. Outermost ring shows light gray sections toward inner device structure. (c) Same as (b), except 20 volts reverse bias applied, causing light gray structure to darken and expand inward. These regions permit small amounts of current to flow under reverse bias (X 300). (d) Ion microprobe image of a channeled transistor shows an image produced by ions of mass 23 (sodium) bombarded from the specimen surface by a plasma. The bright spots represent local surface concentrations of sodium detected in this manner and related by scanning electron microscopy to regions that showed high reverse current leakage (X 200).*

On the basis of these results a failure mechanism was suggested: particulate fallout containing significant amounts of potassium and sodium became deposited on the devices, and in some (but not all) instances combined with the oxide so that alkali diffusion could proceed into the oxide during heat treatment and power aging steps. Most of the alkali remains near the surface of the oxide. When electrical stress is applied some of the alkali atoms become ionized and attract negative charges in the silicon on the opposite side of the silicon oxide dielectric. This causes detrimental reverse current leakages.

## summary

Microscopic methods for microelectronic environmental control studies range from the simplest type of bench microscopes through some of the most elaborate and powerful instruments available, in which magnification and microanalysis are combined.

In considering the analytical and diagnostic instruments discussed in both Chaps. 3 and 4, microscopic methods are unique in that the observer can deduce information from visual clues, frequently without making any measurements. The observer has many techniques at his disposal to obtain information, and as he interacts with data yielded by earlier observations he can decide which techniques will help him the most to form opinions about what he sees. Thus, microscopic methods are often dependent on experience and insight more so than many other techniques. The microscope is frequently a first line of investigation used in conjunction with other techniques.

Quantitative microscopy has become significant in recent years with the introduction of automated scanning systems and computer methods of data handling. This is especially true in measuring particle size distributions or other characteristics of a specimen.

Further advances in microscopy may be anticipated in the areas of automated data handling for light microscopes, and improved resolution and analytical sensitivity for electron beam microscopes. The recent interest in this last field suggests that efforts will be made to push the investigative frontiers to even smaller spots and sensitivities for analysis of microelectronic devices.

## references

1. F. A. Jenkins and H. E. White, *Fundamentals of Optics*, McGraw-Hill, New York, 1950, Chaps. 20, 23.
2. Walter Lang, "Nomarski Differential Interference–Contrast Microscopy," Zeiss *Information* No. 70, p. 114, July 15, 1969 and No. 71, p. 12, June 16, 1969.
3. Charles P. Shillaber, *Photomicrography in Theory and Practice*, John Wiley & Sons, New York, 1944, p. 517 ff.
4. Gareth Thomas, *Transmission Electron Microscopy of Metals*, Wiley, New York, 1962, pp. 1–49 (Chap. 1).
5. P. R. Thornton, *Scanning Electron Microscopy*, Chapmans Hall, 1968, p. 25.

6. D. R. Beaman and J. A. Isai, "A Critical Examination of Computer Programs Used in Quantitative Electron Microprobe Analysis," *Analytical Chemistry*, **42**, No. 13 (November, 1970), pp. 1540–1568.

## journals on methods

Analytical Chemistry
Applied Optics
Applied Spectroscopy
Comptes Rendus Hebdomadaires des Seances de l'Academic des Sciences–Section B–
    Science Physiques (Fr.)
International Journal of Applied Radiation and Isotopes (Brit.)
Journal of Physical Chemistry
Journal of Scientific Instruments (Brit.)
Laboratory Practice (Brit.)
Nuclear Instruments and Methods (Neth.)
Optics and Spectroscopy
Philips Technical Review (Neth.)
Review of Scientific Instruments
Science
Spectrochimica Acta (Brit.)

## journals on problems

Applied Physics Letters
British Journal of Applied Physics
International Journal of Electronics (Brit.)
Journal of Applied Physics
Journal of the Electrochemical Society
Journal of Geophysical Research
Solid State Physics
Zeitschrift fur angewandte Physik (Ger.)
Zeitschrift fur Physik (Ger.)

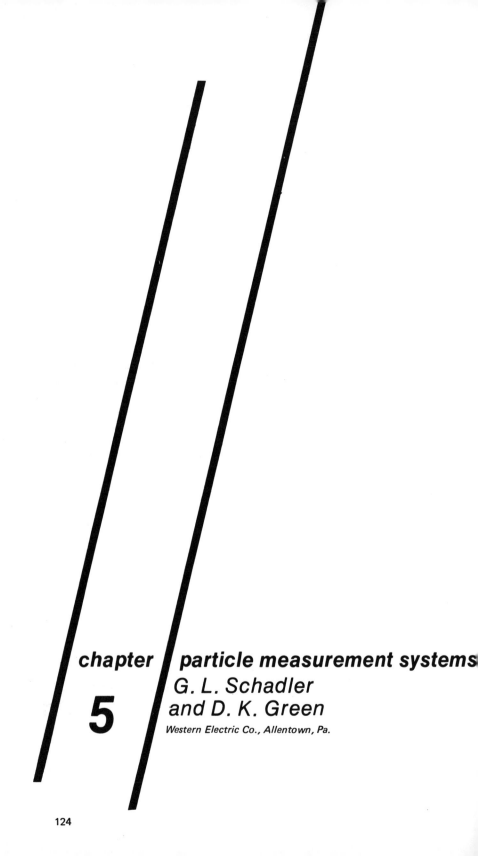

chapter

**5**

*particle measurement systems*
*G. L. Schadler*
*and D. K. Green*
*Western Electric Co., Allentown, Pa.*

A fundamental task in the control of contaminants is the counting and sizing of particles. Some method of measuring particles must be employed to establish the effectiveness (or lack of effectiveness) of a particulate matter control program for product surfaces and their manufacturing environment.

The examination of a product surface for the presence or absence of particles is, of course, the most direct and representative type of measurement. Visual examination with the aid of a microscope can provide not only quantitative information but can often lead to direct identification of the types of particles present. Such identification of particles is the first step in the elimination of their source.

The environment for a sensitive product should be as particle-free as possible. Instruments are available which can give an instantaneous measurement of the particle content of liquids and gases. With these instruments the integrity of fluid handling filters of all types and sizes can be evaluated, and the absolute quality of the resulting environment can be determined. The methods and instruments used in microelectronic manufacturing are discussed in this chapter, along with their advantages and limitations.

## selection of examination techniques

The first consideration in the selection of a particle examination scheme is to establish why the examination is being made. Of interest may be the size distribution of particles in the air or liquid processing environment, the identification of the particles causing a specific product failure mode, or finally, the time occurrences of particle count variations to establish improved manufacturing practices. The reasons for wanting such parameters will determine which technique or series of techniques is most appropriate. To illustrate this point, consider some of the ways in which particles may be collected, examined, and measured.

Particles may be subjected to analysis as they exist in the environment (be it air or liquid), or they may be collected in some manner and transported to a laboratory for subsequent examination. Numerous electronic particle counters exist which are capable of performing in situ measurements. If such equipment is not available or that particular type of analysis is not desired, some form of batch sample must be gathered. This is most commonly done by permitting particles to settle from the environment by gravity or by drawing the test fluid through a suitable filter (impaction or impingement).

If an instantaneous measurement is wanted, then the gravity or impaction method is not applicable. However, these collection processes are satisfactory if one wishes to size or identify particles. Collecting the particle sample by gravity allows the analyst to see how many and what kinds of particles deposit on product surfaces. Realize though, that the number obtained does not represent all the particles existing in the environment air since only the larger ones ($> 5$ microns) tend to settle on surfaces within reasonable time limits. The time required to collect an adequate sample may be anywhere from two hours to several days, depending on the cleanliness of the working environment. Gravity settlement tends to collect particles generated by people and processes in the immediate vicinity of the collection medium. The sample may be more typical of what the product "sees," but it is not necessarily typical of the environmental control surrounding the product.

The vacuum impaction technique generally requires less sampling time than the gravity method since, in a clean environment, the operator can choose a vacuum pump with a high flow rate. Most samples can be collected within a four-hour time interval, although special conditions may dictate longer sampling periods. Moreover, this method permits one to calculate, knowing the time of sampling and pump flow rate, the average concentration of particles which was present in the air. The entire particle size distribution can be obtained without the bias introduced by the gravity method provided that fracture and agglomeration upon collection are minimized. This presupposes a judicious choice of filters as a collection medium. Note that such a sample is an indicator of the environment, not the product surface.

As mentioned, there are instruments to be discussed which instantaneously count and size particles either in an air or liquid environment. They are excellent monitors of the environment and the instantaneous readings allow time correlations between work activity and particle concentrations. When measuring airborne particles, one must keep in mind that particles greater than 10 microns do not remain airborne for significant time periods. Thus, the probability of sampling the larger particles is reduced, which tends to bias the particle size distribution sample in favor of the smaller particles. In terms of total particle count greater than 0.3 micron, the loss of the large particle count is insignificant. However, the risk of electron product failure is greater with large particles than small particles. Also, particle collection for identification is difficult and not usually attempted.

If one is interested in measuring liquidborne particles, then the size information obtained by such counters will be biased in the direction of larger sizes since most instruments have 2 to 5 micron lower thresholds. At the present time, this is not considered a serious deficiency.

It is seen then that in situ sampling by automatic particle counters is far more suitable for measuring environmental particles and their time variations, but is less suited for detecting surface deposits and product failure modes.

## surface measurements

Since this section is concerned with surface measurements, it is assumed that any sample of particulate contamination will have been collected according to one or both of the batch sampling schemes. Once this has been done, one must decide upon the best way to extract the information one desires. For most contamination control applications, it is desirable to know the size, number, and identity of any particles collected. This information is most readily obtained by direct examination of the specimen. Techniques such as gravimetric analysis, which measure some aggregate property of a sample, are not as valuable as those which provide data on individual particles, e.g., microscopy. For this reason the discussion on surface measurements concentrates on the application of manual and automatic microscopy to particle analysis.

### *microscopic methods—manual*

The microscopic method probably remains the most widely used technique for the counting and sizing of collected contaminants. It is attractive to those who wish to obtain additional information over that supplied by gravimetric analysis but who either cannot afford instruments such as light scattering particle counters or who want a direct examination of a surface. Even some of those who can manage the cost of more sophisticated equipment still prefer the microscopic method due to its applicability to samples collected in a variety of ways. The operator can actually see for himself what has been collected and the material is still available for further analysis. The other real worth of the microscopic technique is demonstrated when one wishes to identify the contaminants which have been collected. This ability to identify particles is extremely valuable when trying to locate a source of contamination. With the microscope, information on topography, color, refractive index, crystal structure, etc., can be interpreted and is often sufficient to identify a given material.

**Particle Diameter Definitions.** Even those who are not trained in the techniques of identifying particles by microcopsy can use the microscope to count and size collected particles. Attention must be drawn to the fact that the sizing of particles using the microscopic method requires the definition of a standard dimension of measurement. If one wants to size a regular object such as a sphere or square, there is no ambiguity as to the measurement being a reliable parameter of that body. Consider the situation in which one is confronted with an irregularly shaped particle to measure. For instance, what is the diameter of a particle shaped like a peanut?

In the conventional sense, diameter has little meaning; however, the word has come to apply to a measurement of such particles. Two of the most widely accepted diameters have been those introduced by Martin and Feret. Martin's diameter is the length of a line which divides a particle into two equal areas, as shown in Fig. 5-1. Feret's diameter is the length of a line drawn between two parallel tangents on the particle, as shown in Fig. 5-2. Some constant direction must be chosen in which to measure either diameter. Several other diameters have been employed over the years and appropriate graticles and reticules were fashioned for their estimation.[1,2] Such diameters are often termed statistical diameters because a large number of them must be measured and averaged before they become meaningful.[3]

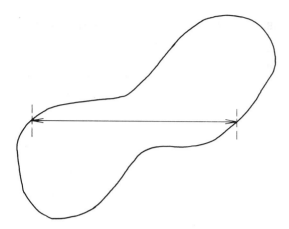

*Figure 5-1 Illustration of Martin's Diameter.*

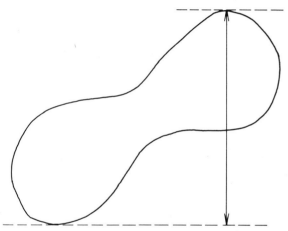

*Figure 5-2 Illustration of Feret's Diameter.*

**Standard Applications.** Once a convention is chosen by which to size particles, a reasonably accurate estimate can be obtained on a given particle. This type of analysis demands that a large number of particles be classified, based on some form of repeatable method. The American Society for Testing and Materials (ASTM) has developed the various standards shown in Table 5-1, which apply microscopy to the counting and identification of particles. Similar procedures may be adopted to suit an individual's particular needs, but it should be emphasized that any technique, once established, should be followed without deviation. The ASTM standards, properly executed, are valuable when comparing results performed by other individuals or companies.

**Limitations.** Microscopic techniques are especially useful to count, size, and identify collected particles when applied to a limited number of samples. This is in spite of the problem of defining the size of an irregularly shaped

TABLE 5-1. Microscopic Particulate Sizing, Counting and Identification Standards

| ASTM Designation | Application |
|---|---|
| D 1357-57 | Sample Planning |
| F 24-65 | Surface Particle Counting |
| F 25-68 | Airborne Particle Counting |
| D 2390-65T, D 2391-65T | Liquidborne Particle Counting |
| F 51-68 | Counting Particles on Clean Room Garments |
| F 71-68 | Identification of Fibers |
| D 2430-65T | Identification of Particles in Fluids |
| F 59-68 | Identification of Metal Particles |

particle. In situations where particle counting and sizing of many samples are required on a routine basis, the limitations of the human operator become apparent. The operator is called upon to make many marginal decisions, and human judgment is far from perfect. Fatigue soon sets in and the operator loses interest in the task at hand. The accuracy of a given measurement may be acceptable, but with many samples examined over a long period of time it becomes questionable. Under these circumstances, precision even with the same operator is commonly placed at ± 30%. Among different operators the variation can be as high as 10 to 1. Even if problems with accuracy and precision did not exist, the time requirements of such an analysis may well exceed one hour per sample—a generally prohibitive consumption of time when many samples are to be evaluated. If one wishes to examine only an occasional sample, this drawback is not too serious. Clearly, it would be desirable to have a technique which would reduce the time requirements, simplify or eliminate the human decision making process, and increase the overall reliability of particle counting and sizing. The automation of microscopy has challenged equipment designers for many years. It now appears that designs coming on the market have a good blend of practicality and versatility for particle analysis in environmental control applications.

### microscopic methods—automatic

Although there are instruments available which lend some degree of automation to the measurement of particles, they will not be considered here. Rather only the more sophisticated image analyzing equipment will be discussed.

The January 1971 issue of *The Microscope* reported six automatic image analyzers to be on the market: the Classimat of Leitz, Digiscan of Kontron, Micro-Videomat of Zeiss, πMC of Millipore, Quantimet 720 of IMANCO, and Telecounter of Schaefer.[4] Fig. 5-3 shows one of these instruments. These systems employ an optical microscope linked to equipment which can electronically analyze the image which is produced and consequently display that image on a television monitor. Each of the units mentioned above carries out

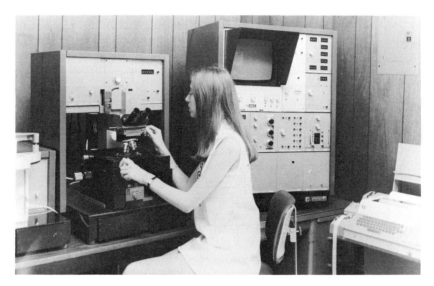

*Figure 5-3 A television imaging system.    (Courtesy Metals Research Instrument Corporation, Monsey, New York.)*

similar measurements, though not exactly in the same fashion. It is intended that comments made here will be general enough to apply to all instruments.

Image analyzers require an input device to obtain the information (image) to be processed. For most applications this is an optical microscope, although it may be a projector system for examining large objects or an electron microscope for looking at very small specimens. After the image has been formed, it passes to a vidicon scanner where it is broken into picture elements and converted to electrical pulses. These pulses are then transferred to the logic circuitry which interrogates the dissected image for the parameter of interest. Proper detection depends upon the image exhibiting suitable gray level or contrast against its background. Some type of threshold control is used to obtain optimum detection. Once the information has been analyzed, it is reconstructed in its original form on a display monitor. Simultaneously, the data may be directed to a calculator, plotter, or other recording/computing device. Options are available which can be employed to automate the entire process from stage movement to data reduction.

Any of several parameters of a particle array may be measured with these instruments. The most common ones are number, area, and some type of defined length. Others (though by no means all) include shape, optical density, and perimeter. Size distributions can be accumulated for any given sample.

Automatic image analyzers are a tremendous aid in the examination of particles. Counts can now be extremely accurate. Particle sizes and areas can be measured with speed and reproducibility impossible by manual means. Perhaps it suffices to say that these instruments represent the merger of a very versatile analytical instrument—the microscope—with the infinite potential of electronic automation and data analysis.

One does not gain such capability without cost. One of the simplest systems costs in the range of $16,000-$20,000. More sophisticated instruments (expanded by adding modular units) may cost as much as $75,000.

## airborne measurements

Although microscopic examination of a product or filter surface is the most direct method of measuring particulate contamination, these methods are tedious and time consuming unless automated. In many instances the feedback of measurement results may be too slow to be useful to a production process. The development of the light-scattering particle counter into a practical form[5] has made it possible to instantaneously measure the particulate matter content of gases down to 0.3 micron and liquids down to 2 microns. Of course, the use of this type of instrument provides only a secondary indication of surface cleanliness. It must be assumed that if a product is exposed only to a clean environment, the number of particles deposited on the product surface from the environment will be zero. Continuous monitoring of the environment surrounding a product gives instantaneous recognition of the environment cleanliness becoming abnormal and the resultant possibility of product surface contamination. Light-scattering particle counters can be set up to provide continuous automatic monitoring of a particular location for indefinite periods of time. Analysis of the resulting data is the only manual chore required. Because of the importance and widespread use of the light-scattering particle monitor, this instrument will be discussed in some detail. Subsequently, a brief commentary will be given on two instruments which measure airborne particle sizes smaller than 0.3 micron.

### light scattering particle counter

**Basic Sensor.** As the name implies, a light scattering particle sensor uses light scattered from particles as the detection mechanism. The concept is based on the Mie[6] theory and experimentation summarized by O'Konski et al.[5] In simple terms the light scattered by a single particle is proportional to the square of the diameter within certain limits. The resulting commercial instruments all use a beam of light from a high intensity lamp which is focused within a dark chamber at a location called the view volume. This volume is traversed by a moving column of air which contains a sample of the aerosol to be studied. The light beam and sample air flow path are usually perpendicular, with the air flow being vertically downward. Beyond the point of intersection with the air stream, the light beam is totally absorbed in a light trap. When particles which are contained in the air stream traverse the light beam, they scatter light in all directions. Light scattering in this case includes both reflected and refracted light.

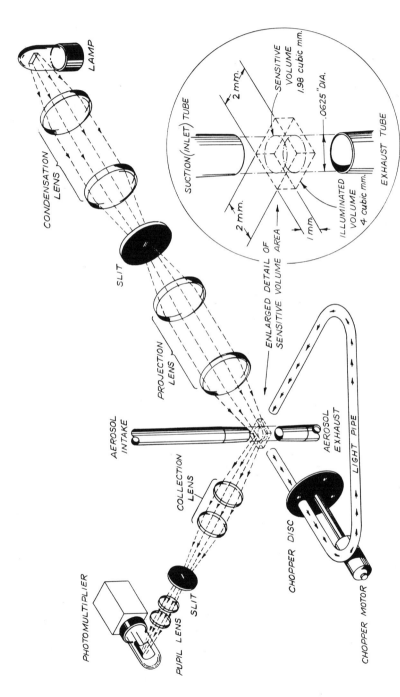

Figure 5-4   Ninety-degree light-scattering optical system.   (Courtesy Royco Instruments, Inc., Menlo Park, California.)

Various optical designs have been used to gather and measure the intensity of this scattered light. In a ninety degree light-scattering instrument a collection lens system is located with its axis perpendicular to both the light beam and the air stream as shown in Fig. 5-4. Any scattered light from the view volume that is within the angle subtended by the collection lens is focused on the light detecting element. Fig. 5-5 shows the optical element arrangement of a near-forward light-scattering sensor. In this particular design the scattered light that leaves the view volume within a small angle surrounding the light trap is collected and focused on the light detecting element. In all of these instruments the light-sensing element is either a photomultiplier tube, photodiode or phototransistor. This tube converts pulses of light to pulses of electrical current and greatly amplifies this current. Within certain limitations it is possible to detect the scattered light from a single particle. The light intensity and therefore the resulting photomultiplier current pulse are proportional to the size of the particle. The time duration of the light pulse and the corresponding current pulse width depends upon the time of residence of the particle in the light beam. This, of course, depends upon the velocity of the sample air stream and the dimension of the view volume in the direction parallel to the stream movement. The view volume is very small so that the possibility of more than one particle being illuminated at one time is small.

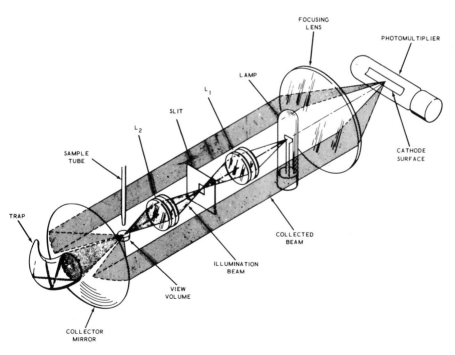

*Figure 5-5 Near-forward light-scattering optical system.    (Courtesy Bausch and Lomb, Inc., Rochester, New York.)*

**Electronic Portion.** Once an electrical signal is produced by the photodetector, a variety of electronic operations can be performed on the signal to accomplish various objectives. The wide range in cost of particle measuring equipment is due to the many available variations in electronic features. The least complex and therefore least expensive instrument is the photometer. Such an instrument simply amplifies and integrates the signal pulses and applies the resulting d-c potential to an indicating meter. A photometer does not numerically count individual particles and is calibrated in terms of apparent particle mass per unit volume of air. A photometer is useful for measuring relatively high particle concentrations such as in leak testing of HEPA* filters and air pollution studies.[7,8]

Particle counters are more complex instruments that actually count and size individual particles. In this type of instrument the photodetector pulses are amplified and applied to a size discrimination circuit, normally a Schmitt trigger. This circuit will provide an output voltage of fixed amplitude whenever the input voltage amplitude exceeds a preset threshold value. As shown in Fig. 5-6, the Schmitt trigger output consists of pulses of fixed amplitude whose width is equal to the input pulse width when measured at the threshold level. These pulses are then applied to a one-shot circuit whose output is one pulse of fixed amplitude and width for every input pulse. At this point in the circuitry a uniform output pulse is present for every photomultiplier pulse whose amplitude exceeds the threshold level. This threshold level is adjusted for proper size sensitivity once the optical system response to known size particles is determined during instrument prime calibration.

The uniform pulses from the one-shot circuit may be used to operate a variety of output devices. They may simply be counted by an electromechanical or electronic counter. They may be integrated and the resulting d-c voltage displayed on a meter (usually logarithmic) reading directly in particles per cubic foot of air. Digital printers or strip chart recorders can be used to provide permanent recording of the measured particle concentrations.

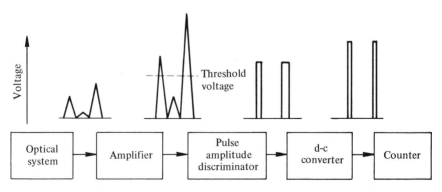

*Figure 5-6 Particle-counter electronics block diagram.*

*High efficiency particulate air filter. See Chap. 10 for further discussion.

*Figure 5-7 Photograph of a multichannel sequential analyzer. (Courtesy Royco In-struments, Inc., Menlo Park, California.)*

Probably the most popular electronic variation that is provided on some particle counters is multi-channel operation. By providing multiple size dis-crimination circuits in parallel, any number of channels that are each sensitive to a different size can be obtained. Figure 5-7 shows a commercial instrument which has 15 size discrimination ranges. A two-channel instrument measuring particles 0.5 micron and larger on one channel and 5.0 microns and larger on the other channel (Fig. 5-8) is very popular since concentration limits of these two size ranges are specified in Federal Standard 209A.*

The counting of particles in a discrete size range can be performed by using an upper as well as a lower size threshold. A particle will be counted only if its size is within the two threshold settings. An instrument with ad-justable thresholds is very valuable for determining particle size distributions.

**Prime Calibration.**    The primary calibration of a particle counting instrument must be established by measuring the electrical pulses produced by intro-

*See Chap. 10 for a discussion on this document.

*Figure 5-8 Simultaneous type particle counter (2 channels). (Courtesy Climet Instruments Corp., Sunnyvale, California.)*

ducing an aerosol of very uniformly sized (monodisperse) spheres of accurately known diameter. The material normally used for this purpose is polystyrene latex. The latex spheres are commercially available as a hydrosol (suspension in water) and are converted to an aerosol by atomization and drying. Available sizes and size distributions range from an average diameter of 0.091 micron and 0.0058 micron standard deviation to 1.947 microns average diameter and 0.0070 micron standard deviation. Larger size particles are available in other materials but with much larger size distributions.

While the calibrating spheres are introduced into the particle counting instrument the amplitudes of the resulting electrical pulses are monitored with an oscilloscope or a pulse height analyzer. The pulse amplitude which is most representative of the average particle size must conform to the instrument manufacturer's specifications. By determining the pulse amplitude for several sizes of particles, a calibration curve can be obtained. The threshold voltage values to which the instrument size discrimination circuitry should be adjusted can be read from this curve.

**Field Calibration.** All light-scattering particle counters are equipped with a built-in secondary calibration system. On most counters this system consists of a bundle of light-conducting fibers which routes a small amount of light from the high intensity lamp to a motor-driven light-chopper disk located outside the main optical chamber. When the system is actuated the light chopper provides pulses of light which are routed by a second fiber optic bundle back into the optical chamber. These light pulses are directed to the photomultiplier where they are detected like any other light pulse. In the calibrate mode the amplified photodetector output is usually applied directly to a d-c meter. At the time of instrument prime calibration, the meter circuit is adjusted to provide a predetermined deflection in the calibrate mode. As a particle counter is used, the lamp intensity slowly decreases due to condensation of tungsten vapor on the glass bulb. By regularly verifying the calibration of an instrument with the built-in system, the photodetector gain is gradually increased to offset the decrease in scattered light intensity.

**Limitations.** The intensity of light that is scattered by a particle depends to some extent on physical characteristics other than size; namely shape, color and index of refraction. An instrument can be calibrated to near perfection with polystyrene latex spheres, but unfortunately, the instrument is probably used to count and size nonspherical particles (even fibers) of various colors whose refractive indices may vary widely from 1.592. The various designs of instruments available will all produce different size information for naturally occurring aerosols. The near forward type of instrumentation is less sensitive to variations in shape, color, and refractive index than are the ninety degree types.[9] Variations in numerical counts of particles larger than a given size can fluctuate significantly simply due to deviations from ideal calibrating spheres.

Some particle counters utilize electronic counters for displaying the total number of particles detected. Other instruments utilize electromechanical counters. Each type of display unit has associated problems. Electromechanical counters have definite speed limitations. While this type of counter may be capable of counting 2400 uniformly spaced pulses per minute, it must be kept in mind that particle arrivals at the detector are entirely random with time. Instantaneous particle arrival rates may easily exceed the maximum counting rate of an electromechanical counter. The result is that some electronic pulses are not counted and a falsely low particle count is obtained. The more concentrated the particulate matter becomes, the worse this condition gets. Most electromechanical counters are equipped with electronic decade counters for this reason. Electronic counters, while capable of high counting rates, can be adversely affected by electronic noise, both in the electrical supply and directly radiated from the source. The result in this case is falsely high particle counts.

It was mentioned previously that the optical system of a particle counter is designed so that normally only one particle at a time is within the view volume. As the concentration of an aerosol increases, the probability of two or more particles being sensed simultaneously increases. This is known as coincidence. The particle counter interprets the resulting pulse as being initiated by only one particle of a size larger than either particle actually present.

The net result of coincidence is a loss in the number of particles recorded and an incorrect shifting of the apparent size distribution toward the larger sizes. Most of the particle sensors available today can accurately count particles in concentrations up to one million particles per cubic foot of air with almost negligible coincidence.

**Sampling Techniques.** Since particle counting instruments measure only the particles in a relatively small volume of air, the statistics of sampling can be applied to the resulting data. For such a small sample the statistical error is equal to $1/\sqrt{n}$, where $n$ is the number of data. According to this formula 100 particles of a given size must be counted in order to achieve an error of $\pm 10\%$. At a class 100* condition if 0.5 micron particles are being counted, this means that a total of one cubic foot of air must be sampled to achieve the $\pm 10\%$ error. With a particle counter sampling at 0.1 cu ft/min, ten minutes would be required to sample one cubic foot of air.

It can be easily seen that the cleaner the environment becomes, the larger the air sample must be to maintain a given statistical accuracy, and the time required to measure the appropriate volume of air with a given particle counter will increase correspondingly. The sample flow rate of a particle counter is a very important characteristic when sampling statistics are considered.

A high flow rate instrument is also desirable in a clean environment for studying particle generation. The maximum flow rate commercially available in a particle counter is 1.0 cfm. Since particle generation sources in clean areas usually supply large numbers of particles during very short time intervals, a particle counter with a high flow rate will have a higher capture probability and will show almost instantaneous response.

On the other hand, a high flow rate ($\geqslant 1.0$ cfm) instrument will suffer coincidence at a relatively low particle concentration; therefore its use is limited to clean environments for all practical purposes. For environments of class 100,000* or higher, lower flow rate instruments should be used. Instruments are commercially available with sample flow rates of 1.0, 0.25, 0.1, 0.01, and 0.0067 cfm.[10-14]

When one samples from an open room or other large volume environment, the air velocity entering the sample probe should not be too different from the average air velocities existing in the area. Since large particles will not follow air streamlines as readily as will smaller particles in areas of streamline curvature, a representative size distribution can be obtained only at nearly isokinetic sampling conditions. If the inlet velocity to the sample probe is much lower than the surrounding air stream velocity, small particles will be diverted around the probe and large particles will impact into the probe, resulting in a sample with the size distribution shifted toward the larger sizes. A sample probe inlet velocity much higher than the ambient will

*See environment class definitions in Chap. 10.

result in a size distribution shifted toward the smaller sizes. If tubing is used to route the air sample to the particle counter inlet, it should be as short as is practically possible and long horizontal runs should be avoided. Clean copper or stainless steel tubing is the best material to use for a sampling tube if accurate size distribution data are required. For most monitoring work, however, accurate size distribution is of secondary importance and flexible plastic tubing such as Tygon* has been found to be satisfactory.

When measuring particle concentrations in small enclosed volumes such as storage dry boxes or environmentally controlled process enclosures, it is rarely possible to place the instrument within the test area. A sample must usually be drawn from the enclosure through connecting tubing. It is extremely important that the sample withdrawal rate does not exceed the rate at which gas is supplied to the enclosure. This will ensure that the enclosed volume maintains a positive static pressure with respect to the outside ambient which should be the normal operating condition. For this type of application, a particle counter with a low flow rate is desirable so that normal conditions in the environment under study are not drastically altered. Ideally, the rate of gas supplied to the enclosure should be increased by an amount equal to the sample withdrawal rate.

The particle concentrations existing in gas cylinders and pipe lines can be determined with a light-scattering particle counter if two precautions are observed: (1) The pressure in the particle counter optical chamber must never exceed approximately 2 psig, and (2) the chemical activity of the gas must be known and the possible consequences of exposing the particle counter to it must be thoroughly evaluated.

Inert gases pose few problems with most instruments. The flow rate indicator should be corrected for the gas being sampled. With certain precautions, some gases considered to be hazardous can be used with a particle counter. Since hydrogen is dangerous only after mixing with oxygen, it can be admitted to a particle counter that is known to be free of leaks. A preflush of nitrogen or other inert gas is required as with any other device exposed to hydrogen. The exhaust from the instrument should be piped to a safe disposal point.

It is not recommended that oxygen or oxygen-rich gas mixtures be admitted to a particle counter since some instruments may contain materials not compatible with oxygen. For oxygen or other gases judged to be incompatible with particle counting instruments, a membrane filter can be used to remove the particles from the gas. The particles on the filter surface can then be counted with the aid of a microscope.

### submicron particle counters

The minimum size particle detectable with light-scattering instruments is about 0.3 micron because of electronic signal-to-noise ratio limitations. Other

*Registered trademark of Norton Inc., Akron, Ohio.

types of counters have been developed to detect even smaller particles. Whitby and Clark[15] have described an electric particle counter which indirectly measures size by determining the electrical mobility of a charged aerosol. This system is reported to measure size distributions in the range from 0.015 to 1.0 micron. Van Luik and Rippere[16] have developed a detector which indicates the presence of very small particles which act as sites for the condensation and growth of water droplets. This instrument is reported to detect particles from 0.001 to 0.1 micron in diameter. With suitable converters which produce nuclei by chemical reactions with the sample gas, this instrument can be used to detect gaseous compounds such as hydrocarbons, sulfur dioxide, and ammonia.

## liquidborne measurements

Unfortunately, not all troublesome particles are suspended and transported in a gaseous medium. Many semiconductor manufacturing operations deal with liquids such as water, photoresist compounds, solvents, acids, and many other chemicals. Quite frequently, these liquids must be as free of particulate matter as possible. Just as with air or other gases, it is desirable to measure the particle content of liquids as a performance check on filtration devices and techniques. Since liquids have much higher viscosities than gases, large and dense particles that would immediately settle out of a gas will remain suspended for relatively long periods in a liquid. It is also much more difficult to keep liquids free of particles after they have been filtered since the viscous drag of a liquid flowing past a settled particle on the wall of a container or a length of tubing results in a much higher removal force than exists for a flowing gas.

### light-scattering and light absorption counters

Light-scattering particle counters are also useful for liquids. In the liquidborne particle counter the optical system contains transparent windows to transmit the light through the liquid. Light scattered from particles contained in the liquid is sensed by a photodetector as in an airborne counter. However, the electrical noise level in a liquidborne particle counter is much higher than in an airborne counter due to interaction of the light beam with the sensor windows and with the liquid medium itself. For this reason the minimum size particle detectable with a liquidborne instrument is usually 2 microns.

Since liquids having a wide range of colors, refractive indices, viscosities, and vapor pressures could conceivably be tested in a light-scattering instrument, many measurement limitations are presented that do not exist with airborne counters. Liquid cavitation, bubble formation, dirt or bubbles forming on optical surfaces, and absorption of light by particles are some of the problems which can occur in counting liquidborne particles. However, these limitations become less significant using the light absorption properties of particles as the sizing mechanism, a slightly different adaptation of the Mie theory.

The light absorption instrument has an in-line optical system similar to the near-forward sensor except that after passing through the view volume the light beam proceeds through a second aperture and impinges on the photodetector.   Instead of the photodetector being normally dark it is normally fully illuminated. As a particle passes through the view volume the light that is absorbed and scattered will no longer reach the photodetector and a signal in the form of a decreased output results. This signal can be treated electronically as with any other type of sensor.

Why light absorption has distinct advantages for measuring liquidborne particles can be easily explained. If the viewing windows in a ninety degree or near-forward light-scattering sensor become contaminated with bubbles or particles, "stray light" is produced in the optical chamber and increases the very small background illumination level of the photodetector. This increases the background noise level of the sensor and reduces the signal-to-noise ratio. A light absorption sensor, however, is very insensitive to contaminated optical surfaces since the detector is normally fully illuminated. The main disadvantage to an absorption type sensor is due to the fact that a photomultiplier detector has a noise level proportional to the illumination intensity. Thus, a large signal is necessary to provide a useful signal-to-noise ratio, affecting the lower limits of particle size detection. The various commercial forms of light scattering or light absorbing particle counters with appropriate data recording equipment will cost around $10,000.[13,14,17,18]

**Sampling Methods.** As previously mentioned it is much more difficult to obtain clean liquids than clean gases. This becomes very evident when sampling supposedly clean liquids with a liquidborne particle counter. The best method of obtaining a representative liquid sample is to install the sensor in-line in the liquid stream to be sampled. This is not always possible since it prevents the sensor from being used in other locations, and the flow rate through the sensor may not be sufficient to satisfy the requirements of the process under study.

It is possible to batch sample a liquid with any of the liquid-borne counters. In batch sampling, a single small container of liquid is emptied through the counter, and all particles exceeding the minimum size threshold are detected. Several precautions must be observed during batch sampling of liquids. All handling containers should be scrupulously clean. The sample should not be agitated to the extent that small bubbles are created since they will appear as particles to the particle counter. This type of batch sampling is particularly suitable for determining particle concentrations in small containers of chemicals as received from the supplier.

Unfortunately neither the in-line nor the pressure batch sampling method is suitable for testing liquids that are in use in open containers. It is extremely difficult to withdraw a liquid from a processing bath, place it in a batch container, and run it through a liquidborne particle counter without adding significant numbers of particles to the sample.

A sampling system used at the Western Electric Company, Allentown Works, overcomes these obstacles by withdrawing a sample via vacuum from an open process tank directly into the particle sensor. The system in its simplest form is shown schematically in Fig. 5-9. Liquids that have been suc-

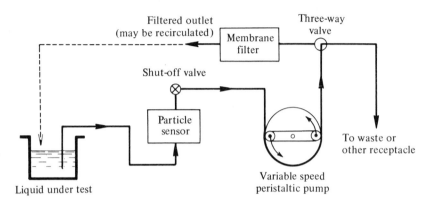

Figure 5-9 Schematic diagram, liquid sampling system.

cessfully monitored with this system include methyl alcohol, isopropyl alcohol, acetone, trichloroethylene, n-butyl acetate, stoddard solvent, deionized water, and several water soluble photographic developing solutions.

The significance of liquidborne particles becomes apparent when one examines particle concentration data obtained from typical process liquids. Concentrations of particles $\geq$ 5 microns in size have been observed from zero in deionized water to 430 particles/ml in a photographic bleach solution. Converting the latter number to units used to express airborne particulate concentrations yields 12,176,310 particles/cu ft. Certainly such gross contamination is unacceptable in either air or liquid environments for microcircuit manufacture. However, a contaminated process bath should be considered relative to the total process chain of events. A subsequent overflowing rinse operation (filtered) may overcome the effects of a previously contaminated liquid bath.

### electrical conductivity particle counter

Another different type of particle counter is available for use with liquids. This type of detector uses an electrolyte solution as a particle carrier. An electrical potential is applied to the electrolyte on both sides of a very small orifice. The electrolyte solution containing the particles to be measured is drawn through the orifice. Under normal conditions a steady-state current flow exists through the orifice. When a particle passes through the orifice it displaces enough volume of electrolyte to alter significantly the electrical conductivity of the circuit. This results in a current pulse which can be electronically detected and analyzed. Since the current pulse amplitude is proportional to particle volume, the electrical output of the detector can be calibrated in terms of either particle volume, area, or diameter. This conductivity type of particle counter can be made sensitive to particles smaller than one micron and larger than several hundred microns depending on the size of the orifice used.[19]

### relative measures of particle contamination

Several other methods of monitoring liquids for particulate matter contamination are available. These methods require a minimum investment in equipment but do require some operator time and skill. The test results are only relative and have no absolute meaning. Size distributions and particle concentrations cannot be obtained. These methods compare liquids on the basis of the time rate of change of flow through a membrane filter under constant pressure.

The silting index test[20] can be performed with very small quantities of liquid. A 10 ml sample of the liquid is placed in a syringe that is equipped with a membrane filter at its outlet. A constant pressure is applied to the syringe by means of a weight. As the liquid is forced out of the syringe, time intervals are measured for 1 ml $(T_1)$, 5 ml $(T_2)$, and 10 ml $(T_3)$ to pass through the filter. The Silting Index is calculated from the formula

$$S.I. = \frac{T_3 - 2T_2}{T_1}$$

This test measures the total mass of particulate matter in the liquid sample with a higher number indicating a greater concentration of particles in the liquid. Liquids with any viscosity gradient must, of course, be measured at a fixed temperature to obtain relative results.

The filter plugging test is similar in principle to the silting index test but uses much more liquid. In this test a membrane filter is placed in a liquid stream and the flow rate through the filter, when new, is determined. After a specified time interval when a definite quantity of liquid has passed through the filter, the flow rate is again measured. The reduction in flow rate or percent decrease is expressed as the filter plugging and is a relative indication of mass of contaminants filtered from the liquid stream. All of the flow measurements must be taken at a fixed pressure and at a fixed temperature if the liquid viscosity is temperature dependent. The equipment needed for either the filter plugging test or the silting index test can usually be purchased for less than $500.

### summary

This chapter has described many different methods of detecting or measuring particles, and it would be unwise to conclude without a few remarks pertaining to instrument accuracy. As has already been pointed out, microscopic particle counting is very subject to human error. Automatic light-scattering instruments are dependent on various physical characteristics of the individual particles, and different optical system designs can interpret similar particles quite differently. The reader is cautioned against interpreting particle count data in an absolute sense. Even comparative data should be critically reviewed. A two-to-one change in the particle concentration of an environment may not be considered significant in many circumstances. Whatever instrument is chosen for a particular job, its capabilities and limitations should be completely understood.

## references

1. H. S. Patterson and W. Cawood, "The Determination of Size Distribution in Smokes," *Trans. Faraday Society*, 32, 1084 (1936).
2. G. L. Fairs, *Chemistry and Industry*, 62, 374 (1946).
3. R. D. Cadle, *Particle Size-Theory and Industrial Applications*, Reinhold Publishing Corp., New York, 1965, p. 4.
4. W. C. McCrone, *The Microscope*, 19, Jan. 1971.
5. C. T. O'Konski, M. D. Bitrun, and W. I. Higuchi, "Light Scattering Instrumentation for Particle Size Distribution Measurement," ASTM Special Technical Publication No. 234, *Particle Size Measurement* (Philadelphia, Pa. ASTM Publication, 1959), pp. 180–206.
6. G. Mie, "Beitrage Zur Optik Trüber Medien Speziell Kolloidaler Metallösungen," *Annalen Der Physik*, 25, 377 (1908).
7. Tentative Standard for HEPA Filters No. CS-IT American Association for Contamination Control, Boston, Mass.
8. "Mass Concentration of Particulate Matter in the Atmosphere, ASTM Standard Test Method D1899-68, ASTM Standards, Part 23, Nov., 1969.
9. J. R. Hopkinson and J. R. Greenfield, "Response Calculations for Light-Scattering Aerosol Counters and Photometers," *Appl. Optics*, 4, 146 (1965).
10. Bausch & Lomb, Inc., Rochester, New York, 14602.
11. Climet Instruments, Inc., 1240 Birchwood Drive, Sunnyvale, Calif., 94086.
12. Dynac Corp., Thompson's Point, Portland, Maine, 04102.
13. Particle Technology, Inc., 734 No. Pastoria, Sunnyvale, Calif., 94086.
14. Royco Instruments, Inc., 141 Jefferson Drive, Menlo Park, Calif. 94025.
15. Kenneth T. Whitby and William E. Clark, "Electric Aerosol Particle Counting and Size Distribution Measuring System for the 0.015 to 1 $\mu$ Size Range," Tellus XVIII, 2 (1966).
16. Frank W. Van Luik, Jr., and Ralph E. Rippere, "Condensation Nuclei, A New Technique for Gas Analysis," *Anal. Chem.*, 34, 1617 (Nov. 1962).
17. High Accuracy Products Corp., 141 Spring Street, Claremont, Calif., 91711.
18. Jet Instruments, Inc., Suite 1717, First National Bank Bldg. East, Albuquerque, N.M. 87108.
19. Coulter Electronics Industrial Division, 2601 Mannheim Road, Franklin Park, Ill., 60131.
20. J. L. Dwyer, "The Silting Index: An Evaluation of Micron and Submicron Contamination in Liquids," ASTM STP No. 342, pp. 69–78.

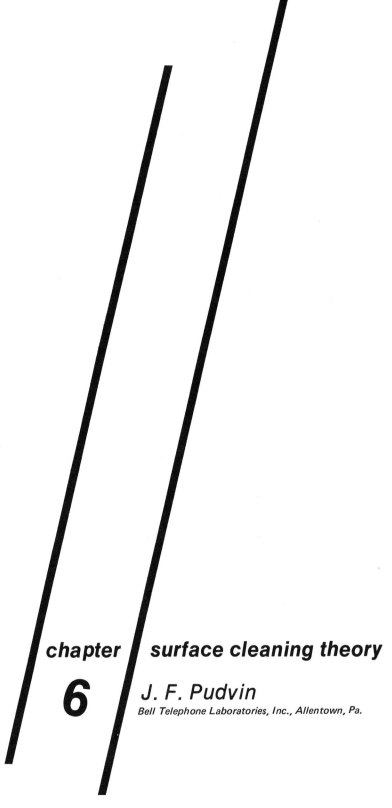

chapter

**6**

surface cleaning theory

*J. F. Pudvin*
*Bell Telephone Laboratories, Inc., Allentown, Pa.*

This chapter will be concerned with the concepts and the principles pertinent to the challenge of getting a surface clean. What is meant by a clean surface in an absolute sense will be described as well as what may be considered a clean surface from a practical engineering point of view. The major physical, chemical, and electrical properties of a surface that are of concern to the fabrication and behavior of electron devices will be considered, followed by a discussion of the nature of the forces operating at a surface which cause an interaction with contaminating solids, liquids, and gases. Since surface forces vary from one material to another, the emphasis will be placed on the semiconductor, metal, or ceramic materials commonly used for microelectronic devices.

The interaction between surfaces and contaminants can determine the approach used to achieve a specific clean surface. The major techniques of surface cleaning will be described in terms of disrupting these bonds. This discussion will set the stage for the subsequent chapter on surface practices utilized in electronic manufacturing.

## surface characteristics

### the "scientific" clean surface

The study of the crystalline nature of matter has resulted in a rather complete understanding of the bulk properties of hundreds of materials. In the interior of a single crystal the atoms behave as though there is an infinite and uniform periodic structure that can be described by mathematical equations. In most cases each atom is surrounded by other atoms in a symmetrical manner so that the electrical forces operating on a given atom are also symmetrical. At a crystal surface, however, the forces acting on a surface atom are no longer symmetrical and result in what has been commonly called a "dangling bond." The abrupt termination of the crystal lattice and the periodic electrical potential that describes it cause the surface atoms to be actually displaced from their ideal lattice position. These two conditions make a surface highly reactive towards any foreign matter outside the crystal. Hence, highly sophisticated methods have been necessary to obtain a perfectly clean crystal surface, or "scientific" surface as we shall call it.

It is appropriate to describe how our best "scientific" surfaces have been obtained along with the limitations of each method. This will be followed by a discussion of "real" or "engineering" surfaces and their chemical and physical properties.

From the instrumental analysis of many crystals it is known that a typical interatomic distance is the order of a few angstrom units. Although the number will vary with different crystal arrangements and depends on the particular interatomic distance observed, a typical number of atoms per square centimeter of surface is of the order of $10^{15}$, and this value is most frequently used in semiquantitative discussions of surface phenomena.

It then follows that if a solid could be purified to the point of only

1 foreign atom per $10^{10}$ there would still be 100,000 of these foreign atoms per square centimeter of surface under the best of conditions. As we shall see later, impurities may be concentrated at a surface, thus making our surface potentially even less pure.

Before discussing any technique for the preparation of a clean surface it is evident that it must be carried out in carefully constructed high vacuum systems where the residual gases are too small to have any effect on the experimental results. Most of the recent work in this field has been performed in systems where the residual pressure is less than $10^{-9}$ torr. The importance of these low pressures is readily seen from elementary kinetic gas theory.[1]

The number of atoms striking a surface is given by the relationship

$$s = \frac{p}{(2\pi mkT)^{1/2}} \tag{6-1}$$

where $s$ = the number of atoms striking a square centimeter per second, $p$ the pressure of the gas, $m$ the atomic or molecular mass of the gas, $k$ the Boltzmann constant, and $T$ the temperature.

For air at $0°C$ and 1 torr of pressure, the value of $s$ is $3.95 \times 10^{20}$ collisions per square centimeter per second. At a pressure of $10^{-7}$ torr, the value of $s$ becomes $3.95 \times 10^{13}$ $cm^{-2}$ $sec^{-1}$. At this pressure, if every molecule of gas that struck the surface remained there (corresponding to a sticking coefficient of 1), in 30 seconds the surface would be covered with approximately $10^{15}$ molecules or one monolayer. At a residual pressure of $10^{-9}$ torr this time for monolayer coverage would be extended to 3000 sec or 50 min—time enough for some experimental measurements on the surface. Sticking coefficients are usually less than unity and in fact may be as small as $10^{-12}$. Hence, the currently available high vacuum systems have been capable of maintaining a newly prepared "clean" surface for adequate times for scientific study.

What then has been actually obtained by investigators, even for short periods of time, as their best achievement of a "truly clean" surface? (We are now neglecting impurities that are in the bulk of the crystal and hence on the surface; rather, we are concerned about "foreign" atoms from outside the crystal that may react with the surface.) The work of H. E. Farnsworth and his colleagues at Brown University is typical of the research attempting to attain atomically clean surfaces.[2] The principal methods employed to obtain a surface with maximum cleanliness follow.

**Heating in a High Vacuum.** The sample is initially cleaned (usually by chemical etching) and then placed in a vacuum system and pumped down. Heating (and pumping) is done at successively higher temperatures until the maximum is reached. Allen and others[3] have shown that a clean surface for silicon can be obtained in this manner by heating at $1280°C$ for two minutes, or 5 minutes at $1380°C$ if greater than 300 Å of oxide is covering the initial surface.

This technique has been most useful for preparing surfaces of the refractory metals such as tungsten, molybdenum, and tantalum.

**Chemical Reactions at High Temperature.** The sample is heated to a suitable high temperature and exposed to an ambient gas that reacts with surface impurities to produce products that are volatile. Both oxygen and hydrogen are frequently used.

**Ion Bombardment and Heating.** The sample is placed in a vacuum system which is pumped and baked to low pressures (usually $10^{-9}$ torr). An inert gas, usually argon, is introduced to $10^{-3}$ to $10^{-4}$ torr. Ion bombardment of the sample is then carried out with accelerating voltages typically in the hundreds and with an ion current density of 10-100 $\mu a/cm^2$.

Damage to the lattice occurs as a result of this bombardment. Lattice atoms are displaced, and argon ions interact with the surface. Hence annealing at elevated temperatures to desorb the argon and heal the lattice defects is part of this method.

The net result of this treatment is the sputtering off of multilayers of the surface to expose a fresh, clean layer. Electron diffraction has shown these kinds of surfaces to have less than 5% of a monolayer coverage by foreign atoms. Materials with melting points as low as 600°C have been successfully cleaned in this way.

**Cleavage and Crushing.** This is a direct method for preparing a clean surface wherein a new surface is created from the bulk structure in a high vacuum by mechanical means. This technique has been widely used with semiconducting materials to prepare surfaces for detailed physical and electrical examination.[4]

**Limitations to a "Scientific" Clean Surface.** Farnsworth[2] cites many instances in which the surfaces prepared by these techniques have had some limitations: 1) bulk impurities may diffuse to the surface during heating and are not removed; 2) surface impurities may diffuse into the bulk (as well as be removed) and although not detected on the surface they may alter the electrical and semiconducting properties of the sample; 3) nonuniform thermal etching may occur due to varying stability of different crystal faces; 4) contamination may occur that is due to the thermionic emitter used in ion-bombardment; and 5) boron which has been transferred by water vapor from the Pyrex[*] glass of the vacuum system may deposit on the surface as a contaminant.

However, when these restrictions on the cleanliness of the surface have been minimized or overcome, these surfaces will serve for the measurement of photoelectric properties, oxidation rates on various crystal faces, and electrical surface properties of materials, especially those with semiconducting properties.

The philosophy that one should adopt on the matter of the degree of surface cleanliness is well put by Dr. Farnsworth: "Although one cannot obtain a completely clean surface, yet for large numbers of applications the

[*]Registered trademark of Corning Glass Company, Corning, New York.

contamination on the surface can be reduced to a point where it has no observable effect on the results."[4]   This is more or less the approach that underlies all cleaning steps employed in fabricating electron devices. The degree of surface cleanliness must meet two criteria: 1) it must be sufficient for subsequent processing, and 2) it must be sufficient to ensure the future reliability of the device or system.

### the "engineering" clean surface

The "practical" or "engineering" surface has been produced by some mechanical (sawing, polishing, lapping) or chemical (etching, sputtering) means and is now subject to oxidation, corrosion, and surface contamination. The surface is no longer in a vacuum chamber at $10^{-9}$ torr, but out in the open. Except for gold, all metals and semiconductor materials will immediately begin to oxidize in air at room temperature or, at least, be quickly covered with an adsorbed gas layer. Other contaminants found on these engineering surfaces can be catalogued as: inorganic residues, organic residues, chemically combined contaminants, and particles.

In recent years instruments and analytical techniques have advanced to a remarkable degree so that in certain instances we can virtually "see" individual atoms. Routine measurements now made on surfaces show that the best surface that can be actually utilized in fabricating electronic devices still contains the order of magnitude of a fraction of a monolayer of foreign atoms other than $H_2O$ or $O_2$, that is, approximately $10^{13}$-$10^{14}$ atoms/cm$^2$. This we will have to consider to be our ultimate "clean engineering" surface, at least at the time of this writing.

A surface, whether it be clean or contaminated, can be characterized by two sets of variables—those relating to its chemical structure and those relating to its physical structure. It is these properties that provide the information we need in order to specify a cleaning procedure for a surface. For example, what is the nature of the tarnish on the metal we must clean, how rough is the surface, are there grain boundaries, etc.? The nature of the surface chemistry will be discussed first.

## chemical structure of a surface

The chemical changes that occur on a surface are many-fold, but basically they fall into two categories: 1) those that depend on the local nature of the surface atoms such as their bonding, relative spacing and size, and the reactivity of the external ambient, and 2) those that depend on the presence and magnitude of free carriers in the bulk of the solid. The latter condition is of course very important in studying semiconducting materials.

### silicon surface chemistry

An example of this complex nature of a surface may be found in the research of Sotnikov and his colleagues on the adsorption of metal ions in dilute solutions onto silicon. To explain his results, Sotnikov[5] proposes the various surface structures for oxidized and hydrated silicon as shown in Fig. 6-1. The initial silicon surface is depicted in Fig. 6-1a as an orderly array (two dimensions only are shown) of silicon atoms with those atoms at the surface having an unpaired electron leading to the descriptive term "dangling" bond. On exposure to air, oxidation quickly occurs giving rise to the structure of 6-1b and on further oxidation to the structure of 6-1c. This surface is known as a siloxane structure.

In the presence of water, the silica surface that has formed can be easily hydrated to give Si–O–H bonds as shown in Fig. 6-1d. The Si–O–H termination is a silanol group and a surface covered by them is called a silanol surface.

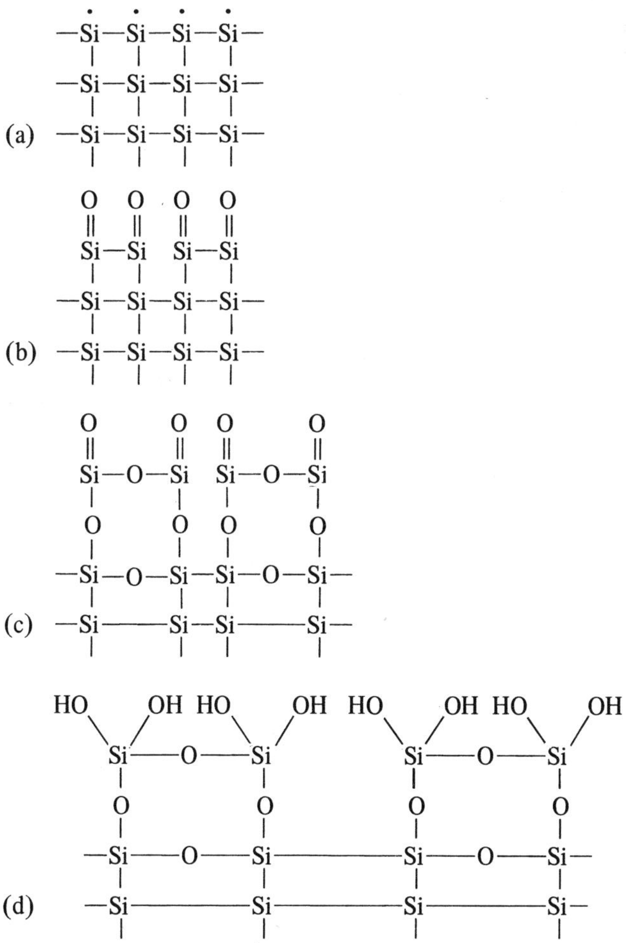

Figure 6-1 (a) Initial silicon surface; (b) oxidized surface; (c) oxidized surface (siloxane structure); (d) silanol surface (hydrophilic).

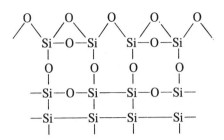

*Figure 6-2 Dehydrated silanol surface.*

If the silanol is heated to temperatures ranging from 200-600°C, dehydration will occur and the surface changes to the structure shown in Fig. 6-2.

The silanol structure is wet by water (hydrophilic) but the dehydrated surface is hydrophobic. In addition, certain dyes such as methyl red will adsorb on the silanol surface but not on the dehydrated surface.

Exposure of a silica surface characterized by silanol groups to hydrofluoric acid or HF vapors causes the OH groups to be replaced by F ions and the surface becomes hydrophobic. Thionyl chloride ($SOCl_2$) can be used to replace the OH group with Cl ions.

### adsorbed ions on surfaces

When a clean surface is placed in an electrolyte solution adsorption of impurity ions in the solution can take place on the surface. As a pertinent example of this adsorption, consider the data summarized in Table 6-1,[6] which shows the extent of adsorption of six elements onto $SiO_2$ film on silicon immersed in very dilute solutions of the contaminating ion. The extent of this is of more significance when presented as relative to the number of silicon atoms per unit area of surface. Since the number of sites on a surface is approximately $10^{15}/cm^2$, these foreign atoms may occupy from one in 10,000 sites to nearly one in 10 sites.

**TABLE 6-1.  Atoms Adsorbed per $Cm^2$ of Surface of $SiO_2$[a]**

| Contaminating Element | Electrolytic Solution | |
|---|---|---|
| | $HF/HNO_3(1:4)$, $20°C$ | 20% KOH, boiling |
| Au | $2 \times 10^{11}$ | $3 \times 10^{12}$ |
| Ag | $2 \times 10^{12}$ | $1 \times 10^{14}$ |
| Cu | $1 \times 10^{12}$ | $2 \times 10^{13}$ |
| Zn | $6 \times 10^{11}$ | $5 \times 10^{11}$ |
| In | $6 \times 10^{13}$ | $2 \times 10^{13}$ |
| Sb | — | $3 \times 10^{12}$ |

[a]For contaminant concentrations of $10^{-5}$ moles/liter in the electrolytic solution.

From V. S. Sotnikov, *Radiokhimiya*, 8, 171 (1966).

Considerable experimental observations[7] tend to confirm the fact that our current processing procedures for fabricating electronic devices consistently leave approximately $10^{13}$ atoms/cm$^2$ of foreign material on a surface. In etching a slice of silicon, for example, the etchant soon contains not only the silicon dissolved but the dopant atoms and diffused impurities that were in that layer as well. That significant adsorption onto the final silicon surface by these ions takes place is verified by Table 6-1. As a result of this adsorption galvanic action on a micro-scale is possible.

### segregation of impurities at interfaces

One of the many phenomena occurring at a surface that is of importance in considering contamination control is that small amounts of an impurity in the bulk material may concentrate to a considerable degree at the surface (or an interface). An excellent discussion of this kind of segregation is to be found in the text by Adamson.[8]   A classic example of this behavior is the case of small additions of ethanol to water which results in a marked reduction of the surface tension, an effect that may be accounted for if the alcohol molecule is selectively adsorbed (or concentrated) at the surface. The Gibbs equation mathematically describes this segregation behavior:

$$\Gamma = - \frac{1}{RT} \frac{d\sigma}{d(\ln c)} \tag{6-2}$$

where $\Gamma$ is the excess concentration of the adsorbing atom or molecule at the interface, $\sigma$ is the interfacial energy (surface tension), and $c$ is the equilibrium concentration of the adsorbing atoms or molecules in the bulk.

Experimentally, one measures the surface tension, $\sigma$, for various concentrations, $c$, of the additive. A plot of $\sigma$ vs $\ln c$ results in a straight line, the slope of which leads to the calculation of a value for $\Gamma$.

It is advisable to keep this phenomenon in mind when looking at various metallizing or brazing systems where impurities might produce this effect or where the molten metal might dissolve small amounts of another metal (such as gold in solder).

An interesting example of this effect is the increased wettability of copper on titanium carbide due to small additions of nickel.[9]   At low concentrations of nickel in molten copper the liquid-solid interface was found to have 24% of the area covered by nickel atoms ($5 \times 10^{14}$ atoms/cm$^2$).

Another example is reported by Kurkjian and Kingery,[10] who made small additions of titanium to liquid nickel wetting a substrate of alumina. A monolayer of titanium was observed at the interface at about 0.01% by weight. This corresponds to $12 \times 10^{14}$ atoms/cm$^2$ and may be compared to $15.3 \times 10^{14}$ atoms/cm$^2$ in the plane of densest packing of oxygen atoms in $Al_2O_3$.

There is still another type of situation where a surface may show a composition significantly different than its bulk composition. The iron-nickel-cobalt alloy (Kovar, Rodar)[*] is frequently used for making metal-to-glass seals and therefore finds considerable usage in the electronics industry. This material is usually lightly oxidized as received and therefore is treated in hydrochloric acid before oxidizing prior to glassing. Later this oxide is removed from the exposed glass-free areas and then gold plated.

The iron and cobalt are oxidized to a greater extent than the nickel so that the iron-cobalt-nickel ratio in the oxide is much different than that in the bulk metal. Table 6-2 shows these data from one experiment. When the oxide is removed, the remaining surface is now (comparatively) nickel rich.

TABLE 6-2.  Analyses of Fe-Ni-Co Surfaces and Oxide

| | Weight % (Approximate) | | |
|---|---|---|---|
| Element | Raw Material | Oxide | Pickled Surface |
| Fe | 54 | 72 | 36 |
| Co | 17 | trace | 16 |
| Ni | 29 | 28 | 48 |

## physical structure of a surface

### semiconductor surfaces

In the real world of manufacture "engineering" surfaces (unlike "scientific" surfaces) are prepared outside of ultrahigh vacuum systems in a relatively contaminated world (minimized of course by the exploitation of clean rooms) using such industrial equipment as saws, grinders, polishers, and metal extrusion and shaping facilities. The materials are not all single crystals but polycrystalline and frequently not even single phase in nature.

Of course, these surfaces are subsequently cleaned to remove contaminants and their effects. But the new surface thus revealed has a physical structure and characteristics that can have profound effects during later processing. A brief survey of the problems associated with the physical structure of a surface follows. Semiconductor surfaces will be discussed first, then the other materials used for fabricating devices such as metals and alloys, ceramics, and glass.

The ingots of semiconductor materials are first sawed into slices which are then lapped and polished to a predetermined thickness. It is well known that such treatment produces a layer of damaged material that has properties far different than those of the bulk. Therefore, this layer is removed by chemical etching to a sufficient depth to eliminate the mechanical damage.

When such surfaces are thus revealed it is found that the single crystal has several kinds of defects and imperfections in it that have been accentuated by the etching operation. Surfaces are not atomically smooth but have a

[*]Kovar is the registered trademark of Westinghouse Electric Corp., Pittsburgh, Pa. Rodar is the registered trademark of Wilbur B. Driver Co., Harrison, N.J.

certain micro-roughness that can affect the perfection of epitaxially deposited layers. It has also been demonstrated that the different crystal faces of semiconductors (and other materials as well) show different etch rates and different oxidation rates.

In addition, the etching step can easily contaminate the surface to the extent described previously. The combination of these impurities and surface defects results in profound electrical effects in the semiconductor. Consult the references for a detailed description of these effects.[11–14]

## metals and ceramics

A wide variety of metals and alloys along with glass and ceramics are used as hardware for encapsulation or as supporting structures for electronic devices. As such they are usually covered with sputtered or evaporated metallized patterns or are electroplated with gold.

Most ceramic materials that have been ground to achieve a certain surface finish have been found to retain a significant amount of residual fine ceramic powder. Experience has shown that it can lead to poor adhesion of metallized patterns and that the powder should be removed before any cleaning and processing by the use of ultrasonic energy.

Surface roughness is a very important property if thin metal films are to be evaporated over the surface. In extreme cases the roughness may be great enough to create holes in the film due to shadowing. On the other hand, a certain degree of roughness may provide additional adherence for the deposited film due to "keying." Microcracks have also been observed in ceramic materials and directly related to faults in subsequent behavior. They can usually be detected by the use of fluorescent dyes that are concentrated in these defects.

Still another type of surface defect which can be detected and which can lead to defects later on is the heterogeneous character of some ceramics and deposited films. An example will illustrate this. High alumina ceramics have been carefully examined for microcracks by the use of fluorescent dyes.[15] In this inspection it is frequently found that large areas (approximately $1/8$ to $1/4$ in. in dimension) appear to lightly retain the dye to an extent that it can be detected while the rest of the surface is completely void of the dye. These areas can probably be explained by a hetergeneous concentration of the fluxing agent in the ceramic.

Metal parts that have been mechanically shaped by machining, deep drawing, or wire drawing may frequently have very poor surface structures. Under several hundred magnification, pits, cracks, and considerable porosity can be seen; in the most obvious cases, ten power magnification is sufficient to detect these defects. These metal parts are usually lightly oxidized so the metal cleaning process is frequently an acid pickle. This usually accentuates the surface condition.

A very porous surface will usually lead to small blisters in electroplated finishes. Both organic contaminants and air may be trapped in these pores and are subsequently covered over by electroplates of low porosity. On heating these parts—usually in air and at a fairly fast temperature rise—the trapped gases and the products of decomposition of the organics expand rapidly and cause blisters in the electroplate before they can completely escape. Cross-sections of blistered materials almost always show some pit or pore as the physical origin of the blister.

A roughness of the surface on a microscale, which is sometimes seen when cross sections are made, can be due to grain boundaries. Some metals, copper being an excellent example, can oxidize in such a fashion as to build up significant amounts of oxides between the crystal grains at the surface. Subsequent pickling to remove surface oxides will also remove the oxides between the grains, leaving microcavities. They too can act as traps for contaminants. A hydrogen brazing operation will also lead to the reduction of these intergranular oxides and open up the surface structure. For copper this condition is minimized by the use of oxygen-free copper as the raw material.

## forces active at surfaces

For the purposes of this chapter it is important to know something of the forces that are acting between the surface and the contamination on it. This contamination may be in the form of a particle of dust, adsorbed gases, an oxide, or a film of some spreading liquid. In all cases it would be desirable to know the nature of the operating forces and an estimate of their magnitude. In the case of weak forces (physically adsorbed gases), heating to room temperature only will break the "bonds," whereas an oxide film can be removed only by active chemical attack.

It will be helpful to describe, qualitatively only, the typical kinds of forces with regard to their origin and magnitude. Then we shall see what forces are of importance with a solid-gas interface and a solid-liquid interface.

### chemical forces

**Chemical Bond—Ionic.** A pure ionic bond arises when there is a complete transfer of an electron creating a positive and a negative ion. The force between these two ions is a Coulomb type force, that is:

$$F_i = \frac{q^+ q^-}{r^2} \tag{6-3}$$

where $F_i$ is the magnitude of the Coulomb force, $q^+$ and $q^-$ are the charges on the ions, and $r$ the distance between them. Because of the inverse square nature of the force of attraction, it decreases with distance but at a rate that is much less than that for other types of forces. Mathematical attempts to calculate the binding energy of an ionic crystal must include a summation over many numbers of neighbors to account for the total energy.

A crystal of NaCl is usually given as an example of ionic bonding and is, therefore, frequently written as $Na^+Cl^-$. The crystal which is most typically ionic in character is CsF since the tendency for a complete transfer of the electron from the Cs to the F is the maximum to be found.

**Chemical Bond–Covalent.** Unlike the ionic bond, there is no transfer of an electron from one atom to another; rather, it is shared between the two. As a result there is a physical accumulation of two or more electrons within the region of overlapping orbitals of the two atomic nuclei and, therefore, there is a directionality to the covalent bond.

The force is given by the following expression:

$$F_c = 6A \left[ \frac{1}{r^7} - \frac{r_0^6}{r^{13}} \right] \tag{6-4}$$

$F_c$ is the magnitude of the covalent force, $A$ is a constant, $r_0$ is the equilibrium distance between the two atoms, and $r$ is the variable distance between the two atoms. It is easily seen that the term $1/r^7$ makes these forces operable over a short range only.

The covalent bond is the type of bond to be found in $H_2$, $Cl_2$, $N_2$, $S_8$, $P_4$, diamond, Ge, and Si and in most organic molecules.

Both experimental and theoretical work have confirmed the fact that most bonds are hybrids and show partial ionic and partial covalent bonding. That is to say, there is both a transfer and a sharing of an electron in the bond. Table 6-3 shows the estimated ionic character between various pairs of atoms calculated from Pauling's data.[16]

TABLE 6-3.   Estimated Fractional Ionic Character of Single Bonds

| Bond | Ionic Character, % |
|------|--------------------|
| CsF  | >90 |
| BeO  | 63 |
| AlF  | 60 |
| CaCl | >50 |
| SiO  | 50 |
| CF   | 44 |
| BeI  | 22 |
| C–H  | 4 |
| C–C  | 0 |

From *Nature of the Chemical Bond*, L. Pauling, Cornell U. Press, Ithaca, N.Y.

**Chemical Bonds–Metallic.** Qualitatively, a metal may be thought of as consisting of an array of "ion cores" surrounded by the freely moving valence electrons. The positively charged cores have a repulsive force operating inversely as the square of the distance. However, there is also an attractive force, referred to as a dispersion force or London force, which is smaller in magnitude than the repulsion Coulomb force but is operable over only very short ranges. This dispersion force will be described later.

The complete picture of metallic bonding is complex, mathematically unsatisfactory, and beyond the scope of this book. The texts by Kittel[17] and Seitz[26] are recommended for further study.

### intermolecular (van der Waals) forces

Dipole-dipole. If a molecule has a nonsymmetrical distribution of charges between its atoms, then it possesses a permanent dipole moment, $\mu$, given by the expression

$$\mu = q \times l \qquad (6\text{-}5)$$

where $q$ is the magnitude of the separated charge, and $l$ is the distance between them. Typical dipole moments (as esu-cm) are: water 1.85, ethanol 1.7, benzene 0, acetone 2.85, and nitrobenzene 3.9. Dipoles can exert forces on one another because of the apparent charge separation. The magnitude of this force is small but finite and is not easily expressed by a simple mathematical formula.

A special case exists for the dipole resulting from the bonding of hydrogen to a strongly electronegative atom such as O, Cl, F, and N. The magnitude of the resulting dipole is so strong in these instances that two molecules are held rather tightly together. An example of this is HF, which in solution behaves as though the dissolved species was $H_2F_2$ or $(HF)_2$. This situation is known as hydrogen bonding.

Dipole-Induced Dipole. A symmetrical molecule in the vicinity of a dipole can become momentarily polarized (that is, there is a slight separation of charges) so that it acts as a temporary, induced dipole. As a result there is a momentary force binding the two species together as though they were two dipoles. These forces are of importance in accounting for the cohesion and adhesion of polymers.

London Dispersion Forces. Pairs of symmetrical molecules have no dipole-dipole forces or induced forces acting on one another, yet they show binding forces nevertheless, as witness the fact that oxygen and nitrogen do liquefy and form solids. The concept of the dispersion force was developed to account for this type of attraction. As Hirschfelder describes it, "At any instant the electrons in molecule $a$ have a definite configuration, so that molecule $a$ has an instantaneous dipole moment (even if it possesses no permanent electric moment). This instantaneous dipole in molecule $a$ induces a dipole in molecule $b$. The interaction between these two dipoles results in a force of attraction between the two molecules. The dispersion force is then this instantaneous force of attraction averaged over all instantaneous configurations of the electrons in molecule $a$."[18]

It may well be seen at first that this type of force is something obscure, rare, and a pure figment of the mind. Yet it has been calculated that the dispersion force is responsible for 40% of the surface tension of mercury and is the principal force acting between Ar, $CH_4$, and $CO_2$ molecules.

## force magnitudes

Table 6-4[19] lists the approximate magnitudes of each of the types of forces mentioned. The information we need on bonding forces can be summarized in the following manner: There are primary bonds (ionic, covalent, and

### TABLE 6-4.

| Type of Force | Energy (kCal/mole) |
|---|---|
| Chemical bonds: | |
| Ionic | 140–250 |
| Covalent | 15–170 |
| Metallic | 27–83 |
| Intermolecular (van der Waals) forces: | |
| Hydrogen bonds | $< 12$ |
| Dipole-dipole | $< 5$ |
| Dipole-induced dipole | $< 0.5$ |
| Dispersion | $< 10$ |

From "Intermolecular and Interatomic Forces," by Good Marcel Dekker, Inc.

metallic) that result from forces arising in the interaction of the outermost electrons of atoms. As a class they are large in magnitude and are the principal factor in the chemical bonds between atoms. There are also secondary bond forces—intermolecular forces—which result from forces arising in the inter-action between permanent and/or temporary dipoles that exist in molecules. They are small in magnitude and are not important to the formation of stable chemical compounds. They do affect the properties of materials such as volatility, viscosity, solubility, miscibility, and surface tension.

## interactions with surfaces

An interface is the boundary arising between two phases (especially between liquids and solids). If the interface is one involving a gaseous phase then the interface is usually called a surface.

It has been aptly put by A. S. Michaels that "All special properties of surfaces or interfaces arise fundamentally from precisely the same forces that are responsible for the existence of matter in condensed (that is, liquid or solid) states."[20] Furthermore, the situation is simplified by the fact that it is not possible for each of the types of forces mentioned to interact at an inter-face with all other forces. As an example, the intermolecular forces of organic compounds cannot interact with the metallic bonds in mercury nor can the hydrogen bonds of water interact with polyethylene.

In dealing with the interaction of contaminating gases, liquids, and solids (particles) with a solid surface we shall make use of these various forces to describe such behavior as binding strength, magnitude of the contamination, and difficulty and method of removal of the contaminant.

The discussion will begin with the interaction of gases with surfaces, that is, oxidation, reduction, physical and chemical adsorption, and degassing. Then we will summarize some of the experimental data concerning the adhesion to and removal of particulate matter from a surface. Organic contamination will then be reviewed.

### oxidation

There are several books[21-23] and papers[24-25] that cover the essentials of the oxidation of materials and discuss the experimental data in the literature. These sources should be consulted for specific information. Only a few of the more important principles can be discussed.

When a clean surface, a freshly etched silicon surface, or a vacuum deposited film is brought out into the air some degree of oxidation will occur immediately. At elevated temperatures the reaction is more rapid, and a few metals may even react with nitrogen. In the presence of water and carbon dioxide a more complicated reaction (corrosion) can occur. Essentially what is happening?

All metals except gold react in air to form oxides at room temperature. The rates vary considerably as one might expect from the differing chemistry of the elements. As the temperature is raised the rate of oxidation will increase rapidly, and at very high temperatures some metallic elements will also form nitrides. Two elements, silver and palladium, form oxides that are relatively unstable so that at several hundred degrees above room temperature these oxides decompose. For most metals the temperature dissociation of the oxide or oxides formed is above the boiling point of the metal.

Oxidation can occur fairly rapidly on certain elements at room temperature. The amount of oxide increases rapidly during the first few hours and then approaches a maximum. For example, at the end of 48 hours silver is covered with an oxide film that is 10-15 Å thick, aluminum with a 25 Å thick film, iron with 35 Å, and copper with 45 Å. Although these are very thin oxides, their presence may easily influence further processing. They may lead to poor adhesion of a covering electroplate if not removed or blistering of the electroplate during a subsequent hydrogen heat treatment or brazing operation due to the reduction of the oxide and the release of water vapor. In general, oxidation is not a simple type of chemical reaction, because as the temperature changes the rate equation describing the reaction may significantly change. If the rate of oxidation is monitored by a weight change, $\Delta m$, then the rate equations take the following forms with respect to the time, $t$:

| | |
|---|---|
| Linear: | $\Delta m = k_l \cdot t$ |
| Parabolic: | $(\Delta m)^2 = k_p \cdot t$ or $= k_p \cdot t + c$ |
| Cubic: | $(\Delta m)^3 = k_c \cdot t$ |
| Logarithmic: | $\Delta m = k_e \cdot \log(at + t_0)$ |
| | $1/\Delta m = A - k_e \cdot \log t$ |

It is very common for an oxidation to be described by two or more of these rate equations. The oxidation of iron, for example, is logarithmic to $200°C$ and then parabolic. Nickel is more complex, being described best by logarithmic to $300°C$, cubic to $400°C$, and parabolic above $400°C$.

These changes are not unreasonable when one considers that at some point the oxide film essentially covers the metal surface completely and that from this point on further oxidation requires the diffusion of some chemical species through the oxide, this being described usually by an entirely different rate.

As an oxide builds up to thickness a change in composition of the oxide is sometimes detectable. An element such as iron can form oxides of different atom ratios because of its multiple outer electronic arrangements (valence). This leads to an oxide film that changes in composition in the following manner. At the metal interface the oxide is richest in metal $FeO$ (Fe:O = 1:1). Next can be found $Fe_3O_4$ (Fe:O = 1:1.3), and at the outside, $Fe_2O_3$ (Fe:O = 1:1.5), which is rich in oxygen. Important points to remember are that these oxides may not be equally soluble in an acid designed to remove this film, or their rates of reaction may not be equivalent. This may lead to an incomplete or spotty removal of the oxide. A specific instance is nickel, where $NiO_2$ is not soluble in hydrochloric acid as is $NiO$.

The forces binding the oxide to the metal surface are truly primary bonds, i.e., ionic and covalent. These forces are quite large for oxides; in fact, they represent some of the strongest bonds. This indicates that the removal of an oxide will require rather vigorous action to overcome these bonds. Indeed, the oxides are usually removed only by strong acids at elevated temperatures.

When a low temperature oxidation occurs in the presence of moisture and carbon dioxide, the final product is best described as a corrosion product. One of the most common examples of such an occurrence is the green coloration which builds up on copper used as building materials.*

At high temperatures water vapor can cause oxidation of certain elements. Use is made of this fact in decarburizing Kovar wherein the alloy is heat treated at $1000$-$1100°C$ in wet hydrogen. The carbon in the surface layers is oxidized by the $H_2O$, but the hydrogen ambient keeps the iron, nickel, and cobalt of the alloy reduced. The conditions—namely temperature, concentration of water vapor, and concentration of hydrogen—necessary to do this sort of selective oxidation have been summarized in thermodynamic charts by Elliot and Gleiser,[28] and Swalin.[29] The principle on which these charts are based is the magnitude and sign of the free energy of the oxidation reaction.

The ability of some of the common metals to oxidize very rapidly when their surface is really clean has to be seen to be fully appreciated. Two

*An interesting example of this type of corrosion is the green colored patina on the Statue of Liberty. Recent analysis showed this green color to be over 95% basic copper sulfate and less than 0.1% copper carbonate. A mild scientific argument in 1962–63 was finally settled by chemical analysis.[27]

common examples are copper and molybdenum. When they have been completely degreased and lightly etched by an appropriate acid to remove the oxide and then washed in high purity water both will tarnish within a few minutes while in the water washing, especially if the water is warm (50-55°C).

It is also common practice to subject substrates to either boiling hydrogen peroxide (15%) or ozone (concentrations as high as 10%) at 100-150°C to ensure the complete removal of residual organic material. Both of these reagents will oxidize most metals and many alloys, hence this treatment must be followed by removal of the oxide film by acid etching or by reduction in hydrogen.

## reduction

The usual practice in fabricating electronic devices is to remove most oxides and tarnishes from metals and alloys by chemical means rather than by hydrogen reduction. This is dictated by the fact that to be effective the reduction must be carried out generally in the 400-550°C range, and this is usually too high for the device or assembly being cleaned.

However, a hydrogen reducing atmosphere is used for most brazing operations and many heat treatments. The presence of a light oxide film on the metals to be brazed is of little consequence since it is usually easily reduced before the braze material becomes molten and flows over the metal.

A simple reduction reaction is that for nickel oxide:

$$2\,NiO + H_2 \longrightarrow 2\,Ni + 2\,H_2O$$

It is seen that the Ni—O bonds are broken in this reaction and that the oxygen atoms are subsequently bound to hydrogen. The water thus formed is evidently a more stable oxide at those temperatures than is nickel oxide. It is important to sweep out of the furnace the water that is formed from this reduction so that it cannot build up in concentration and cause the reaction to proceed in the opposite direction to form an equilibrium condition where both Ni and NiO coexist.

There have been situations where parts placed in a hydrogen furnace for a heat treatment have come out badly oxidized. Two typical causes have been documented. In one instance the furnace muffle was cracked allowing air into the end of the hot zone where the hydrogen and oxygen combined to give high concentrations of water vapor which was carried into a cooling zone. This was sufficient to oxidize the parts. Another condition arose in the following manner: Silicon slices that had been gold plated were placed immediately into the hot zone of a tubular furnace. Due to the very small heat capacity of the silicon they reached the temperature (450°C) of the furnace in a very short time. In this time period the adsorbed air and moisture still surrounded the silicon and the silicon-gold interface was oxidized leaving a nonadherent gold plate. This condition was easily corrected by a slow temperature rise, providing for complete degassing before achieving the hot zone temperature.

A somewhat similar situation can occur (that is, oxidation in an hydrogen ambient) in the following set of circumstances. A copper tubulation, for example, is held in a blind hole in a fixture while being brazed to some other part. If the tubulation fits tightly, the copper may be subjected to oxidation due to adsorbed moisture and air. No subsequent reduction can take place due to essentially no circulation of hydrogen at this surface. Cutting a series of slots in this fixture to give access to the hydrogen corrected this condition.

If it were feasible to design an industrial size facility for the generation and use of atomic hydrogen at levels of 5-10%, then reductions could be carried out at much lower temperatures. As a quick guide, atomic hydrogen is able to reduce common oxides at a temperature of about 150°C below the minimum reduction temperature for molecular hydrogen.[30]

### adsorption of gases onto surfaces

If a gas is allowed to come to equilibrium with a solid surface a concentration of gas molecules occurs at the surface, i.e., the gas concentration on the surface is always greater than the concentration of the gas molecules in the free gas phase. The general process is called adsorption, and depending on the nature of the forces holding the gas molecule to the surface, the process may be further classed as physical adsorption (physisorption) or chemical adsorption (chemisorption). A third kind of behavior may occur, namely absorption, wherein the gas molecules (or atoms) actually penetrate into the bulk material. The first two of these processes will be briefly reviewed: first, as to their similarities and their differences, and then as to the details of each and their importance in obtaining a clean surface.

Thermodynamically, adsorption is a spontaneous process and is therefore characterized by a decrease in free energy of the total system. The orientation in the film of adsorbed gas also reflects a loss in degrees of freedom and hence a decrease in entropy. Consider the familiar thermodynamic equation

$$\Delta F = \Delta H - T\Delta S \qquad (6\text{-}7)$$

where $\Delta F$ is the free energy, $\Delta H$ is the heat content, and $\Delta S$ indicates the entropy of the system. Equation 6-7 dictates that adsorption is always exothermic (heat is evolved) regardless of the type of binding force involved.

In anticipation of the more detailed discussion of each process, consider the criteria that are used to characterize physical adsorption from chemisorption:

1. Physical adsorption is caused by relatively weak van der Waals' (dispersion) forces; chemisorption is caused by electron transfer and chemical bond formation.

2. The heat of physical adsorption is about the same as the heat of vaporization (or heat of liquefaction) of the gas involved, certainly no more than two or three times greater. Chemisorption usually involves heats comparable to those for chemical reactions (except for certain cases that are quite low).

3. Physical adsorption can occur with any gas and any substrate under the proper conditions of pressure and temperature. Chemisorption, on the other hand, is highly selective.

4. Chemisorption is completed with the adsorption of just one monolayer. Physical adsorption, in contrast, can be made to be multilayered and may even occur over a chemisorbed layer.

5. Since physical adsorption is related to liquefaction, it occurs usually at low temperatures and at relatively high pressures. Chemisorption proceeds ordinarily at higher temperatures and at much lower pressures.

6. Desorption is usually much more difficult for chemisorbed layers because of the higher bond strengths and may result in different gases being evolved than were adsorbed.

7. Both physical and chemical adsorption are almost instantaneous. Also complications may enter into the process for both types. This characteristic is not distinguishing.

The literature is rich in details and in both theoretical and practical conclusions. The references given are of special interest to the chemist.[2,31-35]

**Physical Adsorption.** Experiments on physical adsorption systems generate data as to the pressure of the gas phase, the volume of gas adsorbed on the surface, and the temperature. These data are usually summarized in one of two ways. When the volume of gas absorbed is plotted as a function of pressure at constant temperature, the resulting curves are called isotherms. If the measured pressure of gas at a constant volume is plotted as a function of temperature, the curves are called isosteres.

A review of all adsorption data suggested to Brunauer that they fell into one of five typical categories. Curves (isotherms) representing this behavior are depicted in Fig. 6-3. Each type can be qualitatively interpreted in the following way (which affirms the wide variety of situations which can arise in physical adsorption): Type I describes simple monolayer adsorption; Types II and III, forms of multilayer adsorption; Types IV and V are monolayer and multilayer adsorption, respectively, with the addition of capillary condensation.

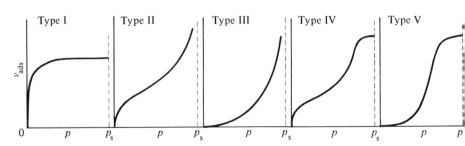

Figure 6-3 The five isotherm types in physical adsorption. (Courtesy American Society for Testing and Materials, Special Technical Publication No. 340, 1963.)

The literature is liberally sprinkled with papers dealing with equations that attempt to describe the experimental data and relate it to physical processes. The equations that seemingly have received the greatest usage over the years have been those associated with Freundlich, Langmuir, and Brunauer-Emmett-Teller (BET).

Much attention has also been given to the nature of the forces between the adsorbant and the gas molecule or atom. The largest forces involved are the previously described London dispersion forces with magnitudes that vary inversely as the 6th power of the distance. To a lesser extent there is the force arising between an electrostatic field at a surface and an adsorbate molecule with a dipole moment or one temporarily induced in it if the molecule is polarizable. Papers that review the latest thinking about these forces are to be found in the references.[35-37]

It has been pointed out that these forces are all relatively weak so that little energy is needed to break the bonds (2–5 cal/mole). For most situations, the gaseous layers held by physical adsorption are easily removed by raising the temperature, thereby providing the molecule with sufficient thermal kinetic energy to overcome the bonding forces. For many situations, this energy is achieved at temperatures well below room temperature. As a consequence, there is little concern about physically adsorbed gases not being eliminated by normal processing. However, if the substrate is porous, some difficulty may arise from the behavior shown qualitatively in Fig. 6-4.[38] The ordinate values for this curve are the weights of water adsorbed per unit

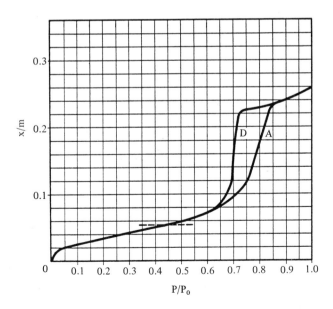

*Figure 6-4  Adsorption isotherm of porous glass at $25°C$. D and A are desorption and adsorption branches.  The dotted line indicates the monolayer.  (After M. J. Rand in Journal of the Electrochemical Society, 109, 402, 1962.)*

weight of dry substrate $(x/m)$ and the abscissa values the relative humidity of the ambient expressed as a fraction.

The hysteresis effect is shown for the desorption cycle and is attributed to the existence of a porous surface structure. This type of behavior can be utilized to measure the extent of the pores and the distribution of pore sizes on a material substrate (see Brunauer and Copeland[33]). A form of porous Vycor* has been used as a getter for water vapor in hermetically sealed enclosures for some kinds of transistors. Its efficacy was based on this type of behavior.[38]

Perhaps our greatest interest in physical adsorption phenomena lies in the use of its experimental techniques to measure the surface area of solids and/or powders (and to measure pore sizes). This sort of information is vital to the selection of materials to be used as an adsorbant substrate, activated charcoal being one of special interest because of its use in removing organic materials from ambient air. Dehydrated silica and activated alumina are others used for their water removal capabilities.

Surface roughness can also be measured as well as the detection of surface cracks and crevices. There is also strong evidence (supported by physical adsorption studies) that surfaces exhibiting these features may even act as a catalytic surface.[39]

The adsorption equation by Brunnauer-Emmett-Teller, known as the BET equation, leads us to a direct measure of surface area. Values of the surface area range from essentially the apparent surface area to values of over 1000 sq. in. per gram for some special carbon absorbents. The references provide more information about this method and, also, how further information may be obtained about the homogeneity of a surface being measured.[31,33,35]

**Chemisorption.** Chemisorption differs from physical adsorption in several significant ways. The energies binding the chemisorbed species are much larger in the range of 15-20 kcal/mole compared to 2-5 cal/mole. Whereas physical adsorption is most prominent at temperatures near the boiling point of nitrogen, chemisorption occurs at room temperature and higher.

The desorption of chemisorbed gases is much more difficult than the desorption of physically adsorbed gases because of the greater binding forces; it is further complicated by the fact that the evolved gases are not always the same as those adsorbed. For example, CO adsorbed on cuprous oxide is desorbed as $CO_2$, $SO_2$ adsorbed on $Cu_2O$ is desorbed as $SO_3$, and $H_2$ adsorbed on chromic oxide comes off as $H_2O$.

It was stated before that chemisorption was highly selective. G. C. Bond[40] has summarized the existing data of chemisorption in Table 6-5. The significant conclusions are that oxygen is almost universally chemisorbed and that the transition metals are outstanding for their broad capabilities for

*Registered trademark of Corning Glass Company, Corning, N.Y.

TABLE 6-5.    Classification of Metals and Semimetals Based on Adsorption
Properties.    (A indicates Adsorption, NA No Adsorption)

| Group | Metals | $O_2$ | $C_2H_2$ | $C_2H_4$ | CO | $H_2$ | $CO_2$ | $N_2$ |
|-------|--------|-------|----------|----------|-----|-------|--------|-------|
| | | | | | *Gases* | | | |
| A | Ca, Sr, Ba, Ti, Zr, Hf, V, Nb, Ta, Cr, Mo, W, Fe[a], (Re) | A | A | A | A | A | A | A |
| $B_1$ | Ni, (Co) | A | A | A | A | A | A | NA |
| $B_2$ | Rh, Pd, Pt, (Ir) | A | A | A | A | A | NA | NA |
| C | Al, Mn, Cu, Au[b] | A | A | A | A | NA | NA | NA |
| D | K | A | A | NA | NA | NA | NA | NA |
| E | Mg, Ag[a], Zn, Cd, In, Si, Ge, Sn, Pb, As, Sb, Bi | A | NA | NA | NA | NA | NA | NA |
| F | Se, Te | NA | NA | NA | NA | NA | NA | NA |

[a] The adsorption of $N_2$ on Fe is activated, as is the adsorption of $O_2$ on Ag films
sintered at $O°$.
[b] Au does not adsorb $O_2$.
( ) Metal probably belongs to this group, but the behavior of films is not known.
From *Catalysis of Metals*, G. C. Bond, Academic Press, London, 1962.

chemisorption.    It is these metals and/or their oxides that serve as the basis
for industrial catalysts.

Studies of catalytic systems lead one to the conclusion that many surfaces are quite heterogeneous in nature and show significant variations in their
degree of adsorption.    We realize that surface atoms have unbalanced forces
but those at edges and peaks are especially unbalanced or unsaturated and
hence of higher chemical reactivity.    These sites are commonly called "active
centers" or "active sites."    Lattice defects can also behave as active sites.
This concept of preferred adsorption on a surface seems to be supported by
experiments in which small amounts of an impurity can "poison" a surface
for chemisorption, especially one that shows catalytic activity.

The type of experimental data which is gathered and the manner in
which it is presented in graphical form is basically the same as for physical
adsorption.    However, one finds that the rate of chemisorption varies with
temperature in a manner to indicate that there is an energy of activation.
This means that of the molecules of gas striking the surface, only a few having
sufficient energy will interact with the surface and result in chemisorption.
Clean metals tend to have small activation energies while oxide surfaces
usually are characterized by appreciable activation energies.

As a molecule adsorbs on a surface at elevated temperatures, several
types of mechanisms may take place.    The molecule may simply be adsorbed
as the entire molecule in a nondissociative process.    In other cases as the
molecule adsorbs on a surface it may be strained by its attempt to "fit"
onto the substrate to the point where its intermolecular bonds are broken

(dissociative). Thirdly, the molecule may react with the surface to form a new chemical compound.

It has been observed that a molecule can apparently be both physically and chemically adsorbed on the same surface. Hair[34] reports the case of $CO_2$ adsorbed on a $NiO-SiO_2$ surface.

Chemisorption also extends to many complex organic molecules. Any manufacturing area is significantly contaminated with organic vapors arising from vacuum pumps, lubricants, solvents, etc. We will describe later the detection of these organic vapors by an increase in the contact angle of a drop of water placed on the substrate initially and then after exposure. A detector consisting of a lightly oxidized sheet of nickel, nichrome, or aluminum is utilized. This is an example of chemisorption, and indeed the best metals to adsorb these impurities are those in the transition series. Use is made of this property for aiding the clean storage of parts and devices.

The references include an excellent summary of studies on the chemisorption of oxygen on semiconductor surfaces.[13]   Sticking coefficients for several semiconductor surfaces are given in Table 6-6. The values simply

TABLE 6-6.  Sticking Coefficients of Oxygen for Zero Surface Coverage

| Crystal Plane | [111] | [100] |
|---|---|---|
| Germanium | $10^{-3}-10^{-4}$ | $10^{-3}$ |
| Silicon | $10^{-1}$ | $10^{-2}$ |
| Gallium Antimonide | $10^{-5}$ | |
| Indium Antimonide | $10^{-6}$ | |

represent the fraction of collisions between the oxygen molecule and the silicon (for example) that result in adsorption. A value of unity would mean that every collision results in adsorption.

A clean surface exposed to oxygen results in a rapid adsorption of about a monolayer (about seconds to minutes in the pressure range of $10^{-1}$ to $10^{-3}$ torr). The next few layers of oxygen are found to adsorb much more slowly. The monolayer is probably held by primary chemical bonds and the next few layers by dispersion forces. Further layers are probably associated with ordinary oxidation processes involving the diffusion of some species through the existing oxide layer.

The mathematical analysis of chemisorption data is well summarized and analyzed in the noted reference[32] which in turn cites the original literature. The same text presents the various attempts to calculate heats of adsorption based on the breaking and formation of bonds. Some of these calculated energies are very close to the experimentally observed values.

### degassing of materials

By various mechanisms considerable amounts of gases can be more or less dissolved in the many materials which are used in fabricating electronic devices. The problem of driving these gases out of materials by heat treatments or vacuum baking is difficult and complex. An excellent summary of degassing experiments on materials of interest to the electronics engineer are included in the references.[1,41-43] The results of the experiments that are liberally quoted in these references can serve only as a guide to your own degassing procedures since the previous history of the material is of utmost importance in determining the extent of dissolved gases.

Some general conclusions concerning the degassing of materials can be drawn from the references cited.

1. The sources of the gas obtained on vacuum fusion are: Gases adsorbed on the surface; gas found in the bulk of the material in true solution, or in the form of a chemical compound, or simply trapped in pores, blowholes, or dislocations; gas from chemical compounds on the surface; gas from the decomposition of lubricants on the surface.

2. The principal gases evolved from materials are water vapor, hydrogen, carbon monoxide, carbon dioxide, nitrogen, and oxygen.

3. As a metal is heated to its melting point the first gases which are released are water vapor and oxygen from adsorbed layers. At intermediate temperatures water vapor, hydrogen, and carbon monoxide are evolved. At the fusion temperature and above, carbon monoxide is a principal gas with smaller amounts of nitrogen, oxygen, and hydrogen.

4. The largest percentage of total gas evolved is at temperatures above the fusion point. The total volume of gas released (at S.T.P.) is usually less than the volume of the metal.

5. The total gas evolved depends on the metallurgical processes the metal has undergone, the kind of anneal it may have seen (and if annealed in hydrogen it will depend on the purity of the hydrogen), and the method of cleaning the surface.

6. Many metals when degassed will continue to remain degassed (except for a monolayer or two of adsorbed gas) if properly handled and stored.

7. Materials that contain finite amounts of carbon can produce CO or $CO_2$ by its reaction with dissolved oxygen or surface oxides. Collins and Turnbull[41] studied this phenomenon in nickel and were able to correlate the evolution of CO and $CO_2$ with the rate of diffusion of carbon in nickel.

8. Degassing materials that contain oxygen, especially in the form of thin oxide films at grain boundaries, in a hydrogen ambient can lead to the rapid formation of water which could cause intercrystalline cracking. This has been reported for copper.[44]

Glass is a special type of material compared to metals from the point of view of degassing. Holland[45] gives a great deal of useful information on this subject. As glass is slowly heated up water is evolved.[46] At first this water is that evolved from surface layers, but this is more or less complete at temperatures below 150°C. At higher temperatures water is still evolved in large amounts. This comes from the diffusion of water to the surface from the bulk of the glass; it is known as "water of constitution."

The amount of water attributed to surface layers varies with the age of the glass and its chemical treatment. Weathering (aging) of glass usually produces a complex hydrate on the surface of glass, sometimes to an extent that makes it appear hazy. This layer will produce significantly more gas. If this layer is removed by etching in HF the gas evolved is comparable to the amount evolved from freshly drawn glass.

Glass is also outgassed due to the effects of electron bombardment[47] or interaction with plasma.[48] These phenomena can be of importance when metals are being evaporated or sputtered onto glass in making thin film type devices or metallization patterns.

### particulate matter on surfaces

The problem of contamination due to the presence of particles on a surface is particularly serious since it is extremely common and because so little of the theoretical understanding we have of the problem has been translated into sound working techniques for their removal to the extent needed for electron devices.

Despite the capabilities of HEPA filters and achievements in clean room design it is virtually impossible to keep a surface free from dust and other particulate matter. Even in a laminar flow hood one can see particulate matter adhering to the jet black finish of a microscope after only a few days of exposure. Similarly, it has been demonstrated that silicon slices stored in a typical hood for a week will show a significant increase in surface sodium atoms due to the pick-up of particles.[49]

Two common methods for removing particles from a surface are blowing with a jet of air and a rapid flush with water or inert solvent. We shall see in this section how successful these techniques can be. The difficulty of removing this particulate matter can be better appreciated when translated into everyday terms—cleaning an automobile. Even when driven at high speeds only a small fraction of the particulate matter on the finish of the automobile is "blown-off." Perhaps this comparison may not seem fair because organic films are also deposited on the automobile that would tend to increase the adhesion, but an organic film is also deposited on the silicon slice. If an attempt is made to wash the dust off the automobile by running water from a hose (no soap), experience again says that it cannot be done even with the highest pressure from a hose.

What are the forces that contribute to the so-called adhesion of particles to a substrate? A theoretical discussion of the forces involved in adhesion is possible. However, for a practical case the relative contribution of these forces is usually unknown. Nevertheless, numerous measurements have been made of adhesion forces and are so reported. An excellent treatise[50] has been published that covers the literature extremely well.

Three types of forces may contribute to the total observed adhesive force:

1. Intermolecular Interaction. This type of force is the London disper-

sion or van der Waal's force previously discussed. It is of significant value when the particle is within approximately 100 molecular diameters of the surface. The factors entering into the calculation are the particle size, the true contact area, surface roughness, and chemical composition of both particle and surface.

2. Electrical Forces. When the particle makes contact with the substrate, an electric force of attraction arises from the contact potential difference; the greater this difference, the larger is the adhesive force. If one of the surfaces is a semiconductor, then these forces are further influenced by the ratio of donor-acceptors. They are also affected by the area of contact. If moisture condenses on the substrate and penetrates between the surface and particle this type of force is reduced to zero.

Furthermore, if the particle is charged electrically, as it approaches the substrate surface there is an "image interaction" that varies as $1/r^2$. This type of force exceeds in magnitude both of the above-mentioned molecular interactions and electrical forces. If the substrate is conductive the charge on the particle may leak away and reduce the adhesive force. The presence of moisture will also result in a reduction.

3. Capillary Forces. If a condensed liquid fills the space between the particle and substrate (this is possible above 65% relative humidity), a capillary force occurs which exceeds the magnitude of the other three types of forces. A change in degree of hydrophobic character of the surface will significantly alter the magnitude of this force due to the change in wettability.

Other factors which contribute to the magnitude of the force of adhesion are the chemical composition, shape, size, and plasticity of the particle. The surface roughness of the substrate is of importance as well as its freedom from organic deposits. It is a matter of experimental observation that the adhesive force changes with the duration of the contact and that the force of adhesion for a particle arriving on a surface by free fall is different from one being left on the surface by the evaporation of water or a solvent. Examples of this behavior are given later.

What is the magnitude of the force of adhesion in a typical situation? As a rough guide the adhesion force is the order of 1 dyne for a 100 micron particle and about 0.1 dyne for a 1 micron particle. Specifically, for glass spheres, 40-60 $\mu$m in diameter, adhering to a steel surface, the force of adhesion was found to be 0.2 dyne.

A practical problem to consider is the effectiveness of an air blast in blowing dirt off a surface. Table 6-7 shows typical experimental data for glass spheres on a metal substrate.[50] The effectiveness of this removal method is rather disappointing. Also, the previously mentioned observed greater adhesion arising when particles are deposited through the evaporation of water is quite evident. The effects of vibration on removing dust particles are summarized in Table 6-8.

Whitby[51] has given an example of the relative magnitude of the adhesion forces, the particle masses, and the force exerted on the particles by air flow

TABLE 6-7.

| Particle Size Range, μm | % Particles Removed Air velocity in meters/sec | | |
|---|---|---|---|
| | 2.8 | 5.6 | 11.2 |
| (Deposited on surface by free fall) | | | |
| 100–150 | 27 | 70 | 82 |
| 50–100 | 9 | 17 | 23 |
| <50 | 5 | 5 | 9 |
| (Deposited on surface by evaporation from water) | | | |
| <100 | 0 | 0 | 0 |

Tables 6-7, 6-8, and 6-9 from *Adhesion of Dust and Powder*, A. D. Zimon, Plenum Press, 1969

TABLE 6-8.

| Vibrational Force in g-units (Surface vertical) | % Particles Removed Air velocity = 11.2 m/sec | |
|---|---|---|
| | Free settling | Evaporation |
| 0 | 23 | 0 |
| 5 | 52 | 9 |
| 15 | 64 | 9 |
| 20 | 69 | 0 |
| 60 | 75 | 9 |

(drag forces).  At a very high velocity (10,000 cm/sec) it is found that the drag force tending to move a 1 micron diameter particle is smaller than that of the force of adhesion by at least an order of magnitude.

When flowing water is used to dislodge particles either the adhesive force must be overcome or the weight of the particle must be overcome. A diameter of 100 μm is about the cross-over point for this changeover. Thus, particles < 100 μm in diameter will have adhesive forces greater than their mass while those > 100 μm will have a mass greater than their adhesion. For the latter case, > 100 μm, we can neglect the influence of adhesion, and the minimum flow of water needed to dislodge the particle need only be great enough to set the particle in motion. For a particle 50 μm in diameter the adhesive force

TABLE 6-9.

| Particle Size Range in microns | Velocity for 100% removal cm/sec |
|---|---|
| 500 | 10.8 |
| 250 | 8 |
| 100–80 | 7 |
| 40–30 | 17 |
| 10–5 | 41 |

is approximately five times the mass of the particle, and for 7.5 $\mu$m it is 45 times larger.

Table 6-9 summarizes data from Zimon[50] concerning the water velocity needed to obtain 100% removal of particles—glass spheres from steel. Unlike the case for removal by air flow, the fact that the particles adhere to the surface from the evaporation from water does not significantly change the velocity of water flow necessary for their removal.

### organic contaminants

Organic contaminants are universally present in a manufacturing area. Nearly all of them are hydrophobic in nature, meaning that surfaces with these residues covering them are not wet by water. This behavior is the basis for the tests employed to detect surface organics.[52–54]

In surface chemistry studies it has been customary to measure wettability by measuring the contact angle of water on the surface. This contact angle, $\theta$, is the angle that a drop of liquid makes with a flat surface, measuring through the liquid from a tangent drawn to the edge of the drop. A low angle denotes a wettable surface and a high angle a nonwettable surface.

The apparatus described by Fort and Patterson[54] is capable of making measurements of contact angles on very small areas using a drop of water as small as 4 microliters. The reproducibility of the measurement is about $\pm 2°$, although more precise values can be obtained at low angles.

Numerous experiments[55–59] have been conducted on the wettability of a wide variety of surfaces. It is a well founded conclusion, and an extremely important one, that the presence or absence of hydrophobic contaminants is not the only factor contributing to wettability. The chemical nature of the surface is also important, as Table 6-10 summarizes.

In view of the information given in Table 6-10 the common substrates that are easily wettable are metals with a light oxide film, glass, mica, and ceramics. These surfaces could thus be used for the detection of organic

TABLE 6-10.   Wettability of Surfaces

I. *Surfaces that Wet*
 1. Oxides.
 2. Metal plus metal oxide film.
 3. Metal plus foreign oxides (alumina, silicates).
 4. Metal plus adsorbed chemical films (wetting agents).
 5. Hydrated silica (Si−OH surface bonds).
II. *Surfaces that Do Not Wet*
 1. Gold
 2. Metals free of oxides
 3. Organic polymers or surfaces with organic films.
 4. Surfaces with organics *in* the surface.
 5. Metals plus certain adsorbed ions ($F^-$).
 6. HF etched silicon (Si−F bonds).
 7. Strongly heated silica (Si−O−Si surface bonds).

contaminants by their increasing contact angle observed as the result of exposure to a contaminating ambient. Most ceramic materials have a rough surface texture, due to the techniques of their fabrication, which can act to increase the wettability of a surface and thus lead to questionable results. Glass is difficult to clean reproducibly but might be used. The best substrates are mica and oxidized metals.

To prepare test specimens small squares of metal sheet are first cleaned, polished, and lightly oxidized. Contact angle measurements on these surfaces are usually 3-5°. To detect organics, these samples are exposed to the ambient to be tested for an appropriate period of time. A second contact angle measurement of a drop of water on this new surface is then made.

White[60] has exposed various metal samples to laboratory air to compare their sensitivities to organics. The results are shown by the curves of Fig. 6-5. The rapid increase in contact angle for oxidized nichrome (chromium oxide film) and aluminum is most striking. It is readily apparent that hydrophobic contaminants are present in laboratory (and plant) ambients to a degree that a clean surface can be quickly contaminated in only a few hours time. Furthermore, the use of mica as a practical test piece is not warranted in view of its lowest sensitivity.

Numerous experiments have led to the conclusion that these hydrophobic contaminants are chemically adsorbed on the metal oxide. (It may be no coincidence that the metal oxides that show the greatest contact angle in-

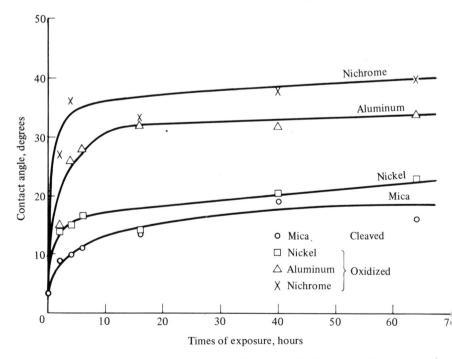

*Figure 6-5 Contamination of surface exposed to laboratory air.*

creases are also frequently used as catalytic surfaces in organic reactions.) Both aluminum and nichrome as oxidized metal surfaces have been used to detect the presence of minute amounts of organics in various gaseous ambients. Tables 6-11 and 6-12 summarize some of these types of data.[61]

**TABLE 6-11.** Evaluation of Gaseous Ambients for Hydrophobic
Contaminants

*Detecting Surfaces: Lightly oxidized Al or Nichrome; initial contact angle < 8°*

| Ambient | Exposure Time | Increase in Angle | Degree of Contamination |
|---|---|---|---|
| Laboratory Air | 2 hr | $13°-22°$ | Moderate |
| Mineral Oil Vapors | 2 hr | $>25°$ | High |
| Forming Gas: (10% $H_2$, 90% $N_2$) | | | |
| At Source | 2 min | $8°$ | Moderate |
| Thru 6″ Tygon | 2 min | $19°$ | High |
| Thru 6″ Copper | 2 min | $10°$ | Moderate |
| Polyethylene at 50°C | 5 min | $26°$ | High |
| Oil-Diffusion System | 2 hr | $26°-49°$ | Very High |
| Vac-Ion System | 2 hr | $0°-2°$ | Very Clean |

**TABLE 6-12.** Contamination in Vacuum System

| Preheating of Chamber | Heating Temp. During Test | Contact Angle On Nichrome |
|---|---|---|
| None | $25°$ | $6.5-9.5°$ |
| None | $120°$ | $22-24°$ |
| 1 hr at 300°C | $120°$ | $8.5-11°$ |

M. L. White[60] also describes the use of contact angle measurements in evaluating the retention time of DOP* in HEPA filters after leak testing and in evaluating various types of storage containers. Special attention is directed to his use of the "gettering" properties of aluminum oxide for designing storage containers.

Organic contaminants are rather strongly held on device surfaces and on supporting structure surfaces as well. The forces are largely the dispersion forces that we have discussed several times before. We may overcome these forces by heating to elevated temperatures where the thermal kinetic energy is sufficient to drive off the organic molecule. However, at elevated temperatures, in air, many of the organic contaminants also undergo thermal decomposition or react with oxygen (and/or moisture) to give volatile products. Hence, a rather common method for the removal of these organics is to heat in air at 400–450°C for 10–30 minutes to completely oxidize them.

*Dioctylphthalate.

However, for many common organic contaminants, their removal can be effected through their solubility in solvents (other than water). Vapor degreasing in chlorinated organic solvents is a very common method and is widely employed in the electronics industry. However, not all organic contaminants are soluble in any one solvent (either hot or cold) so that more than one must be employed to achieve complete removal.

Alkaline cleaners have also been successfully employed, as has ultrasonic energy, to remove organics.

## summary

In this chapter, the point has been made that contaminants that are found on a surface are held there by a variety of forces having a rather broad range of magnitude. Removing these contaminants can be thought of as a process in which these chemical bonds are broken by subjecting the system to stresses that are sufficiently energetic to overcome the holding forces.

The methods that are usually employed can be categorized as either mechanical or chemical in nature. In their net result these methods sometimes attack the substrate (and the contaminants as well) or may simply remove the contaminant(s) effectively from the surface. In the latter case, this may occur by the solubility of the foreign material in an organic solvent or water, or it may result from a chemical attack by acids, bases, etc.

To establish the optimum cleaning cycle the contaminants (or at least their type) must be identified. Once their chemical character is established one course of action is to trace their origin and eliminate it. Usually, however, methods are developed to remove all traces of the contaminant. It is at this point that knowledge of the type of bonding existing between the contaminant and the substrate may be of utmost importance in selection of cleaning methods.

The nature of the many contaminants that must be considered and their bonding types and energies have been discussed with this in mind. The basic techniques of contamination removal were also presented. The details of execution of these cleaning methods as well as specific solutions are to be presented in the next chapter.

## references

1. S. Dushman, *Scientific Foundations of Vacuum Technique*, Chap. 1, 2nd Ed., J. M. Lafferty, Ed., J. Wiley & Sons, New York, 1962.
2. H. E. Farnsworth, "Atomically Clean Solid Surfaces—Preparation and Evaluation," Chap. 13, *The Solid-Gas Interface*, E. Alison Flood, Ed., Marcel Dekker, New York, 1967.
3. F. G. Allen, J. Eisinger, H. D. Hagstrum, and J. T. Law, *J. Appl. Phys*, *30*, 1563 (1959).
4. E. Drauglis, R. D. Gretz, and R. I. Jaffee, *Molecular Processes on Solid Surfaces*, p. 629, McGraw-Hill, New York, 1969.
5. V. S. Sotnikov, A. S. Belanovski, and F. B. Nikishova, *Radiokhimiya*, *4*, 725 (1962).

6. V. S. Sotnikov, *Radiokhimiya*, *8*, 171 (1966).

7. W. Kern, *RCA Review*, *31*, 207–264 (1970).

8. A. W. Adamson, *Physical Chemistry of Surfaces*, 2nd Ed., Interscience Publishers, New York, 1967.

9. T. J. Whalen and M. Humenik, Jr., "Surface Tension and Contact Angles of Copper-Nickel Alloys on Titanium Carbide," *Trans. Met. Soc. of AIME*, *218*, 952 (1960).

10. C. R. Kurkjian and W. D. Kingery, "Surface Tension at Elevated Temperatures. III. Effects of Cr, In, Sn, and Ti on Liquid Nickel Surface Tension and Interfacial Energy with $Al_2O_3$," *J. Phys. Chem.*, *60*, 961 (1956).

11. N. B. Hannay, *Solid-State Chemistry*, Prentice-Hall, Inc., Englewood Cliffs, N.J., 1967.

12. N. B. Hannay, Editor, *Semiconductors*, Reinhold Publishing Corp., New York, 1959.

13. A. Many, Y. Goldstein, and N. B. Grover, *Semiconductor Surfaces*, North-Holland Publishing Co., Amsterdam, 1965.

14. P. J. Holmes, Editor, *The Electrochemistry of Semiconductors*, Academic Press, New York, 1962.

15. W. C. Lo, Unpublished Data, Bell Telephone Laboratories, Allentown, Pa.

16. L. Pauling, *Nature of the Chemical Bond*, 3rd Ed., Cornell U. Press, Ithaca, N.Y., 1960.

17. C. Kittel, *Introduction to Solid State Physics*, 4th Edition, J. Wiley, New York, 1971.

18. J. O. Hirschfelder, C. F. Curtiss, and R. B. Bird, *Molecular Theory of Gases and Liquids*, J. Wiley, New York, 1954.

19. R. J. Good, "Intermolecular and Interatomic Forces," Chap. 2, in *Treatise on Adhesion and Adhesives*, Vol. I, R. L. Patrick, Editor, M. Dekker, New York, 1967.

20. A. S. Michaels, "Fundamentals of Surface Chemistry and Surface Physics," Symposium on Properties of Surfaces, ASTM Special Technical Publication, No. 340, Amer. Soc. for Test. and Materials, Philadelphia, 1963.

21. Karl Kauffe, *Oxidation of Metals*, Plenum Press, New York, 1965.

22. O. Kubaschewski and B. E. Hopkins, *Oxidation of Metals and Alloys*, Academic Press, New York, 1962.

23. U. R. Evans, *The Corrosion and Oxidation of Metals*, Edward Arnold, Ltd., London, 1960.

24. A. U. Seybolt, "Oxidation of Metals," *Advances in Physics*, Vol. 12, No. 45, 1 (1963).

25. K. Kauffe, "On the Mechanism of the Oxidation of Metals", *The Surface Chemistry of Metals and Semiconductors*, H. C. Gatos, Ed., J. Wiley & Sons, New York, 1960.

26. Frederick Seitz, *The Physics of Metals*, McGraw-Hill, New York, 1943.

27. Staff Report, "The Truth About Miss Liberty," *Materials in Design Engineering*, p. 80, June 1963.

28. J. F. Elliott and M. Gleiser, *Thermochemistry for Steelmaking*, Addison-Wesley, Reading, Mass., 1960.

29. R. A. Swalin, *Thermodynamics of Solids*, J. Wiley and Sons, New York, 1962.

30. A. A. Bergh, *Bell System Tech. J.*, *44*, 261 (1965).

31. D. M. Young and A. D. Crowell, *Physical Adsorption of Gases*, Butterworths, Washington, 1962.

32. D. O. Hayward and B. M. W. Trapnell, *Chemisorption*, 2nd ed., Butterworths, Washington, 1964.

33. S. Brunauer and L. E. Copeland, "Physical Adsorption of Gases and Vapors on Solids," Symposium on Properties of Surfaces, ASTM Special Technical Publication No. 340, ASTM, Philadelphia, 1963.

34. M. L. Hair, *Infrared Spectroscopy in Surface Chemistry*, Marcel Dekker, New York, 1967.

35. E. A. Flood, Editor, *The Solid-Gas Interface*, Vol. 1, 2, Marcel Dekker, New York, 1967.

36. F. M. Fowkes, "Calculation of Work of Adhesion by Pair Potential Summation," in *Hydrophobic Surfaces*, F. M. Fowkes, Ed., Academic Press, New York, 1969.
37. A. C. Zettlemoyer, "Hydrophobic Surfaces," in *Hydrophobic Surfaces*, F. M. Fowkes, Ed., Academic Press, New York, 1969.
38. M. J. Rand, "The Adsorption of Water Vapor by Porous Glass," *J. Elchem. Soc.*, *109*, 402 (1962).
39. A. Wheeler, "Reaction Rates and Selectivity in Catalyst Pores," *Catalysis*, Vol. II, Chap. 2, Reinhold Pub. Co., New York, 1955.
40. G. C. Bond, *Catalysis by Metals*, Academic Press, London, 1962.
41. R. H. Collins and J. C. Turnbull, "Thermal Degassing of Tube Materials," Advances in Electron Tube Techniques, Proceedings of the 5th National Conference, D. Slater, Ed., Pergamon Press, 1960.
42. R. H. Collins and J. C. Turnbull, "Degassing and Permeation of Gases in Tube Materials," Advances in Electron Tube Techniques, Proceedings of the 5th National Conference, D. Slater, Ed., Pergamon Press, 1960.
43. W. Espe, "Methods of Technology of Degassing Metals," *Vakuum Technick*, *5*, 34–72, 78–81 (1956).
44. E. Mattson and F. Schückler, "An Investigation of Hydrogen Embrittlement in Copper," *J. Inst. of Metals*, *87*, 241 (1958).
45. L. Holland, *The Properties of Glass Surfaces*, J. Wiley & Sons, 1964.
46. B. T. Todd, "Outgassing of Glass," *J. Appl. Phys.*, *26*, 1238 (1955).
47. B. J. Todd, J. L. Lineweaver, and J. T. Kerr, "Outgassing Caused by Electron Bombardment of Glass," *J. Appl. Phys.*, *31*, 51 (1960).
48. T. W. Hickmott, "Interaction of Atomic Hydrogen with Glass," *J. Appl. Phys.*, *31*, 128 (1960).
49. George Schneer, Unpublished Data, Bell Telephone Laboratories, Allentown, Pa.
50. A. D. Zimon, "Adhesion of Dust and Powder," (translated by M. Corn), Plenum Press, New York, 1969.
51. K. J. Whitby, "The Starting Point for Contamination Control," Second Annual Convention, Amer. Assoc. For Contamination Control, Boston, Mass., April 30–May 3, 1963.
52. ASTM F-21-62T Hydrophobic Surface Films by the Atomizer Test. ASTM F-22-62T Hydrophobic Surface Films by the Water-Break Test.
53. I. Langmuir and V. J. Schaeffer, *J. Amer. Chem. Soc.*, *59*, 2405 (1937).
54. T. Fort, Jr., and H. T. Patterson, *J. Colloid Sci.*, *18*, 217 (1963).
55. M. L. White, *J. Phys. Chem.*, *68*, 3083 (1964).
56. R. A. Erb, *J. Phys. Chem.*, *69*, 1306 (1965).
57. F. M. Fowkes, *ASTM Spec. Tech. Pub.*, *360*, 20 (1963).
58. S. Anderson and D. D. Kempton, *J. Amer. Ceram. Soc.*, *43*, 484 (1960).
59. T. H. Elmer, I. D. Chapman, and M. E. Nordberg, *J. Phys. Chem.*, *67*, 2219 (1963).
60. M. L. White, "The Detection and Control of Organic Contaminants on Surfaces," *Clean Surfaces*, G. Goldfinger, Ed., Marcel Dekker, New York, 1970.
61. M. L. White and J. F. Pudvin, Unpublished Data, Bell Telephone Laboratories, Allentown, Pa.

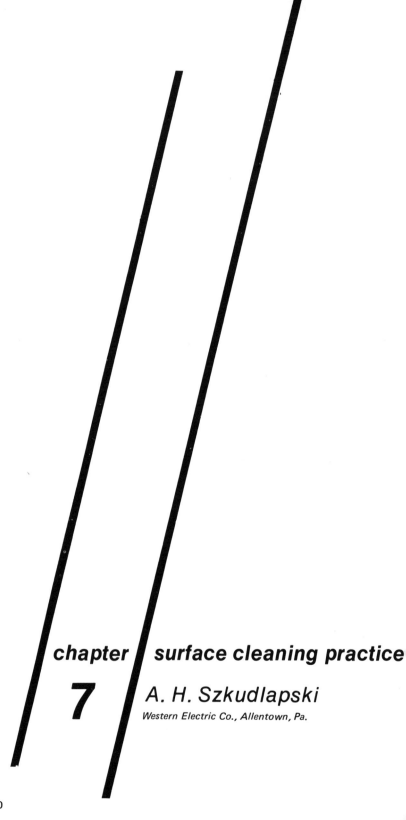

**chapter** / **surface cleaning practice**

**7** / *A. H. Szkudlapski*
*Western Electric Co., Allentown, Pa.*

Although there are numerous methods available, the basic purpose of all surface cleaning techniques is to remove all contaminants—physical, organic, or inorganic—without damaging the device or part. Such contaminants must be removed to reduce or avoid rejects during subsequent process steps.[1,2]

This chapter explores the cleaning methods which are available and commonly used in the manufacture of electronic devices. The initial sections discuss the use of various chemicals in solvent cleaning and vapor degreasing. Inorganic chemical compositions used for alkali and acid cleaning are then discussed along with the use of electric power for some applications. The use of special cleaning methods, involving ultrasonic energy and high temperature cleaning, are presented, as well as the post cleaning methods of rinsing and drying. The chapter closes with a general discussion of properly engineered cleaning systems.

## organic solvents

Organic solvents are used preferably to remove soils organic in origin, and each solvent should be analyzed in detail before making a selection. The physical properties of the solvents are well known and are available from chemical handbooks. Of special importance is their flammability and toxicity. Boiling points are also important; the wider the spread of boiling points between solvent and contaminant, the more effective the separation in a solvent system.

### solvent immersion

Immersion cleaning is one of the simplest cleaning techniques available and involves the use of a liquid to remove foreign matter by simple immersion of the part. The cleaning action is dependent on the solvency of the fluid, which may be an organic solvent, an aqueous solution of a single agent, or a mixture of suitable cleaning chemicals. In the latter category a large number of commercially available products can be used, primarily in removing photoresists. The cleaning can be performed in open containers, such as beakers, and the liquid may be heated or agitated. In many cases a series of containers are used with each subsequent cleaning performed in a purer solvent to improve cleaning efficiency. Often at least two different solvents are employed since organics of both animal and mineral origin may be present.

Cleaning costs are generally low when using this method since only simple facilities are required; however, in many cases, considerable chemical waste results with no provisions for redistillation or recovery of the solvents. These cleaning techniques are used less frequently than vapor degreasing, electrolytic cleaning, ultrasonic cleaning, detergent solutions, or specially formulated cleaning compositions.

When evaluating solvent cleaning methods, solvency or solvent power is most commonly measured by the kauri-butanol system. The solvent power is expressed as the amount of solvent which produces a definite turbidity when added to a standard kauri gum solution in butanol as compared to the amount of benzene used in a similar titration. More precisely, the value is the number of milliliters of solvent which must be added to 20 grams of standard kauri-butanol solution at 25°C to produce sufficient gum precipitate to make a printed paper appear blurred and illegible when viewed through the flask containing the solution. At the prescribed conditions 100 milliliters of benzene will produce the described condition, and hence, the kauri-butanol number for benzene is 100. The benzene value is the arbitrarily accepted standard for this test.

Kauri-butanol values for some commonly used solvents are listed in Table 7-1. These values serve only as guides, and other factors such as temperature, time, and materials must be considered in selecting the proper solvent.

#### TABLE 7-1.

| Solvent | Kauri-Butanol Value at 25°C |
|---|---|
| Trichlorotrifluoroethane | 31 |
| Stoddard Solvent | 34 |
| Mineral Spirits | 39 |
| Perchloroethylene | 90 |
| 1, 1, 1-Trichloroethane | 124 |
| Trichloroethylene | 130 |
| Chloroform | 208 |

### vapor degreasing

This cleaning method removes solvent soluble contamination such as oil, greases, and organic matter from surfaces by a two step action. Initially, the solvent vapors dissolve the contaminants. Secondly, the condensate provides a liquid flow which carries away the organic soils by gravity. Contaminants which are held to the surface by the solvent soluble material are also generally removed by the cleaning action. This method is not effective in removing nonorganic particles unless the operation is performed with ultrasonic energy.

Vapor degreasing is most commonly used as the first step of a cleaning sequence to break the adhesion of particles caused by oil and grease films, but sometimes it is sufficient to provide the degree of cleanliness required. This is especially true when used with an ultrasonic unit where contamination levels between 1.0 and 0.1 monomolecular layer can be achieved. This cleanliness range is sufficient for miniaturized parts which are subsequently subjected to electrical, chemical, or electrochemical processes.

Physically, a vapor degreaser consists of a tank containing heat elements at the bottom and water-cooled condensing coils around the top perimeter of the tank. The heating elements provide the heat required to maintain the solvent at the boiling point, thereby producing a hot, high-density vapor zone. To prevent significant solvent losses the cooling coils, located at the top of the tank, condense the vapors. Many types of degreasers are commercially available and selection among them is dependent upon the treatment cycle desired. Figure 7-1 shows several typical variations. The treatments may include exposure to various solvent phases, e.g., vapor only, spray-vapor, vapor-spray-vapor, vapor-immersion-vapor, boiling solvent–cool solvent-vapor, and other combinations.

Selection of the proper degreasing solvent has been in practice narrowed

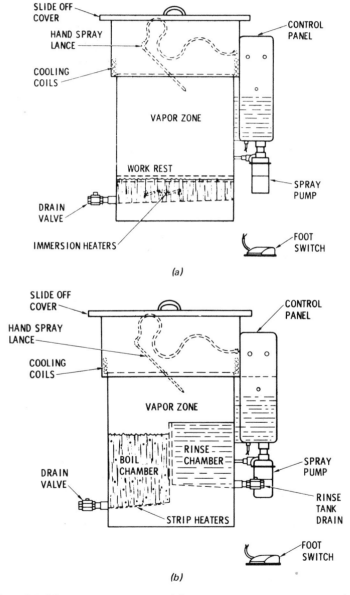

Figure 7-1 (a) Vapor spray degreaser; (b) vapor immersion spray degreaser; (c) ultra-sonic vapor degreaser. (Courtesy Baron-Blakeslee, Inc., Chicago, Illinois.)

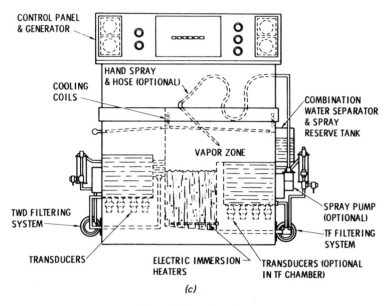

Figure 7.1 continued

to a few agents. Trichloroethylene (ClCH:CCl$_2$), perchloroethylene (CCl$_2$ : CCl$_2$), trichlorotrifluoroethane (CCl$_2$FC · ClF$_2$), methylene chloride (CH$_2$ · Cl$_2$), and 1,1,1-trichloroethane (CH$_3$ · CCl$_3$) are mostly used. Some azeotropes (constant boiling point mixtures) based on Freon TF and some suitable degreasing solvent (e.g., methylene chloride) are also used. Blends of Freon TF and isopropyl alcohol are available also.

The properties of the individual degreasing solvents are presented in Table 7-2. These degreasing solvents are characterized by lack of flammability, excellent solvency, high vapor density relative to air, and relatively low heat of vaporization and specific heat. They also possess a boiling point within a range to facilitate effective vapor condensation and chemical stability when used for degreasing and are generally noncorrosive to metals of degreaser construction or piece parts treated.

One drawback of the chlorinated solvents is their high toxicity. Ventilation is necessary since repeated inhalation of even small amounts of these vapors may have a narcotic effect on the operator. Proper protective clothing should be worn to prevent skin contact and subsequent dermatitis due to loss of natural oils. Also, chlorinated solvents can hydrolyze in the presence of water and active metals (aluminum, zinc) to yield hydrochloric acid.[3] To prevent this harmful decomposition which would attack the materials of construction, solvents are blended with minor quantities of stabilizers, such as diisopropyl amine. Trichloroethylene and perchloroethylene are the most widely used of the chlorinated solvents primarily because of their solvency,

TABLE 7-2. Properties of Commercially Available Vapor Degreasing Solvents

| | Trichloro-ethylene | Perchloro-ethylene | Methylene Chloride | Trichloro-trifluoro-ethane Freon TF | Methyl Chloroform (1,1,1-tri-chloro ethane) |
|---|---|---|---|---|---|
| Boiling Point, °F | 188 | 250 | 104 | 118 | 165 |
| Flammability | Nonflammable under vapor degreasing conditions | | | | |
| Surface Tension at | 32° | 32.3 | 28.2 | 19.6 | 25.5 |
| 68°F. (Dynes/cm) | (77°F) | | | | (74°F) |
| Latent Heat of Vaporization (b.p.) | | | | | |
| Btu per lb. | 103 | 90 | 142 | 63 | 105 |
| Saturation in Air, 100% Relative Humidity at 68°F., lb. Solvent per | | | | | |
| Cu. Ft. Air | 0.03 | 0.008 | 0.19 | 0.27 | 0.06 |
| Specific Heat (Liquid), | | | | | |
| Btu per lb. per °F | 0.23 | 0.21 | 0.28 | 0.21 | 0.24 |
| Specific Gravity: | | | | | |
| Vapor (Air = 1.00) | 4.53 | 5.72 | 2.93 | 6.75 | 4.60 |
| Liquid (Water = 1.00) | 1.464 | 1.623 | 1.326 | 1.514 | 1.327 |
| Threshold Limit, Values ppm in Air | | | | | |
| by Volume | 100 | 100 | 500 | 1000 | 350 |
| Kauri Butanol Value | 130 | 90 | 136 | 31 | 124 |

Source: T. J. Kearney, *Metal Finishing Guidebook Directory*, p. 183, Metals and Plastics Publications, Inc., Westwood, N.J. (1970).

cost, and good stability. The prime difference between these two is their boiling points.

The fluorinated solvents are characterized by lower toxicity, are very stable with respect to heat, water, and metals, and do not attack many types of plastics and elastomers. However, they are expensive compared to chlorinated solvents. Thus, while ventilation is not normally required, facility design will include distillation to minimize loss of the fluorinated solvents.

In brief, it can be stated that vapor degreasing is a highly effective means of removing soils, particularly those in the oil and grease family. Adequate protection can be provided against personnel hazards connected with the toxic vapors and skin contact. It can be said that vapor degreasing, in properly designed equipment and operated according to safely established routines, is a simple, relatively safe, rapid, and economical method of cleaning surfaces.

## inorganic cleaners

Aqueous, inorganic-based cleaning can be divided into acid and alkaline systems. Acid type systems are used to remove rust, scale, corrosion products, and soldering or welding flux residues. The second category, alkaline cleaners, is used to remove greases, oils, drawing compounds, fingerprints, and other organic-based contaminants. The alkaline cleaners differ from organic solvents in that instead of dissolving organics, they either emulsify or saponify them.

## alkaline cleaners

This method of cleaning is used to a very limited extent in the manufacture of electron devices as compared to acid cleaning. An alkaline cleaner is a mixture of alkaline salts (e.g., phosphates, carbonates, silicates, etc.) with one or more surfactants. A variety of compositions is available from commercial producers, some of which are proprietary formulations, for both general or specific cleaning formulations. One example of the use of a commercial alkaline cleaner, Alconox,* is the pre-sputter cleaning of ceramic substrates where tantalum nitride is the sputtered film. Ceramic substrates after ultrasonic degreasing in trichlorethylene are blown dry by air and immersed in a boiling aqueous solution of Alconox (3 g/l.) for 5 minutes. The parts are spray rinsed using hot deionized water followed by immersion rinsing in boiling water (15–20 minutes) of the same quality. Drying is accomplished using dry nitrogen. Glass substrates before sputtering are also cleaned using an Alconox solution.

This method of alkaline soak cleaning will lift the soils from the surface, disperse them, and prevent their redeposition. Advantages of this simple method of cleaning using low concentrations of special formulated alkaline cleaners are numerous: low cost, effective removal of water-soluble soils, a resulting hydrophilic surface, and low toxicity.

Besides cleaning, alkaline solutions with complexing agents (e.g., KOH solution with n-butanol) are used for etching silicon. An anisotropic etch is used, removing silicon 30 times more rapidly along the (100) crystal plane where the atoms are less densely packed than along the (111) plane. At the same time such etchants practically do not attack silicon dioxide or silicon nitride. This has significance in integrated circuits design; air isolation can be used in place of junction isolation of electrical components, resulting in higher device packing per area of silicon and increased production yields.

## acid cleaners

Acid cleaning and the related areas of chemical polishing and etching are among the more important chemical processes in microelectronic product fabrication. Besides such substrate materials as silicon and high alumina ceramics, there also exists a variety of piece parts (headers, cans, lead frames, etc.) which require cleaning. Also, materials used as product handling equipment such as glass (Pyrex) and quartz require a variety of acids and acid mixtures for cleaning. Some acids or mixtures of acids are used solely for cleaning; other formulations are intended to remove material during chemical

*Registered trademark of Alconox, Inc., New York, N.Y.

etching or polishing. Etching is used in device fabrication (e.g., separation etching of discrete or integrated circuit beam lead devices) and to evaluate semiconductor crystal properties (e.g., to obtain dislocation densities in silicon crystals).[4]

The design of an acid process for cleaning, etching, or polishing involves consideration of surface preparation, chemically compatible process equipment, chemical purity, process material etching characteristics, and means of stopping the chemical reaction at the proper moment. These points will be outlined and then illustrated with common microelectronic examples.

As a rule acid processes require materials or parts to be relatively clean from dust particles, residual abrasive particles, grease, and oils. For parts which underwent some kind of heat treatment prior to acid cleaning (e.g., Kovar oxides from a glassing operation), mild surface cleaning operations such as effective rinsing in water with ultrasonic agitation may be all that is necessary. Of importance is the cleanliness of the etching containers. Since most of the semiconductor etching solutions contain hydrofluoric acid which attacks glass, chemically inert containers made of polyethylene, polypropylene, or Teflon* are commonly used.

Microelectronic substrate materials require reagent grades of chemicals for the etching or cleaning mixtures since chemical purity is of extreme importance to the success of the processes. Cleaning of piece parts and chemical polishing (bright dipping) are usually performed using commercial grade acids. In most cases these operations are related to header or lead frame cleaning prior to electroplating. The purity of chemicals and their procurement will be discussed in Chap. 9.

The reagent grade acids and other chemicals (e.g., halogens) are specified by the percent concentration by weight and the amount of allowable impurities. Reagent grade chemicals conform to the standards of the American Chemical Society, and the term ACS grade is used interchangeably. Thus, hydrofluoric acid (HF) is 48–51%, nitric acid ($HNO_3$) is 69–71%, and glacial acetic acid ($HC_2H_3O_2$) is 99.7% by weight. These three acids form some of the most important etchants for silicon. One composition of $5:3:3$ by volume of $HNO_3:HF:HC_2H_3O_2$ is used both in the initial cleaning of polycrystalline silicon charges before single crystal Czochralski pulling technique and also to etch apart beam lead integrated circuits manufactured on a silicon wafer.

The commercial-grade acids used for cleaning piece parts before plating are: nitric acid ($HNO_3$-42°Bé), hydrochloric acid (HCl-20°Bé), sulfuric acid ($H_2SO_4$-66°Bé). and glacial acetic acid $HC_2H_3O_2$ at 99.5% by weight. The degrees Baumé used by the chemical industry approximate the concentrations of ACS grade acids.

The etching of semiconductor materials, especially silicon, has been reviewed by many authors. The mechanism of silicon etching in the $HNO_3$-HF system is documented.[5-7]   The same is true for the etching rates for

*Registered trademark of E. I. du Pont de Nemours and Co., Wilmington, Del.

Environmental Control in Electronic Manufacturing

$HF-HNO_3-H_2O$ mixtures[8] and the same system with addition of $HC_2H_3O_2$.[9] Halogens are often part of the etchants, including bromine[10] and iodine.[11] Iodine monobromide (in glacial $HC_2H_3O_2$) is also used.[12]

Of great importance is the post-etching treatment which must remove the etchants from the various surfaces in the shortest possible time. This is especially true when etching semiconductor surfaces; otherwise, staining and further etching may result. Some techniques specified for etching silicon recommend flooding of the etchant with a diluent such as water or acetic acid ($HC_2H_3O_2$) to stop the reaction. If used on a small laboratory scale this method is satisfactory. In other cases it becomes very uneconomical and generates considerable quantities of waste etchants. Rinsing practices will be discussed later.

**Silicon Cleaning.** An example will show the use of single and mixed acids coupled with other techniques in cleaning epitaxial silicon wafers. Before the p-n junction formation is accomplished through diffusion an oxide is grown on the silicon wafer to serve as a selective diffusion barrier in some areas of the wafer. The quality of this oxide is very important, and an elaborate cleaning technique before its growth is required.

The epitaxial silicon wafers are placed in a wafer-holding quartz rack and immersed sequentially in single and mixed acids held in quartz beakers kept at 80–90°C for a period from five to fifteen minutes. A typical silicon wafer cleaning sequence using acids of ACS grade is shown in Table 7-3.

**TABLE 7-3. Acid Cleaning Sequence for Silicon Epitaxial Wafers**

(1) Immerse rack in $H_2SO_4$ for ten minutes.
(2) Transfer the beaker into an ultrasonic tank with deionized water and ultrasonically clean for five minutes.
(3) Cascade rinse the rack in deionized water.
(4) Immerse rack in 1 : 1 by vol. mixture of $H_2SO_4$ and $HNO_3$ for fifteen minutes.
(5) Cascade rinse as under (3).
(6) Immerse in HF at room temperature for one minute.
(7) Cascade rinse as under (3).
(8) Immerse rack with silicon wafers in a fresh solution of mixed acid ($H_2SO_4 : HNO_3 = 1 : 1$ by vol.) for fifteen minutes.
(9) Cascade rinse as under (3).
(10) Dry wafers in nitrogen tunnel.
(11) Grow oxide on wafers in a furnace.

**Piece Parts Cleaning.** In the preparation of glassed Kovar headers an important step is the removal of the mixture of black oxides formed during glassing. The headers are placed in a polyethylene basket and immersed with occasional agitation for 20 minutes in a 1 : 1 by vol. solution of 20°Bé HCl at 70°C. Next they are rinsed in tap water and immersed one minute in HF at room temperature. This removes the small amount of glass along the header's leads which may break later during tumbling. The final step involves Freon drying of the parts. The same Kovar headers are prepared for gold plating using a bright dip as one of the preceding steps. The parts (in the same baskets) are immersed for a period from 5-7 seconds in a mixed acid at 70°C. The composition of the commercial grade acids is by volume: 3 parts $HC_2H_3O_2$ : 1 part $HNO_3$. Additionally, 15 ml of the same grade HCl is added to 1000 ml of the mixed acid. Thorough rinsing with water follows the bright-dip step. On the average about 0.5 mil of the surface is removed during this step.

Copper parts are prepared for plating by using a bright dip formulated from acids of commercial grade in the mixture 7 : 7 parts of $H_2SO_4$ : $HNO_3$, and 1 part of water by vol. Heat is generated on mixing these chemicals; thus, the bright dip is prepared in advance and allowed to reach room temperature before use. Parts in polyethylene baskets are bright dipped in bulk for about ten seconds, depending on the geometry of the individual parts. When bright-dipping Kovar and copper piece parts, the baskets must always be agitated to ensure uniform polishing.

**Operating Considerations.** Acid cleaning is very useful in altering the surface of a material and is a relatively easy operation to perform and control. The one major exception would be the various chemical polishing operations (bright dipping) which require skill and experience to be performed properly. Due to the corrosive nature of the chemicals involved special care has to be taken to ensure proper choice of materials of construction for such items as processing equipment, containers, and handling fixtures. Of all the methods used in cleaning surfaces this one needs the greatest attention of the operator to prevent accidents. Finally, special attention should be given to efficient and economical utilization of the acid solutions because of their concentration and the resultant problems in waste treatment.

### *electrolytic cleaning*

Electrolytic cleaning is used primarily as a final aqueous cleaning of parts prior to electroplating. As a rule it is preceded by one or more of these steps: solvent cleaning (vapor degreasing), pickling and/or bright dipping, and soak cleaning.

The principle of operation for this method is based upon the electrolysis of water in an aqueous electrolyte. Cleaning is achieved by the scrubbing action and upward agitation of bubbles of hydrogen and oxygen released by the decomposition of water. These solutions are usually alkaline although acid type (activating) solutions are used in special cases. The solutions are usually weak, thus limiting chemical attack on the parts without electric current. If the parts to be electrocleaned are made the cathode they are said to be cathodically or direct current cleaned. If the parts are made the anode they are said to be anodically or reverse current cleaned.

During cathodic cleaning, hydrogen is liberated at the surface of the work (the cathode). The volume of hydrogen liberated at the cathode is twice that of oxygen which is discharged at the anode for a given current density. Thus more scrubbing is achieved at the cathode. But this characteristic should not be the only one considered in selecting the type of electrocleaning to use. The choice of cathodic or anodic cleaning is, as a rule, dictated by the type of metal or alloy being cleaned, and to a degree, by the type of cleaning solution (strongly or mildly alkaline) and the operating conditions. Direct current cleaning is used to clean metals such as chromium, tin, lead, brass, magnesium, and aluminum which would be dissolved or etched by anodic cleaning.

Cathodic cleaning may lead to metallic smut being plated out if metallic impurities find their way into the cleaning solution. Conversely, reverse current cleaning provides a slight deplating action which prevents smut formation and deposition of nonadherent metallic films resulting from drag-in impurities. Also, oxygen discharged on the metal does not cause embrittlement as hydrogen does.

Many of the piece parts used in electronics manufacture are either copper or Kovar based. The copper parts are usually nickel plated before final gold plating. The purpose of the nickel deposit is to act as a barrier to prevent outdiffusion of copper into gold. The Kovar piece parts are glassed headers which are bright dipped as a step in the cleaning cycle before gold plating. These nickel plated and Kovar examples and other parts of solid nickel (lead frames) are cathodically cleaned or activated in an alkaline cyanide solution prior to gold strike plating.

The cleaning solution contains 80 g/l. of sodium cyanide (NaCN), and 2 g/l. of sodium hydroxide (NaOH). It is operated at 65°C and approximately 6 volts. Times of activation vary depending on whether the parts are rack plated or barrel plated and the size of the lots. In the latter instance the time may extend to several minutes, but with racked parts it may be only thirty seconds. Other electrocleaning solutions in wide use are based on such chemicals as phosphates, silicates, carbonates, and combinations of these.

## nonchemical cleaning methods

Heretofore the discussion has centered on the use of chemicals for surface cleaning. However, the mechanical forces of ultrasonics and the effects of high temperature are important tools available to achieve high orders of surface cleanliness. Ultrasonics cleaning is often used in combination with many different organic or inorganic chemicals as a cleaning booster, hence the reason for treating it as a separate subject. On the other hand, high temperature cleaning or heat treatment is often used as an alternative to chemical cleaning.

### ultrasonic cleaning[13-22]

Agitation plays an important role in efficient cleaning, especially in the removal of soils composed of fine particles. It can be accomplished in many ways, the simplest example being to boil the solution. However, for various reasons not all cleaning solutions can be heated to boiling. An effective technique to achieve good agitation on a microlevel is through the use of ultrasonic energy.

The human ear can usually respond to frequencies ranging from about 20 to 18,000 vibrations per second (18 kHz). Sound vibrations above 20 kHz are considered ultrasonic. Sound behaves differently passing through various media. In an elastic medium like a metal, both the sound energy and the medium vibrate at the same rate. When sound passes through a liquid the liquid ruptures or cavitates. Small vacuum bubbles are created which almost immediately collapse. This cavitation occurs only at or above the sonic intensity level of the threshold of cavitation. The rapid implosion of thousands of these bubbles occurs thousands of times a second and provides a rapid and vigorous cleaning action throughout the tank area. Consequently, this cleaning action takes place even in blind holes, crevices, and other difficult-to-reach surfaces. The cleaning action is enhanced by the use of a suitable cleaning solution as the conducting medium.

An ultrasonic unit consists essentially of a generator, transducers, and a cleaning tank as explained in Fig. 7-2. The transducer transmits the ultrasonic energy into the cleaning tank, and while vibrating at the ultrasonic frequency it is alternately moving toward and away from the liquid in the tank. A compression wave is produced when the transducer moves towards the liquid, while a rarefaction wave results when it moves away from the liquid. If the amplitude of the transducer vibration is sufficiently high the rarefaction ultrasonic wave will lower the pressure on the liquid in the tank below its vapor pressure, at the temperature of operation. When this occurs, thousands of minute vapor bubbles are formed in the low-pressure region. One-half cycle later the compression wave moves through this region, thereby causing the

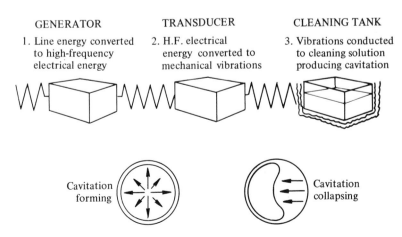

*Figure 7-2 How ultrasonic cleaning works. (Courtesy Turco Products, Division of the Purex Corporation Ltd.)*

vapor to recondense to liquid, and thus forming tiny partial vacuum bubbles. These implode with great force and cause the ultrasonically-induced cavitation, i.e., the formation and implosion of vapor bubbles.

Cavitation can occur only as stated when the local pressure on the liquid is less than its vapor pressure; thus, the amplitude of the ultrasonic waves generated by the transducer must be great enough to satisfy this condition. This minimum amount of power is referred to as the "threshold of cavitation" and is different for various liquids, but the threshold must be exceeded to achieve ultrasonic cleaning.  It is only that ultrasonic energy above the threshold that causes the formation of cavitation bubbles and thus ultrasonic cleaning.

As one can infer, the liquid used for ultrasonic cleaning will have an important effect on the cleaning efficiency.  The four principal physical properties of the liquid which have the greatest influence on cleaning efficiency are vapor pressure, surface tension, viscosity, and density.  Since temperature influences these properties it also has an influence on the effectiveness of cavitation.  A detailed discussion on these properties and their interdependence is given in the *Contamination Control Handbook.*[13]

Although increasing temperature usually improves the chemical action of the cleaning solution, one may reach a temperature at which, despite better chemical action, there is a marked decrease in the cavitation intensity and thus, in the overall cleaning.  This temperature is called "maximum effective operating temperature."  For water this temperature is about $180°F$; for ethyl alcohol and methylene chloride it is about $85°F$. Cavitation effectively ceases when a cleaning fluid reaches its boiling point.

Several cleaning solutions can be used in ultrasonic cleaners.  These include deionized water (with and without some form of wetting agent), mild acids, mild alkalies, chlorinated or fluorinated hydrocarbon solvents, and low molecular weight alcohols.  Although most solvents cavitate very readily it has been reported that acetone will not support cavitation sufficiently to be used as an ultrasonic cleaning medium.[14]

Among the types of soils removed by ultrasonic cleaning are small particulate matter, light oils and greases, solder flux, microbial contaminants, finger prints, and even oxides from metallic surfaces.  A major application for ultrasonic cleaning is in the manufacture of silicon wafers for integrated circuits.  Typically these wafers are in the order of 1½ in. in diameter and 0.006 in. thick.  They are quite delicate but can be cleaned quite vigorously using this technique.  Results of the effectiveness of ultrasonic cleaning of silicon wafers using deionized water and wetting agents are presented in Table 7-4, based on the experiment that follows.

In the experiment silicon wafers with contamination on their surfaces were cleaned using a nonionic wetting agent in deionized water with ultrasonic energy.  These wafers and two without the wetting agent treatment (the control sample) were analyzed on a spark source mass spectrometer on both

TABLE 7-4.  Effectiveness of Wetting Agent Use In Ultrasonic Cleaning

| | Polished Side | | Lapped Side | |
| | Control | Cleaned | Control | Cleaned |
|---|---|---|---|---|
| Zirconium | – | – | 30 | – |
| Copper | 10 | – | – | – |
| Manganese | – | – | 2 | – |
| Chromium | 3 | – | – | – |
| Calcium | 30 | 1 | 10 | 3 |
| Potassium | 30 | – | 30 | 2 |
| Chlorine | 5 | – | 2 | – |
| Aluminum | 30 | 10 | 300 | 3 |
| Magnesium | 50 | – | 100 | 3 |
| Sodium | 1000 | 3 | 300 | 50 |
| Fluorine | 5 | – | 20 | – |
| Boron | 2 | – | 1 | – |

the polished and the lapped sides of the wafers for surface cleanliness. The results are expressed in units of the limit of detection for each element such that "1" was just detectable, "10" was ten times that much, etc. A value of "10" represents approximately one atomic monolayer or its equivalent for most elements. These results, which provide a relative measure of surface contamination, show that the treatment (0.25% concentration of wetting agent in 0.2 micromho/cm water with 15 minutes ultrasonic agitation followed by cascade rinsing and nitrogen tunnel drying) removed most of the contamination.[15]

TABLE 7-5.  Typical Ultrasonic Cleaning Sequence for Silicon Wafers

1. Place wafers to be cleaned into a clean, fused silicon rack.  This is important because the cleaning procedure will remove any soil from a used rack and may redeposit the soil on the wafers.
2. Immerse the wafers in 0.25 per cent solution of non-ionic wetting agent in deionized water (0.2 micromho/cm max.) and ultrasonically agitate the wafers for fifteen minutes min.
3. Rinse wafers and rack in free flowing deionized water (0.2 micromho/cm max.) for two minutes min.
4. Boil rack with wafers for a minimum of five minutes in a solution of 15 per cent by vol. hydrogen peroxide (ACS grade), and 85 per cent by vol. of deionized water (purity as above).
5. Rinse as under (3).
6. Immerse rack with wafers in 10 per cent by vol. (ACS grade) hydrofluoric acid and 90 per cent by vol. of deionized water (purity as above) for one minute.
7. Rinse in cascade rinser for two and five minutes respectively in the three compartments until the 0.2 micromho/cm conductivity of deionized water is reached.
8. Dry rack with wafers in a nitrogen tunnel.

A typical ultrasonic cleaning procedure for silicon wafers is given in Table 7-5. The process aims at destroying traces of organic matter remaining from previous manufacturing steps and the removal of any formed silicon oxide.

While ultrasonic cleaning of about 40 silicon wafers (1 sq. ft in area) seems to be a very simple operation, many factors have to be taken into account to achieve maximum cleaning effectiveness. For instance, consider the proper choice of the cleaning solution and its proportions for the cleaning job to be done. "A polar soil is removed best by a polar solvent and a nonpolar soil is removed best by a nonpolar solvent. Also, for a given polarity of solvent, the lower its molecular weight the easier it will penetrate the contaminant. The solvent also needs a low surface tension to wet the surface of the material."[16]

Degassing of the cleaning solution is an important step, the purpose of which is to remove the air dissolved in the cleaning solution. Degassing is accomplished by cavitation. The time for complete degassing will vary with the type of equipment and cleaning solution. This may range from seconds to minutes (up to thirty minutes) although partial degassing (80–85%) usually takes place within 5 minutes. To make degassing most effective, parts must be slowly immersed and removed from the solution during cleaning to reduce aeration. The ultrasonic unit should be operating continuously during removal and immersion of parts.

Parts loading must be also considered. In general, parts to be cleaned should be placed as close as possible to the radiating surface although neither the parts nor the parts basket should rest on the bottom of the tank. The most difficult surface to clean should face the transducer, and one should not forget that the weight of the parts being cleaned should not exceed 35 lb/kW of transducer output.

Incorrect container design or one having too high a mass can greatly reduce the effectiveness of the system. Fine mesh baskets (300 mesh) are frequently used for cleaning small parts. Wire mesh baskets are less efficient than solid sheet metal baskets unless the mesh is coarser than ¼ in. or finer than 200 mesh.[17] Parts are frequently supported on racks, jigs, and in beakers. Glass beakers are especially useful when cleaning very small parts in acid solutions. The cleaning tank is filled with water, the beaker immersed in the water bath, and the ultrasonic energy transmitted through the glass and into the solution with little or no loss. However, if a polyethylene plastic beaker were used for the same purpose it would be found that the cavitational energy would be attenuated by 25–50%. Elastomers and nonrigid plastics will similarly absorb ultrasonic energy and will produce a shadowing effect. Insulated parts may have to be specifically oriented. Critical surfaces to be cleaned should face the transducer, but racked parts should be positioned vertically rather than being stacked one on top of the other. All surfaces including blind holes should be wetted, and the level of the cleaning fluid should be at least ¾ in. above the parts.

Other factors to be taken into account are the two kinds of noises which may be produced by the cleaning system. The first, a buzzing or hissing sound, is produced by cavitation bubbles. The second kind, resembling screeches or squeals, should be avoided. The latter is usually caused by the improper use of the equipment, low solution level, moving parts in the tank too fast or too violent agitation of the liquid, two or more nonsynchronized generators, the use of chemicals which do not form clear solutions, or by an excessive accumulation of greasy soils.

In precision cleaning particulate matter must be removed from the cleaning system by cautious filtration and recirculation. The rate of filtration should not cause turbulence as this would interfere with the cavitation process. One should guard also against possible introduction of air by the filtration system, which would defeat the purpose of degassing the cleaning solution.

As with many other manufacturing processes, one likes to measure its efficiency since it is known that ultrasonic cleaning efficiency depends on the ultrasonic power transmitted to a cleaning liquid. Although a close relationship exists between cleaning efficiency and cavitation intensity the two are not identical. Among the many techniques used to monitor ultrasonic cleaning there are several which are commonly used. Each of these measures different parameters and is based on various assumptions. The methods are foil erosion, heat rise, sound probe, and transducers-amplitude measurements. The first two are highly empirical but simple and probably as accurate as the other two mentioned. The details of each test are included in the references.[18-20]

Ultrasonic equipment is quite safe to use, assuming proper installation and use. It must be properly grounded and shielded. Any prolonged contact of any part of the body with ultrasonically activated equipment and/or solutions could cause bone or tissue damage. However, the body part would usually be removed from the solution by reflex action due to heat and pain before any damage could occur. Any prolonged exposure to subultrasonic (15-18 kHz) frequencies should be avoided. Special care should be taken when working with certain toxic solvents such as benzene or chloroform because of their low threshold limit values.

The advantages of using the ultrasonic cleaning technique are many. The method is fast and effective, removing soil even from minute cracks, pores, and indentations. As a rule it requires less heat. In fact, on prolonged use the ultrasonic energy will heat the solution. There is no harm to surface finishes and no need to dismantle assemblies. Delicate parts (e.g., silicon wafers 0.006 in. thick) can be cleaned vigorously yet gently.

Among the disadvantages is the noise which usually accompanies this cleaning method. This method is more complex than other methods, and unless used with a proper cleaning solution it may not achieve the desired results. There are reports of damage done during cleaning to transistors and circuit boards. In the first case a transistor failed mechanically at the junction between the disk and the 10.5 mm nickel wire after five minutes immersion in Arklone P* (trichlorotrifluoroethane) ultrasonically agitated at 25 khz, while a similar device (in which the nickel wires were gold plated) was cleaned without damage.[21]    In the other example circuit board damage by resonance effects or dissipation of ultrasonic energy was minimized by maintaining the surface of the cleaning fluid by float control to within 1 to 1½ in. of the transducers.[22]

## surface heat treatment

Many materials, both metal piece parts and semiconductor materials, are exposed to heat treatments for various reasons. The purpose may be not only to clean the surface but also to remove inclusions of unwanted solids and gases. Heat treatment is also used to obtain the required grain structure of a metal and to diffuse desired impurities as in the case of *p-n* junction formation.

An example of heat cleaning practice is the reduction of silicon dioxide, almost always present on the surface of a silicon wafer, to silicon. This treatment is performed in an epitaxial reactor using hydrogen at $1200°C$ for about 10 minutes. In addition, this procedure will also stabilize the thermal conditions within the reactor for subsequent reactions. Immediately after $SiO_2$ reduction a more thorough cleaning step is used, exposing the silicon wafer to about 3% by volume of anhydrous hydrogen chloride in a stream of hydrogen. This treatment takes place at the same temperature for about five minutes. Generally, one micron of the highly polished silicon wafer surface is removed; thus, any entrapped polishing particles or metallic impurities are changed to chlorides and are swept away with the excess of hydrogen, leaving a nearly perfectly clean and polished silicon surface upon which an epitaxial film will be grown. This is perhaps the single most important step to obtain a good epitaxial layer. It is important that the anhydrous HCl be of the highest purity. Otherwise, traces of water, nitrogen, or methane can cause the formation of oxides, nitrides, or carbides of silicon, which can initiate the formation of defects in the epitaxial layer. A similar technique is used to clean epitaxial susceptors. A higher proportion of anhydrous HCl in a stream of hydrogen is used at $1200°C$ until the silicon layer is volatilized and the susceptor surface of silicon carbide is exposed.

Alkaline potassium cyanide–potassium gold cyanide $KCN–KAu(CN)_2$ plating solutions used heavily in industry can produce a porous gold deposit having large grain structure. This type of deposit, as has been shown by gas analysis, contains a major amount of hydrogen, carbon monoxide, carbon dioxide, and also traces of other gases such as nitrogen, hydrocyanic acid, or

---

*Registered Trademark of Imperial Chemical Industries, Ltd., London, England.

cyanogen. These, during the life of an electron device, would slowly diffuse out and lower its reliability. To prevent this it has been the practice to out-gas critical piece parts plated with cyanide gold in a vacuum. This operation is performed at $10^{-5}$ torr between 350–400° for 2½ hours.

Iron-cobalt-nickel alloy (Kovar or Rodar) is widely used in glass-to-metal seals, especially in headers for transistors. A typical analysis of the alloy would show the following percentages: Ni = 29; Co = 17; Mn = 0.5 max.; Si = 0.2 max.; C = 0.06 max.; Al, Mn, Zr = 0.1 max.; Ti = 0.02 max.; and iron the remainder. The removal of carbon from these alloys is a prerequisite in the production of good glass-to-metal seals. After the seal is made, the presence of carbon in the metal causes bubbles of $CO_2$ and CO to form at the glass metal interface. The presence of these bubbles will cause blistering and, therefore, poor seal strength. To remove the carbon from the sealing alloy the metal parts are usually decarburized in wet hydrogen. After degreasing and chemical cleaning, the alloy piece parts are annealed and decarburized in wet hydrogen at about 1000°C for about thirty minutes. An oxide is formed later at about 650°C in a slightly oxidizing atmosphere, and the glass is applied to this oxidized alloy surface.

As indicated, heat treatments in the microelectronics industry perform a variety of purposes without which we could not reach the present state-of-the-art. However, the processes are not without disadvantages. The major liabilities are the costs for expensive equipment, lengthy heat-up and shut-down cycles, and the gases used to provide the right atmosphere. Additionally, hydrogen atmosphere furnaces present a hazard to the work area because of the explosive properties of the gas. Nonetheless, surface heat treatments will continue as an important part of the microelectronic processing.

## post cleaning practices

An integral part of surface cleaning are the techniques of stopping the chemical reaction at the proper moment, the removal of all chemical solutions, and the return of the product surface to a dry state for further processing. Rinsing and drying have been included in all of the illustrations of cleaning routines; what is implied in these two operations deserves explanation.

### rinsing methods

Rinsing is important to prevent immediate corrosion and surface discoloration, minimize contamination of successive treating solutions (including expensive precious metal baths), and enhance the adhesion of electrodeposits, films, or organic coatings of subsequent processes. After wet chemical treatment of metal surfaces for scale oxide removal, pickling, bright dipping, etching, alkaline cleaning, or cyanide activation, these relatively concentrated solu-

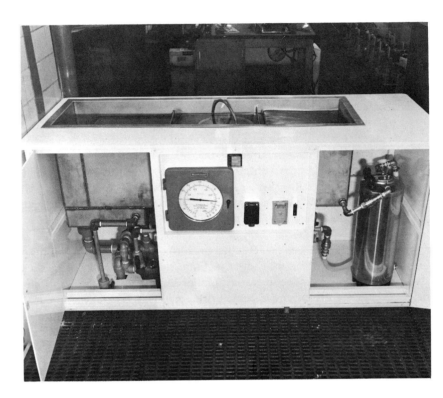

Figure 7-3 Cascade rinsing tank for plating operations.

tions must be removed quickly and effectively from the surface of the work. This is especially true in automated plating systems where, for example, poor rack design, blind holes on parts, or processing of cup-shaped parts would soon result in plating rejects if inefficient rinsing were used.

The problem of good rinse tank design is one of economy and efficiency in "diluting," that is, finding the required minimum rinse flow rate and the minimum permissible time between successive rinses.[23-26] The level of concentration of contaminants in the rinse tank is the important factor in rinsing. The rinsed work leaves the tank with a solution concentrate upon it. This constitutes the drag-in to the next process tank. The lower this equilibrium concentration is, the fewer rinsing steps are required to accomplish the desired result.

A typical cascade or counter current rinse tank used in plating shops is shown in Fig. 7-3. The parts are rinsed counter current to the flow of water so that the purity of each chamber increases as one proceeds through the

TABLE 7-6.   Comparison of Rinsing Methods[a]

| Type of Rinse | Single | Series[b] | | Countercurrent[c] | |
|---|---|---|---|---|---|
| Number of Rinses | 1 | 2 | 3 | 2 | 3 |
| Total Water Feed, gpm | 10 | 0.61 | 0.27 | 0.31 | 0.10 |

[a]Total rinse water volumes required to maintain equilibrium concentration $C_e$ in final rinse at 0.001 × drag-in concentration $C_o$ ($V_o$ = 0.01 gpm).
[b]Separate Water Feed
[c]Same Water Used in Tank 2 from Tank 3, etc.
From A. K. Graham, *Electroplating Engineering Handbook*, p. 759, Van Nostrand Reinhold Co., New York (1971).

tanks. The multiple tank system makes possible an efficient utilization of water by having the same water flow from the third to the second to the first tank (Table 7-6).

Efficient rinsing and cascade rinse systems are affected by the variety of device shapes and quantities which are processed at one time. It has been found that the drag-out volume per unit area is smallest with well drained vertically racked flat parts and may be up to fifty times greater with cup-shaped parts which have been very poorly drained. For example, in a gold plating operation a single lot of 60,000 Kovar cans in a 8 in. × 18 in. plating barrel is handled at one time, a very difficult draining and rinsing operation. In another case a lot of 10,000 molybdenum buttons is loaded into a 3 in. × 5 in. barrel. These parts tend to adhere to each other, and it is very difficult to break up the lot for efficient rinsing.

Rinsing is also used to remove residual ionic material from semiconductor surfaces or metal and ceramic piece parts following other compatible manufacturing steps (e.g., organic solvent cleaning). In this final rinse the various electronic parts or subassemblies are usually submerged in moving, distilled, or deionized water of very high purity (10–18 megohm-cm resistivity or conductivity of 0.1–0.055 micromho/cm). The water washes away many of the absorbed ions, especially when used in conjunction with ultrasonics. Many times the water contains a wetting agent.

The detection of water soluble contaminants can be accomplished by monitoring the change in the conductivity of ultrapure water. Sensitivity depends on the ratio of parts area to solution volume and on the speed with which equilibrium is reached. For example, a cascade rinser in a clean bench as shown in Fig. 7-4 is used to process silicon wafers of various sized lots. The wafers are stacked individually in a vertical position in special quartz handling trays. The total surface area of a 40 wafer (1.25 in. dia.) lot including the tray may be slightly over 1 sq. ft. These wafers are rinsed in high purity deionized water for thirty seconds in the first and second successive cascade compartments, and five minutes in the last compartment, or until the water reaches a minimum of 8.0 megohm-cm.

Suitable specifications and tests of rinsing efficiency are a necessary part of rinsing routines. A typical specification used to evaluate the effectiveness

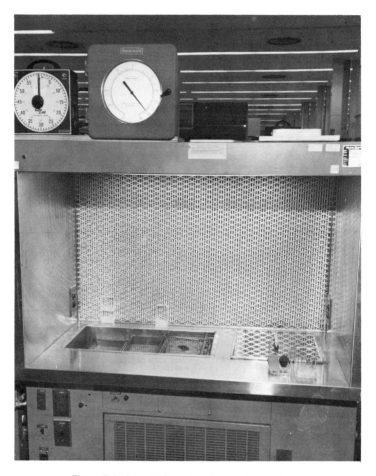

*Figure 7-4  Cascade final rinse for semiconductors.*

of rinsing a freshly plated piece part is the following:  The rinsed part shall not cause an increase in conductivity of more than 0.2 micromho/cm when one square inch of surface area is immersed in 100 ml of pure water (0.4 micromho/cm max.) for 30 ± 2 seconds.  Prior to the test the water shall have come to equilibrium with the carbon dioxide of the atmosphere.  An equivalent ratio of surface area to water volume may be used.

A test commonly used to measure the effectiveness of rinsing residual plating salts consists of checking for sodium and potassium ions ($Na^+$, $K^+$) using flame photometry.  In this test piece parts amounting to 6 sq. in. of total area are immersed in 10 $cm^3$ of high-purity water for twenty minutes minimum.  The final alkali ion concentration should not exceed its initial level (blank test) by more than 0.25 ppm.  This test is usually preceded by a rapid, qualitative test where a plated part is examined under ultraviolet illumination. Visible fluorescence indicates the possible presence of plating salts and usually requires rewashing the plated lot of piece parts.

### drying methods

Drying is almost always a part of the overall cleaning cycle. The one exception is vapor degreasing, since the parts dry almost immediately on removal from the vapors. Emphasis should be placed on clean environment conditions during drying and prior to the next operation. Otherwise, the cleanliness level achieved by the cleaning operation will be lost. If superclean surfaces are required as is usual with most semiconductor wafers, both the final rinsing and drying operations should be performed in a clean bench or clean room environment.

The drying methods used in the fabrication of various electron devices can be generally divided into the following categories:

1. Mechanical, which includes wiping, air blowoff, and dry nitrogen blowoff.
2. Thermal methods including ovens, vacuum, infrared and heat light banks, and hot air or nitrogen drying.
3. Chemical displacement, which uses chemicals such as alcohol or acetone, or the relatively new method of using Freon T-DA35 followed by Freon TF, to displace the water from the parts.

Wiping is used mainly with larger assemblies where the drying action is achieved by absorption, using lint-free cloth. This method, a gross or predry operation, is normally not preferred. In both the air and the dry nitrogen tunnel techniques an appropriate sized container is flooded with air or nitrogen heated to between 100-150°C. The physical force of the gas stream and heat displaces the cleaning solution or water. Before use the gas supply is filtered and often dried or purified in catalytic units of the types described in Chap. 12. The air drying method is used mainly with piece parts. The nitrogen drying tunnel is used mainly to dry semiconductor wafers and is often included with a cascade water rinse in a clean bench as a final drying position for precision work.

One method used extensively in the plating shop with metal parts which are not easily damaged is the use of centrifugal spin drying. When used alone (without electric heat) it can be included under the mechanical categories as the drying action is performed by air which not only displaces water but also aids its evaporation. Usually electric heaters are used to increase the evaporation rate. The direction of spin of the dryer can be reversed, which also aids drying. Thus, the centrifugal spin dryer is an example of a mixed mechanical-thermal method.

An interesting example of performing final surface cleaning of polished silicon wafers after their detachment by solvent soaking from the polishing plates is presented in Fig. 7-5. This machine[27] is placed in a laminar flow hood and is completely automatic other than loading and unloading of the

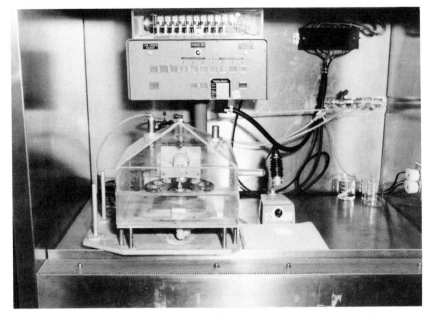

(a)

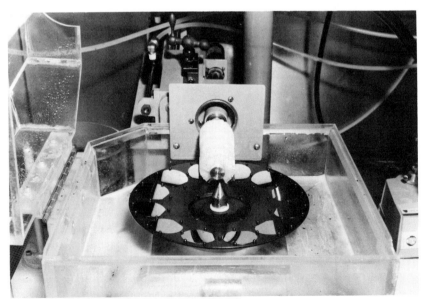

(b)

Figure 7-5 Precision cleaning for silicon wafers.

supporting plate. Different size disks can support up to 12 silicon wafers (from 1¼ to 3 in. diameter). Once the wafers are positioned with the polished face up on the motor driven wafer-holding plate the apparatus cover is closed, and an automatic cleaning cycle begins by scrubbing using a dilute detergent solution. The scrubber is a cylinder-shaped nylon brush which advances over the slices and rotates while a nozzle sprays a wetting agent–water mixture onto the brush. In a preset time sequence the wetting agent nozzle shuts off, a second nozzle begins to spray deionized water (10–18 megohm-cm resistivity) for rinsing, the rotating brush retracts from its contact with the wafers, and the rinse water shuts off. The wafer holding plate which has been rotating at a slow velocity accelerates to a much higher rpm to centrifugally fling residual water off the wafers which may cause stains to form. The clean, dry wafers are removed from the machine, inspected, and stored for further processing. A typical cycle of scrub-cleaning, rinsing, and drying takes two minutes although the controls allow it to be shortened or lengthened.

The use of thermal methods, such as ovens and light banks, causes evaporation of the liquid by heat. Generally, the use of ovens for drying is limited to the drying of ceramic substrates (e.g., 2 hours at 100°C ± 10°C) since many in-process semiconductor materials are heat sensitive. Infrared and heat light banks are not usually used if the control of contaminants is a factor.

The use of chemicals such as alcohols and acetone in drying was, until recently, fairly widespread in the plating room operations. It speeded up drying because of rapid evaporation of the solvent, although for drying bulk loads it still required additional heat—usually hot air. Adequate precautions had to be taken when using this method because of the highly flammable nature of the solvents mentioned.

Flammability worries have been completely eliminated when using Freons or other commercial forms of halocarbons for drying. This displace-

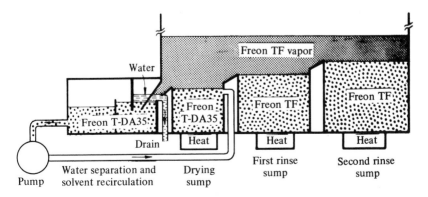

Figure 7-6 Freon T-DA 35 drying system. (Courtesy E. I. du Pont de Nemours & Company, Wilmington, Delaware.)

ment drying system is illustrated in Fig. 7-6. This system depends on water displacement by Freon T-DA35, removing water from metals, plastics, and material combinations. It is used in conjunction with Freon TF in specially designed vapor degreasers. In drying a cup-shaped part or parts with small crevices some agitation is necessary to bring Freon T-DA35 in contact with all surfaces. The experience with this system indicates that it is effective and has many advantages. Among them are:

1. Relatively low operating temperature ($118°F$) is possible.
2. Drying cycle can be completed in 4 to 7 minutes.
3. The chemical is nonflammable and of low toxicity (TLV of 1000).
4. Freon T-DA35 and water are immiscible; therefore, the water is separated and discharged without contaminating the solvent.
5. Spot-free and contamination-free surfaces are obtainable.

Although the initial cost of Freon is high when compared to solvents such as acetone or alcohol, the recycling design of the Freon drying equipment makes this method as economical as the other solvent methods of drying. The latter are not only dangerous from a flammability standpoint but also tedious and cumbersome because of the large volume of dirty solvents that must be transported, collected, and disposed of.

## *summary*

Cleaning processes play an indispensable role in the manufacture of microelectronic products, particularly in the preparation of surfaces before each succeeding treatment or processing step. Ineffective cleaning may lead to costly rejects in subsequent processing steps. The removal of contaminants in the preparation of a surface can be effective only in a properly engineered cleaning system.

The selection of the proper cleaning agent is complex. It will include, among other considerations, the reasons for cleaning, which may be dictated by proper surface preparation, reliability, function, appearance, or all of these. In the selection of the proper cleaning agent, one should know the type of soil or contaminant being removed, whether it is of an organic (oil, grease) or inorganic (salts, rust, fine particles) nature. Of course, the cleaning agent itself, in conjunction with materials of construction of the cleaning tanks and auxiliary equipment, should not leave a different kind of impurity in place of the contaminants it removes. Finally, the choice of the cleaning agent is dictated by the material to be cleaned. Ferrous or nonferrous metals, combinations of both, alloys, plastics, ceramic substrates, glass, and semiconductor materials will have to be cleaned using different cleaning agents or different mixtures.

In making the proper choice of equipment for effective cleaning, whether it is a tray, rack, fixture, basket, or a barrel, one should not forget the configuration of the parts cleaned or the work load. Flat surfaces, especially fragile ones with high intrinsic value in the later stages of manufacture (exemplified by silicon wafers), should be cleaned in a special rack and in a different manner than metal piece parts containing crevices or blind holes. Of importance also is the work load typified by the part size, quantity, and type of soil. As was stated in the ultrasonic cleaning example, there must be a power intensity sufficient to clean but not enough to fracture a delicate silicon wafer.

Finally, any combination of cleaning methods chosen from immersion, soak, dip, wipe, electrolysis, vapor degrease, ultrasonic, heat treatment, or combinations of the above incorporating compatible rinsing and drying methods is acceptable provided it will produce the desired or required degree of surface cleanliness.

## references

1. F. J. Biondi, *Bell Lab. Record,* 36, 289 (1958).
2. J. F. Pudvin, *The Institute of Electrical Engineers* (London), 1125 (1960).
3. J. J. Demo, *Corrosion,* 24, 141 (1968).
4. E. Sirtl and A. Adler, *Z. Metallk*, 52, 529 (1961).
5. D. R. Turner, *J. Electrochem. Soc.,* 107, 810 (1960).
6. B. A. Irving, *The Electrochemistry of Semiconductors,* P. J. Holmes, Ed., Academic Press, London, (56 Refs.) 1962.
7. H. C. Gatos and M. C. Lavine, *Progress in Semiconductors,* Vol. 9, pp. 1–46, A. F. Gibson and R. E. Burgess (Ed.), Temple Press Book Ltd., London, 1965.
8. H. Robbins and B. Schwartz, *J. Electrochem. Soc.,* 106, 505 (1959).
9. H. Robbins and B. Schwartz, *J. Electrochem. Soc.,* 107, 108 (1960).
10. R. D. Heidenreich (to Bell Telephone Laboratories, Inc.), U.S. Patent 2, 619, 414 (Nov. 25, 1952).
11. P. Wang, *The Sylvania Technol.,* Vol. II, 50 (1958).
12. A. H. Szkudlapski (to Western Electric Co., Inc.), U.S. Patent 3, 272, 748 (Sept. 13, 1966).
13. *Contamination Control Handbook,* NASA CR-61264, Clearing House for Federal Scientific and Technical Information, Springfield, Va., 1969, pp. III 42–60.
14. L. R. Jeffrey, Jr., *Metal Finishing Guidebook Directory,* 208, Metals and Plastics Publications, Inc., Westwood, N.J. (1970).
15. T. H. Briggs, Unpublished Data, Western Electric Co., Inc. Allentown, Pa. May 13, 1965.
16. W. J. Colclough, *Ultrasonics* 6, 21 (1968).
17. C. Glickstein, *Basic Ultrasonics,* J. F. Rider, New York (1960), pp. 89–96.
18. J. A. Farris, "Cleaning and Material Processing for Electronic and Space Apparatus," ASTM Special Technical Publication No. 342, Nov. 1963, p. 103.
19. T. J. Bulat, *Air Engineering,* June 1962.
20. A. E. Crawford, *Ultrasonics,* 2, 120 (1964).
21. G. C. M. Byrd and W. M. Clay, *Ultrasonics,* 4, 195 (1968).
22. P. J. Bud, *Electronics,* 36, 86, (1963).
23. H. L. Pinkerton, *Electroplating Engineering Handbook,* 2nd Ed., Chap. 34, A. Graham, Ed., Reinhold Publishing Corp. (1962), p. 705.
24. U. F. Marx and S. D. Cashmore, "Symposium on Pretreatment for Metal Finishing Processes," Institute of Metal Finishing, London (1965), p. 11.
25. A. F. Mohrnheim, *Plating,* 56, 715 (1969).

26. J. B. Kushner, *Metal Finishing Guidebook,* Metals and Plastics Publication, Inc. (1970), p. 502.
27. D. R. Oswald (to Bell Telephone Laboratories) U.S. Patent pending.

## *bibliography*

1. S. K. Ghandhi, *The Theory and Practice of Microelectronics,* John Wiley and Sons, Inc., New York, 1968.
2. *Contamination Control Handbook,* NASA CR-61264, Clearing House for Federal Scientific and Technical Information, Springfield, Va. 1969.
3. F. L. Dwyer, *Contamination Analysis and Control,* Reinhold Publishing Corp., 1966.
4. "Cold Cleaning with Halogenated Solvents," ASTM Special Technical Publication No. 403, July 1966.
5. *Handbook of Vapor Degreasing,* ASTM Special Technical Publication No. 310, April 1962.
6. "Cleaning of Electronic Device Components and Materials," ASTM Special Technical Publication No. 246, March 1959.
7. "Cleaning and Material Processing for Electronics and Space Apparatus," ASTM Special Technical Publication No. 342, Nov. 1963.
8. S. Spring, *Metal Cleaning,* Reinhold Publishing Corp., 1963.
9. "38th Annual Metal Finishing Guidebook Directory," Metals and Plastics Publication, Inc., 1970, Westwood, N.J.

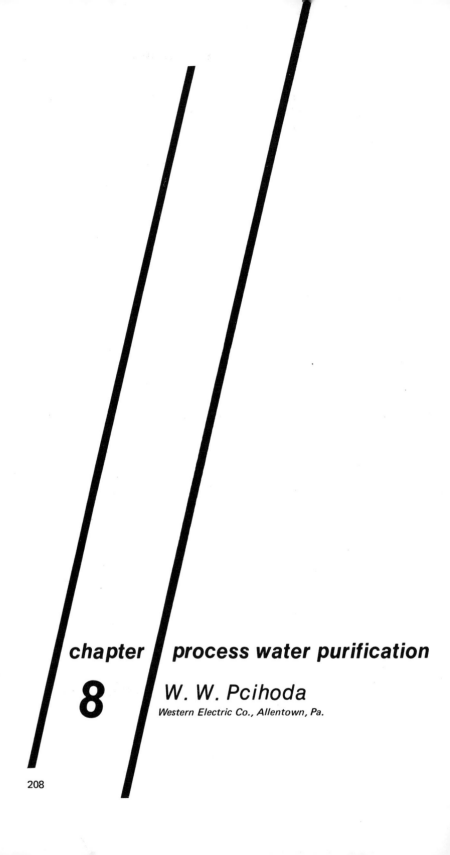

**chapter**

# 8

# process water purification

## W. W. Pcihoda
*Western Electric Co., Allentown, Pa.*

Water is one of the most widely used raw materials in the operation of a microelectronics manufacturing plant.    It serves both as a medium and vehicle in a variety of manufacturing processes and plant related operations. In addition to its use in general applications (boiler feed water, air conditioning, etc.), process water is associated with manufacturing operations indirectly as a cooling medium and directly in solutions, baths, and for cleaning, etching, and rinsing purposes.    In each case some method of water treatment is generally required to control contaminants to acceptable levels.    From a product contamination control position the quality of water used directly in the manufacturing processes is of major concern.

This chapter presents criteria for high purity water used for direct microelectronic processing applications in which contamination control is of prime interest.    Methods of water treatment are described as well as the reasons for using these treatments and the benefits gained from them.

## raw water

Most people would be surprised to learn that the water which they drink is often unsuitable for many applications in the microelectronic industry. Although the water is potable it may contain significant amounts of inorganic and organic impurities as well as particulate contaminants and dissolved gases. Table 8-1 shows some of the impurities typically encountered in raw water, i.e., water to be treated for some specific use.

The source of water supply will give some indication of the type of impurities one can expect to find in raw water.    The two main sources of raw water are ground water supplies obtained from wells and surface water supplies obtained from reservoirs, lakes, rivers, etc.    Well water usually contains high amounts of hardness ($Ca^{++}$, $Mg^{++}$, etc.) as a result of contact with the mineral deposits in the vicinity of the well.    There is a tendency for the hardness or mineral content to increase as well depth increases.    Natural filtration of well water through adjoining sandy areas usually provides water which is relatively clear (free from turbidity) and low in organic content.

Surface water, by contrast, usually has considerably lower concentrations of dissolved minerals.    However, turbidity and organic content are usually higher than in well water because of the presence of silt and clay particles,

TABLE 8-1    Typical Impurities Found in Water

A. Inorganic
   $Ca^{++}$, $Mg^{++}$, $Na^+$, $K^+$, $Fe^{++}$, $Mn^{++}$, $HCO_3^-$, $CO_3^=$, $SO_4^=$, $Cl^=$, $OH^-$, $NO_3^-$
B. Organic
   Humic acid, fulvic acid, alkylbenzenesulfonates (used in detergents), oils, hydrocarbons
C. Particulate
   Silt, sand, oxidized iron or manganese, colloidal silica, microorganisms, organic matter
D. Dissolved Gases
   Nitrogen, oxygen, carbon dioxide, hydrogen sulfide, and others

dust settling from the atmosphere, decaying vegetation, and—in some instances—industrial wastes and human sewage. Surface water supplies fluctuate quite rapidly in particulate concentration, varying with the season and weather conditions. Deep well supplies, on the other hand, remain relatively constant in composition. The reader interested in typical analysis of surface and ground waters is referred to Lange's *Handbook of Chemistry.*[1]

## water quality requirements for microelectronic production

For direct use in the manufacture of microelectronic components, high purity water is necessary with control exercised on inorganic, organic, particulate, and microbial contaminants. Depending on the application (general plating baths and preliminary rinses at one end, to precision cleaning and final rinsing of finished devices at the other end) the level of purity required may vary within the total range of "high purity" water. Table 8-2 shows tentative

TABLE 8-2   Typical Range of Specifications for High-Purity Water

| Parameter | Range |
|---|---|
| Resistivity (megohm-cm) | 15–18 |
| Total electrolytes (ppb as NaCl) | 25–30 |
| Particle count (no./ml) | 150 |
| Nom. max. size (microns) | 0.5 |
| Organics (by $CO_2$ formation) (ppm) | 1.0 |
| Dissolved gases (ppm) | 200 |
| Living organisms (no./ml) (max.) | 8 |

Source: "Ultrapure Water for the Semiconductor and Microcircuit Industries," G. P. Simon and C. Calmon, *Solid State Technology*, February 1968.

specifications for high purity water used by some integrated circuit and semiconductor manufacturers. The stringent requirements placed on the quality of the water used by these manufacturers is a necessity because of the performance requirements of the components being produced. Many of these have been detailed in previous chapters and will not be elaborated at this point. Suffice to say that there is a need to produce, in large quantities, water of a purity which made it a laboratory curiosity not too many years ago.

## water purification processes

Selection of the most suitable water treatment installation for a particular manufacturing activity will depend primarily on evaluation of such factors as the raw water supply and the purity and quantity of water required. To better understand water purification and help evaluate these factors the major steps necessary to produce high purity water from raw water will be discussed. These are shown schematically in Fig. 8-1.

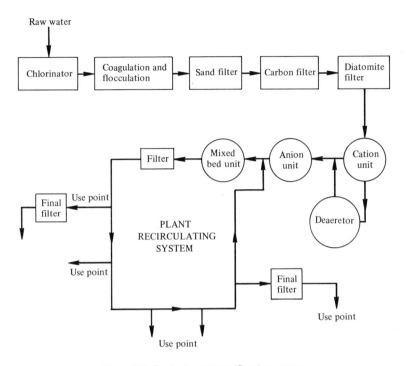

*Figure 8-1 Typical water purification system.*

## chlorination

Municipal water supplies are chlorinated to prevent disease. Microbial contaminants can cause problems in water treatment schemes, as will be discussed later, and must be destroyed by chlorination. Chlorine, when added to water, hydrolyzes according to Eq. (8-1) to form hypochlorous acid:

$$Cl_2 + H_2O \rightleftharpoons HOCl + HCl \qquad (8\text{-}1)$$

The hypochlorous acid thus formed provides the disinfecting properties required. Water to be used in producing high purity water is chlorinated to oxidize or "burn-up" organic matter in the water, to oxidize iron or manganese to enable easier removal by settling, and to reduce color and odor producing materials.

Most installations requiring large amounts of chlorine will use chlorine gas for economic reasons. On smaller installations calcium or sodium hypochlorite may be used as a source of chlorine.

## coagulation and flocculation

To remove the very finely divided particles found in water, it is necessary to use some form of coagulation and flocculation. Coagulation is the neutralization of the electrical charge on these finely divided particles by introducing a material (particle) having an opposite electrical charge. Coagulation is most effective if pH is controlled between 5.5 to 8.0 and a strong anion such as chloride or sulfate is present.

Aluminum or iron salts such as alum (aluminum sulfate) or ferrous sulfate are generally used in coagulating. They form a floc of aluminum or iron hydroxide. The precipitates of the hydrous oxide are positively charged and will neutralize the negative charge of the sulfate or chloride anions in the water, resulting in the coalescence of the fine particles into a porous, gelatinous precipitate.

The flocculation process, which normally follows coagulation, brings the small particles together by gently mixing the water. This tends to entrap smaller particles in the flocculent forming mass. More rapid settling occurs as the agglomerates increase in size. The water is generally passed to a retention basin where the agglomerates settle out. Retention times of two to six hours are common although this can be reduced to one or two hours through the use of upflow clarifiers.

Depending on the incoming raw water the turbidity of most waters can be reduced through coagulation and flocculation to 5 ppm or less. For electronic processing, water of consistently high quality under 1 ppm turbidity is necessary, and filtration must be included in the water treatment scheme.

### *filtration*

Filtration complements the coagulation and flocculation processes by removing small particles which have not been removed by these processes in a limited retention time, i.e., those particles or agglomerates which are too light to settle out in a reasonable time. Filters and filter media for use with water are generally classified into two broad categories: depth type filters and screen or surface type filters. Depth type filters remove particulates by providing layers of filter media to trap and remove the particulates. Examples of depth filter media include sand, coal, urethane foams, and wound cartridges of various natural and synthetic fibers.

A screen filter is a medium having a number of perforations or holes oriented in a prescribed pattern. The perforations or holes are generally well defined with regard to size and tolerance. Particles are removed from a fluid by screen filters by virtue of the specific size of the hole. Particles larger than the hole diameter are retained on the screen while particles smaller than the hole diameter are passed through the screen. The household window screen is the most common example of this type of filter. The screen filter which has come into wide use in the electronic industry is the membrane filter. A membrane filter is a microporous polymeric sheet of material characterized by extremely small pores of a precisely controlled size. Membrane filters are manufactured with pore dimensions as large as 10 microns and as small as 0.01 micron. They are installed at final points of use to ensure freedom from microscopic particles.

Depth filters can be characterized as follows:
a. have relatively high contaminant handling capacity;
b. remove some very fine particles;
c. are not absolute;
d. have low pressure drop;
e. have a tendency for migration of filter media because of the nonhomogeneity.

By contrast, screen filters can be characterized in the following manner:
a. have relatively low contaminant handling capacity;
b. remove only those particles larger than hole diameter;
c. are absolute;
d. have high pressure drop.

In actual use depth filters and screen filters are used to complement each other. A depth filter having high dirt handling capacity will precede a screen filter of definite hole size. In this manner one can economically achieve high dirt handling capacity in addition to absolute particle retention. Specific types of depth filters will be discussed at this point because of their common use in the water purification cycle.

**Sand Filters.** Sand filtration is generally employed to remove turbidity and suspended solids from water. It is also used following coagulation and flocculation to remove the "light" or small particles not removed by these processes.

Sand filters consist of a bed of graded sand with coarse grades used on the bottom layers and built up to a top layer having a grain size approximately 0.5 mm in diameter. The small particles in the water along with some of the floc from the preceding coagulation and flocculation process are trapped by the sand. In addition, a gelatinous film is formed around the sand grains through biological action. These mechanisms result in removal of finer particles than one could expect by the sand alone. Most of the suspended matter is removed in the top few inches of bed depth. Normal bed depth is approximately 15 to 30 in.

In some instances anthracite coal may be substituted for sand. This medium provides higher filtration rates, extended filter life between backwashing, and the use of less water for backwash purposes. There is some sacrifice of water quality when anthracite coal is used. It does have the additional advantage that no silica is added to the water if the water should be alkaline and hot. Often a combination of sand and coal is used to make use of the advantages of each medium.

Sand filters may be one of two types: gravity or pressure. In the former the water flow is, as the name implies, by gravity. Pressure type filters have the advantage of operating in-line without additional pumping. Flow rates of approximately 3 gal/sq. ft/min are normally encountered. Higher flow rates have been achieved through the use of polyelectrolyte filter aids (high molecular weight electrolytes).

**Diatomaceous Earth Filters.**  A somewhat more complex system than sand filtration which removes even finer particles from water is the diatomaceous earth filter.  This system uses diatomaceous earth, the skeletal remains of microscopic one-celled aquatic plants known as diatoms which range in size from 5 to over 1000 microns.

A typical filtration cycle using the diatomite consists of three steps: (a) precoat, (b) filtration and body feed addition, and (c) removal of the filter cake.  In the precoat step a thin layer of the filter aid (diatomite) is deposited on a filter septum such as vertical filter leaves or hollow porous ceramic tubes (Fig. 8-2).  To reduce the tendency of the particles in the water to plug the precoat, additional diatomite (called body feed) is added with the water being filtered.  Thus new layers of filter media are being continually built-up. Porosity of the cake is maintained in this manner and permits long economical cycles on raw water.

Filtration is continued until the resistance of the filter cake has increased to a point where either the pressure drop, the filtration rate, or a combination of the two has reached the economic limit.  Then the cake can be removed mechanically or it can be blown off using air pressure.

Diatomaceous earth filtration can remove particles smaller than the 10-20 micron range.  The grade of diatomaceous earth which is used is dic- tated by the quality of the effluent desired.  In some instances asbestos or

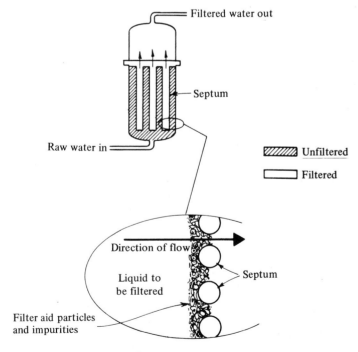

*Figure 8-2 Diatomaceous earth filtration.*

cellulose fibers can be added to the earth to increase the mechanical integrity of the cake.

**Carbon Filters.** Following the filtration steps previously described, it is an accepted practice to pass the water through a bed of granular activated carbon. The carbon bed or carbon filter serves a two-fold purpose. It removes most of the soluble organics and excess chlorine residuals remaining from the chlorination process. In carbon filtration a bed depth of approximately five feet is used. Flow rates of 2 to 3 gal/sq. ft/min are nominal in this process.

## deionization

At this point the water has had the largest portion of the particulate burden removed and the organic content has been similarly reduced. However, little has been done to remove the dissolved minerals or hardness producing elements. This is done most effectively and economically by ion exchange techniques.

Ion exchange deionization, or demineralization as it is sometimes called, can remove all but very slight traces of minerals generally found in water. The purity of water, i.e., freedom from ionic impurities (mineral content), is generally expressed in terms of specific conductance or its reciprocal, specific resistance. Specific conductance is the conductance between the opposite faces of a 1 cm cube of water. It is generally expressed as micromho/cm. Pure water has a specific conductance of 0.0546 micromho/cm at $25°C$ (or conversely, a specific resistance of 18.3 megohm-cm).[2] Deionization is capable of producing water having a specific conductance approaching these values, whereas using distillation, specific conductances of 1 or 2 micromho/cm can generally be achieved.

Ion exchange is a reversible equilibrium reaction. In water treatment there is an exchange of ions between a solid (ion exchange resin) and the water. The ion-exchange resins are crosslinked polymers in the form of highly porous spherical beads having a large number of ionizable groups attached to them. The ions exchanged in the water are the result of the ionization of the inorganic materials dissolved in the water. Aside from the active ionizable groups attached to the resin, the resin itself is insoluble, chemically and physically stable, and permeable to diffusion of ions from the water.

There are two general categories of ion exchange resin: cation and anion exchange resin. Cation exchange resins are used to replace metallic cations (e.g., $Mg^{++}$, $Ca^{++}$) by hydrogen ions. A typical reaction involving cation exchange is shown in Eq. (8-2):

$$2R{\cdot}SO_3H + CaCO_3 \rightleftharpoons (R{\cdot}SO_3)_2\,Ca + H_2CO_3 \qquad (8\text{-}2)$$

In this equation $R{\cdot}SO_3$ represents that portion of the cation exchange resin which is insoluble.

Anion exchange resins are used to replace anions such as $Cl, SO_4^=$ and $HCO_3^-$, which are generally found in water. These resins are available in forms containing active amine or quaternary amine groups and are referred to as weakly basic or strongly basic anion exchangers. Weakly basic anion exchangers are generally used when silica removal is not a problem. When silica removal is a necessity the strong base anion resin must be used with some loss of regenerant efficiency. The reactions involving weak base and strong base anion exchangers are shown below:

Weak base anion:    $RNH_3 \cdot OH + HCl \rightleftharpoons RNH_3 \cdot Cl + H_2O$    (8-3)

Strong base anion:  $2R_4N \cdot OH + H_2CO_3 \rightleftharpoons (R_4N)_2 \cdot CO_3 + H_2O$  (8-4)

During the anion exchange residual hydrogen ions, $H^+$, and the hydroxyl ions, $OH^-$, combine to form water leaving no ionic material.

As the deionization process progresses through its cycle, the resin gradually becomes exhausted ($H^+$ and $OH^-$ have been replaced by other cations and anions) and must be regenerated. The cation exchange resins are regenerated by sulfuric acid or hydrochloric acid. Sulfuric acid is most commonly used for economic reasons. Some care must be taken during regeneration with sulfuric acid because of the possible problem of precipitation of calcium sulfate in the resin bed. This can be overcome by initially using dilute regenerant and gradually achieving the strength of acid recommended by the resin manufacturers. Hydrochloric acid can be used to remove the calcium sulfate precipitation should it occur. Anion exchange resins are regenerated using sodium hydroxide.

In the deionization process the sequences for producing high quality water can be (1) two-step or separate bed deionization; (2) multiple-step deionization (variation of (1)); or (3) mixed bed (monobed) deionization. These are shown schematically in Fig. 8-3.

In the two-step or separate bed deionization, water is introduced to the cation exchanger and then to the anion exchanger. Usually the anion exchange medium is of the strong base type so that silica removal can be more effective. The multiple step employs an additional anion exchanger. This is of the weak base type and is inserted between the cation and the strong base anion unit. The weak base anion, having a higher capacity, will remove free mineral acidity more economically than the strong base anion. The latter can then remove the silica. By utilizing this system one can take advantage of increased capacity and regenerative efficiency of the weak base anion and still obtain the silica reduction experienced in the two-step operation. The specific conductance of water deionized by processes (1) or (2) is generally in the range of 1 to 20 micromho/cm, the value being determined by the raw water, resins used, and the operating conditions.

If it is desired to obtain a specific conductance of 0.2 micromho/cm or less ($>$ 5 megohm-cm), mixed bed deionization is required. The mixed bed or monobed deionizer consists of a single vessel containing an intimate mixture of cation and anion exchange materials. This then represents an infinite series of separate bed deionizers. The advantages of this system are more complete removal of dissolved solids, shorter rinse cycles, and less rinse water.

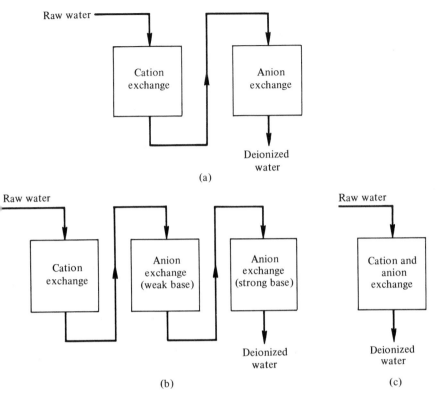

Figure 8-3 Processing sequences for deionizing water. (a) Two-step or separate bed deionization; (b) multiple bed deionization; (c) mixed bed (monobed) deionization.

Disadvantages are more complicated regeneration and often higher operating costs. Also, effluent conductivity is easily affected by resin fouling and hence, water pretreatment as described previously is a necessity. Mixed beds are most effective when used to polish the effluent from the separate bed deionization process.

In some instances, particularly where the bicarbonate hardness of water is high, a deaerator can be used to remove carbon dioxide from the water and lighten the load on the anion resin. In cation exchange we have the following:

$$Ca(HCO_3)_2 + H_2R \rightleftharpoons CaR + 2H_2CO_3 \qquad (8\text{-}5)$$

In the vacuum deaerator carbon dioxide formed according to Eq. (8-5) is removed from the water by application of a vacuum supplied by steam jet eductors or vacuum pumps:

$$H_2CO_3 \rightleftharpoons H_2O + CO_2\uparrow \qquad (8\text{-}6)$$

The resulting reduction in the amount of $H_2CO_3$ in the water conserves ion

exchange capacity of the anion resin.  Aeration, or intimate mixing of air
with the water, will also accomplish carbon dioxide degassing.

### reverse osmosis

The preceding discussion on water purification is based on well established
techniques which have been in common use for many years.  More recently
the reverse osmosis process is being given consideration for high purity water
applications.  This method, developed as part of the government desalinization
studies, can be explained as follows.  If a concentrated solution is separated
from a dilute solution by a semipermeable membrane there will be a movement of solvent from the dilute solution to the more concentrated solution.
This movement of solute continues until equilibrium is achieved.  At equilibrium a differential head (known as the osmotic pressure) exists which in the
case of water with dissolved solids is approximately one psi for each 100 ppm.
If one can reverse this situation, i.e., pressure sufficient to overcome the
osmotic pressure is applied, solvent flow will be in the opposite direction;
hence, the name reverse osmosis.  By providing a high water pressure and a
method for removing the concentrate, reverse osmosis can be used to provide
a continuous supply of pure water.  A system of water purification using
reverse osmosis is shown schematically in Fig. 8-4.

The membranes for the process are generally cellulose acetate on a suitable support material.  Pressures used in this method approach 400 psi in
some instances.  Reverse osmosis units are compact as can be seen in Fig. 8-5,
which shows a 40,000 gpd unit with dimensions approximating 12 ft. by 4 ft.
by 4 ft.

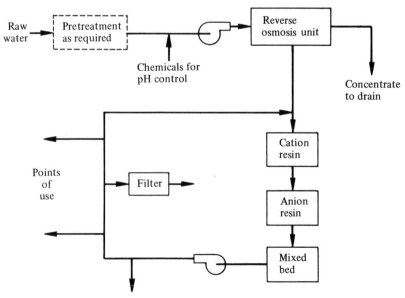

*Figure 8-4 Water puification using reverse osmosis.*

*Figure 8-5 Reverse osmosis system. (Courtesy Polymetrics, Division of Technical Equities Corp., Inc.)*

On most raw waters it is believed that reverse osmosis can replace much of the pretreatment such as coagulation and filtration prior to deionization. Some manufacturers claim the following benefits for the reverse osmosis process: 1) ultra filtration (0.05 micron); 2) mineral removal (90%+); 3) organic removal (95%+); 4) bacteria removal; 5) low cost; 6) simple, infrequent maintenance.

For raw waters of high turbidity, pretreatment is still necessary prior to reverse osmosis processing. For improved membrane lifetime and performance of the units pH control is required and design temperatures and pressures must be maintained at specified levels. Recent case histories[3,4] indicate that reverse osmosis units are playing an active role in new water treatment systems, in some instances replacing more conventional components.

## pure water distribution and control

### distribution planning

If the water is to be used in quantities of more than a few gallons per minute and at several locations consideration must be given to distribution of the water to the points of use. Because of the purity of the water most piping materials have a tendency to contaminate the water to a certain degree. Platinum and platinum group metals would be the most corrosion resistant piping materials, but the cost of such a distribution system is obviously prohibitive. Pure block tin may offer sufficient corrosion resistance at a more moderate cost. Distribution systems today find extensive use of polyvinyl chloride, polyethylene, and stainless steel. The latter is inclined to contribute trace metallic impurities while the plastics have a tendency to generate particles and contribute organics from filler materials and lubricants used in forming processes. The effects of these trace impurities can be minimized by limiting pipe lengths and by employing point-of-use scavengers.

A typical water purification system is included in Fig. 8-1. In this system water is distributed by means of a recirculating grid system from point of generation to points of use. Water which is not used is returned to the deionizers for repolishing. At points of use small systems may be employed to "polish" the water, i.e., remove trace impurities that may have resulted from the distribution system itself (e.g., due to repairs, particulate generation by valves being turned on and off, or introduced during initial assembly of the system). These small systems may consist simply of a membrane filter and mixed bed resin column or may be a water treatment system in miniature, i.e., a depth type prefilter, a carbon bed, mixed bed resin column, and a membrane filter (Fig. 8-6). In the latter case, the prefilter will remove any gross contaminants which might have entered the system, while the carbon bed and resin column are used to remove soluble organics and inorganics, respectively. Assurance of particle-free water is provided by the membrane filter to the degree associated with the pore size of the membrane.

*Figure 8-6  Point-of-use water treatment system (Courtesy Millipore Corp., Bedford, Massachusetts.)*

In some instances small stills are used to polish the water taken from the distribution systems. These stills are often quite elaborate and produce limited quantities of water. They find application where a sterile water free of organic traces is required.

## microbial control

Sterility of the process water for electronic processing is not an absolute necessity as it is, perhaps, in the pharmaceutical industry. However, bacteria and algae represent a source of particles and residues which can contaminate product. To ensure such particulate removal many manufacturers are using membrane filters in the range of 0.45 and 0.22 micron pore sizes which will remove bacteria. Unusually rapid clogging of these filters can be caused by bacteriological growth. Thus, bacteria in distribution systems and treatment facilities can be a serious problem. Since bacteria multiply at a geometric rate, one bacterium in a system, if properly nurtured, can multiply to millions in a relatively short period. Carbon beds and ion-exchange beds are most easily affected by contamination with bacteria. The many impurities in the water trapped in the carbon and resin beds often contain the nutrients required to sustain and nourish the bacteria, thereby acting as a culture medium. The bacteria tend to coat the carbon granules and the resin beds, drastically reducing their capacity.

Steam has been used to kill the bacteria in a carbon bed. However, the entire bed must attain the temperature of the steam in order to ensure killing all the bacteria. If all bacteria are not killed, the survivors will multiply and the original conditions will soon be re-established. Chlorine has also been used to disinfect carbon beds. A drawback to the use of chlorine is the large quantity of chlorine required because of absorption of the carbon and the breakdown of the carbon granules. Following chlorine treatment, carbon beds do not generally revert back to their original activity.

Formaldehyde solutions (2%) have been more successful for sterilizing carbon than steam or chlorine. The solution is introduced to the bed and held there for one to two hours. It can then be flushed with water and re-activated with steam.

With the ion-exchange resin a mild germicide such as Sterimine* is used. These germicides are based on trichloromelamine and are generally formulated with salts, wetting agents, buffers, and inert ingredients. They are strong enough to kill the bacteria but do not harm the resin as would a strong chlorine treatment. Formaldehyde can be used in the manner described for treating carbon beds. The bacteriological problem is generally more severe in the anion bed than in the cation bed. This can be attributed to the use of acid for regenerating the latter.

In the distribution system itself bacteria can grow in dead end runs of piping which can become stagnant over a period of time. Sterilization of a grid system can be accomplished using a weak solution (approximately 0.5%

*Wallace and Tierman, Division of the Penn Walt Corp., Bellville, N.J.

or less) of sodium hypochlorite. This solution should not be introduced to the resin system. Recirculating grid systems, because of their continuous movement, are helpful in maintaining low bacteria counts once a system is sterilized.

Detection of bacteria in a system can readily be accomplished through the use of the Standard Plate Count Methods of the American Public Health Association[5] (for further discussion see Chap. 16). Counts of one or two colonies are normal. When bacteriological contamination has occurred counts will soar to hundreds or thousands of colonies, sometimes becoming too numerous to count. Proper sterilization methods should reduce this to the normal levels.

## *monitoring*

To ensure that a water system is functioning as intended, monitoring the water quality is necessary. Sampling should be done at the point of generation to ensure that the purification equipment is operating satisfactorily. Samples should be taken at scattered points along a distribution system to determine if the grid is contaminated or degrading the water to a degree more than normally expected. Sampling of the return line from a grid system is logical because by the time water reaches this point it should have been "contaminated" if a problem area exists in the grid.

Conductivity and pH are relatively easy to measure and can be done continuously in-line. These tests will indicate the effectiveness of the ion-exchange process and when regeneration is necessary. If mixed bed deionization is used, specific conductances in the range of 0.055 to 0.10 micromho/cm (10-18 megohm-cm) should be expected. The acid-alkali balance should remain within a few tenths (of a pH unit) of pH 7.

Indications of filter effectiveness can be monitored using the filter plugging test[6] which measures the decay of flow rate through a filter. An 0.45 micron filter is inserted into a stream of liquid being monitored. The initial flow rate is measured as well as the flow rate after a specified time interval. The decay in flow or percent plugging of the filter can be calculated, and a relative indication of contaminant level can be noted.

The three tests mentioned above can quickly give an indication of system conditions. Most other tests are rather time consuming and must be performed in a laboratory. For instance, there are various particle counting instruments available for monitoring particulate levels in the water (see Chap. 5). Combustion methods combined with infrared spectrophotometry are used to measure organics in water (see Chap. 3). Because of the time and money involved with this type of test it is recommended that they be used on a periodic sample basis. These, along with other process monitoring techniques, are discussed in Chap. 16.

## *summary*

The source of water for an electronic manufacturing plant and its end use will determine what treatment facilities are necessary to produce high quality water necessary for manufacturing operations. The repeated use of water for various steps in the electronic manufacturing cycle makes it imperative that the best possible water be provided. Emphasis must be placed on removal of particulate and organic contaminants as well as the traditional removal of inorganic contaminants using ion-exchange. No longer is high purity water a laboratory curiosity limited to small volumes. It must be produced in large quantities and be available at the opening of a spigot. With new, more sophisticated devices becoming available, the engineer will be challenged to improve on the schemes described herein and provide even higher quality water.

## references

1. N. A. Lange, *Handbook of Chemistry—Revised Tenth Edition*, McGraw Hill, N.Y., 1967.
2. G. P. Simon and C. Calman, *Solid State Technology*, February, 1968, pp. 24–25.
3. Radovan Kohout, "Case History of the 324,000 GPD Reverse Osmosis Ultimate Water Plant at Microsystems International," presented at the Industrial Water and Pollution Conference and Exposition, Chicago, Ill., March 14–16, 1973.
4. R. T. Skrinde, W. M. Steeves, L. S. Shields, Jr. and T. L. Tang, "Economic and Technical Evaluation of Reverse Osmosis for Industrial Water Demineralization," presented at the Industrial Water and Pollution Conference and Exposition, Chicago, Ill., March 14–16, 1973.
5. *Standard Methods for the Examination of Water and Wastewater*, Twelfth Edition, 1965, American Public Health Association.
6. Charles W. Baldwin, "TI Gets Super Clean Water for Semiconductor Work," *Plant Engineering*, May, 1963. pp. 144–147.

## bibliography

1. Robert L. Howe, *Applied Chemistry for Water Purification and Waste Treatment*, Caglayan Scientific Publisher, Istanbul, Turkey, 1967.
2. *Betz Handbook of Industrial Water Conditioning*, Betz Laboratories, Inc., Philadelphia, Pa., 1962.
3. James L. Dwyer, *Contamination Analysis and Control*, Reinhold Publishing Corp., New York, 1966.
4. *Contamination Control Handbook*, NASA CR-61264, Clearing House for Federal Scientific and Technical Information, Springfield, Va., 1969.
5. E. D. Driscoll, "Industrial Water Treatment Process," presented at the Fourth Annual Liberty Bell Corrosion Course, Philadelphia, Pa., 1966.

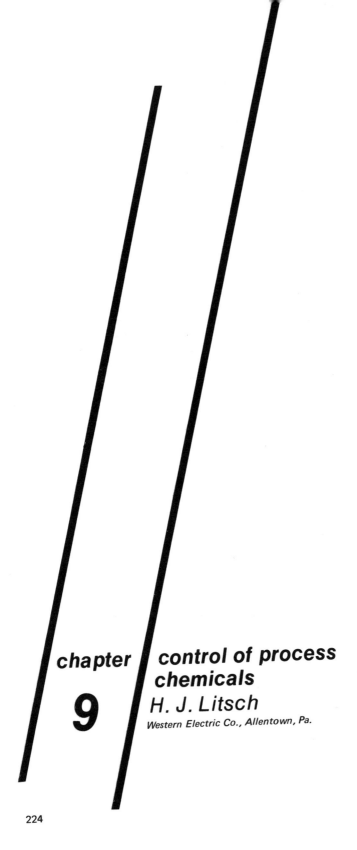

chapter

9

**control of process chemicals**

*H. J. Litsch*

*Western Electric Co., Allentown, Pa.*

In the manufacture of microminiature electronic devices it has become apparent that process chemistry and the associated chemicals are now intimately integrated into the manufacturing system. In the past these chemical processes were considered adjunct but not primary to the preparation of reliable products. The obvious change of product design, such as integrated circuits, has directed control efforts into the process chemical area to achieve the optimum reliability with these near-microscopic devices. It is the understanding of these process chemicals, their relative significance, and their need for control that will be described in this chapter.

## control by specification

A process chemical can be described as any substance of a chemical nature and of specific purity needed for a chemical process function. Typical process chemicals associated with semiconductor and thin film manufacture fall into the general categories of: (1) semiconductor raw materials; (2) etches and polishes; (3) dopants; (4) high purity metals; (5) plating solutions; (6) solvents and cleaners; (7) encapsulants; (8) specialized products; and (9) materials for containers and equipment.[1] Each of these materials contains impurities in some degree. The specific application in each instance must be used to determine which levels of impurities can be tolerated and what kind of specifications are most appropriate in each category.

### semiconductor raw materials

The purity of semiconductor starting materials is of prime importance since the electrical behavior of solid state devices is so greatly dependent on impurity gradients in the raw material. Thus, high purity semiconductor materials such as silicon, germanium, gallium arsenide, and indium antimonide require high purity $SiHCl_3$, $SiCl_4$, $GeCl_4$, Ga, As, In, and Sb as input ingredients. In addition, high purity acids and gases such as $H_2$, HCl, HF, A, and He are required as part of the raw material preparation.

As a prime example of this group, consider the material requirements of $SiCl_4$, a main source of high purity silicon. In the early 1960's $SiCl_4$ could only be purchased in an impure "as manufactured" state. To produce silicon useful for semiconductor device manufacture it was necessary to further process the material by distillation, adsorption, and associated treatments.[2] In this processing method the "as manufactured" $SiCl_4$ was first refluxed with copper turnings to remove sulfur. It was then preferably treated with $AlCl_3$ and $Cl_2$ and passed through an adsorption column containing either activated alumina or silica gel. The $SiCl_4$ was then cleaned of $Cl_2$ by refluxing in a stream of dry $N_2$ or $H_2$. By reducing the purified $SiCl_4$ with $H_2$ in Ta tubes, polycrystalline Si rods were obtained. After removal of the tantalum the Si rods with several zone passes in dry $H_2$ became n-type with a resistivity of 50-100 ohm-cm. Further refining to remove donor impurities, such as phosphorus, produced p-type silicon with resistivities above 8000 ohm-cm. This technique as described was required to produce a useful process chemical. To provide a supply of "good" $SiCl_4$ a specification was generated which suitably described the input raw material. This specification placed control at the 1 ppm level on the elements (other than silicon) which could be detected by emission spectrography. A use test was included as part of the material acceptance criteria which currently requires that epitaxial grade $SiCl_4$ be capable of producing noncompensated epitaxial layers of 25 ohm-cm, n-type. The original "as manufactured" $SiCl_4$ has to be further processed similar to the described refining method in order that the material can meet the specification. Through these efforts effective control of the raw material quality has been achieved and can now be purchased routinely since inspection per a specification is possible.

In this case the specification was developed by the user to meet his own needs. In other cases standard industry specifications may be appropriate.*

## etches and polishes

Etches and chemical polishing solutions for the several semiconductor materials are generally comprised of either one or a mixture of high-purity HF, $HNO_3$, HAc, $H_2SO_4$, $H_2O_2$, NaOH, and KOH with appropriate additives. These materials are critical to the manufacture of semiconductor devices since adsorption of metal ions present in the etchants as impurities can seriously affect the device performance.

Sotnikov and Belanovskii[3] evaluated the adsorption of ions on Ge, Si, and quartz during etching in appropriately doped etchants. The etchants studied were 49% aq. HF and 65% aq. $HNO_3$ in a 1:4 volume ratio (com-

*ACS Specifications: These specifications are prepared for industry by the American Chemical Society, and they generally cover the reagent type material where impurities as well as minimum assay are specified.

ASTM Specifications: The American Society for Testing and Materials is an international, technical, scientific, and educational society devoted to the promotion of knowledge of the materials of engineering and the standardization of specifications and the methods of testing. ASTM is involved primarily in the standardization of methods of test and definition of terms relating to materials.

NF Specifications: These specifications are prepared by the Committee on National Formulary under the direct authority and supervision of the Council of the American Pharmaceutical Association. Although the medical field is the major benefactor of these specifications, special use manufacturers may find specific application of the detailed description of materials covered.

USP Specifications: Prepared primarily as a legal document, these specifications provide U.S. Government regulatory agencies with enforceable standards of purity and strength. Their main application centers in the medical field although it is not uncommon to find a material adequately specified for microelectronic product manufacture.

Military Specifications: Generated for the Army, Navy, and Air Force, these specifications often are more selective than other standards in the control of certain impurities. It is possible that military contract needs in electronic manufacture may have generated specifications which can be useful. Because of the broad scope of military documents detailed examination of each specification is required before its applicability can be properly determined.

monly abbreviated as CP), 20% KOH, and 30% $H_2O_2$. The concentration of Ag, Au, Cu, In, Sb, or Zn ions adsorbed during etching was determined as a function of the ion concentration in the etchant and the etching time. Their studies covered the range of concentrations from $10^{-6}$ to $10^{-1}$ atomic percent and an etching time of $\leqslant 15$ min, with etchant temperatures of $20°$, $107°$, and $104°C$ for CP, KOH, and $H_2O_2$, respectively. The experiments showed that the adsorption isotherms were mostly Freundlich type ($\log n \propto \log c$), and the maximum amount adsorbed on Ge or Si was $10^{17}$ atoms/cm$^2$, much higher than that adsorbed on quartz ($10^{15}$ atoms/cm$^2$). The kinetic curves showed rapid equilibration ($< 5$ min) of the etchant. For 6–8 ohm-cm Ge or Si the conductivity type was found not to affect the adsorption. Further studies by Sotnikov, Dyner, and Belanovskii[4] established that with NaOH as the etchant and using radioactive cobalt, $Co^{60}$, the number of cobalt atoms adsorbed on a 1 cm$^2$ surface increased sharply when the NaOH concentration was increased from 0.0 to 0.5%.

Keiler[5] studied the physical adsorption and the electrochemical deposition of cations such as Au, Hg, Cu, As, Sb, In, Ga, Zn, and Fe from etching and nonetching solutions such as $H_2O_2$-HF, $HNO_3$-HF, dilute HCL, 10% HF, and water. Working mainly with Ge and Si using tracer techniques he found that differences in surface contamination were of several orders, depending on whether the cation deposition was electrochemical or by adsorption. Electrochemical deposition was found to depend on diffusion. In addition, when surface contamination by As, Cu, Ag, or Au reached $10^{14}$–$10^{16}$ atoms/cm$^2$, the reverse voltage values of diodes decreased.

Kern[6] showed that contamination of Si by metallic impurities from reagent solutions is particularly severe in aqueous HF since the protective layer of natural oxide is stripped without etching the semiconductor layer. Kern[7] further demonstrated that etching of Si results in physical adsorption of up to $10^{17}$ F$^-$/cm$^2$ of silicon. This chemisorbed fluoride was removed only with hot deionized water.

Each of these examples illustrates the vital significance of the purity of the etching materials. With adequate specification and inspection these materials can be controlled. Proper selection of the chemical specification must be made with impurities in mind. Tables 9-1 and 9-2 illustrate the variety of specifications available for typical acids. In the case of the ultrapure HCl, note the significant increase in the cost of the materials which accompanies the increase in purity. These examples are indicative of the specification variations available in industry which usually provide sufficient latitude of choice for the process engineer. With extreme sensitivity to impurities prevalent in the microminiature semiconductor field the choices of etchant are critically related to the device reliability. There are many varieties of suitable specifications available for these etching materials as well as an ample number of chemical suppliers. In the case of etching mixtures, however, it may be necessary to develop a detailed specification if the etchant is to be purchased from a chemical manufacturer.

TABLE 9-1   Typical Grades of Hydrochloric Acid

|  | Technical | ACS Reagent | Ultrapure HCL Min. 30% |
|---|---|---|---|
| Approx. price/pound | $0.40–$0.50 | $0.65–$0.75 | $22.00–$26.00 |
| Assay | – | 36.5–38.0% | 30% min. |
| Specific gravity | 20°Be' | 1.19 | – |
| Color | – | A.P.H.A. 10 max. | – |
| Residue after ignition | – | 0.0003% max. | – |
| Impurities |  |  |  |
| Extractable organics | – | (0.0005%) T.P.T. | – |
| Copper (Cu) | – | 0.0005% max. | Max. 0.005 ppm |
| Nickel (Ni) | – | 0.0005% max. | Max. 0.005 ppm |
| Cobalt (Co) | – | – | Max. 0.005 ppm |
| Zinc (Zn) | – | – | Max. 0.005 ppm |
| Aluminum (Al) | – | – | Max. 0.01 ppm |
| Calcium (Ca) | – | – | Max. 0.05 ppm |
| Potassium (K) | – | – | Max. 0.05 ppm |
| Sodium (Na) | – | – | Max. 0.05 ppm |
| Sulfate ($SO_4$) | – | 0.00008% max. | – |
| Sulfite ($SO_3$) | – | 0.0001% max. | – |
| Free chlorine (Cl) | – | 0.00005% max. | – |
| Ammonium ($NH_4$) | – | 0.0001% max. | – |
| Arsenic (As) | – | 0.0000005% max. | Max. 0.005 ppm |
| Heavy metals (as Pb) | – | 0.00001% max. | Max. 0.005 ppm |
| Iron (Fe) | – | 0.00001% max. | Max. 0.02 ppm |
| Bromide (Br) | – | Approx. 0.005% | – |

TABLE 9-2   Typical Grades of Hydrofluoric Acid

|  | Technical | ACS Reagent |
|---|---|---|
| Approx. price/pound | $0.53 | $1.28 |
| Assay (HF) | 52–55% | 48.0–51.0% |
| Fluosilicic acid ($H_2 SiF_6$) | – | 0.01% |
| Residue after ignition | – | 0.0005% |
| Chloride (Cl) | – | 0.0005% |
| Phosphate ($PO_4$) | – | 0.0001% |
| Sulfate and sulfite (as $SO_4$) | – | 0.0005% |
| Arsenic (As) | – | 0.000005% |
| Copper (Cu) | – | 0.00001% |
| Heavy Metals (as Pb) | – | 0.00005% |
| Iron (Fe) | – | 0.0001% |

## dopants

These substances are generally highly purified compounds which are introduced into semiconductor metals as an impurity in controlled amounts by deposition or diffusion methods. Since the ultimate purity is desired in almost all cases, the highest purity available "as manufactured" is normally specified. Typical specifications for these materials are: boron, 99.99999$^+$% pure; arsenic, 99.999$^+$% pure; and boron in silicon at 2.4 X $10^{18}$ boron atoms/cm$^3$ of silicon. Phosphine, PH$_3$, in helium at 10 and 200 ppm is also commonly used.

The capability of these materials to affect the resistivity in semiconductors is typified by boron. With high purity boron it is possible using the Czochralski process to vary the resistivity of silicon from 20 to .005 ohm-cm by varying the boron concentration from 6.5 X $10^{14}$ to 4 X $10^{19}$ atoms/cm$^3$ of Si. If impurities are present in the boron dopant a change in concentration of dopant would be required in order to produce the desired resistivity. More significantly, a serious decrease in carrier lifetime would correspondingly result, making the doped silicon valueless.

Further use of dopants can be found in epitaxial processing. Herein the dopant is added in the gaseous state to be appropriately deposited in the epitaxial layer. Diborane, B$_2$H$_6$ (20 ppm in ultra-high purity hydrogen), and phosphine, PH$_3$ (20 ppm in ultra-high purity hydrogen), are two of the more common dopants used for this processing technique. Process sensitivity to impurities in the dopant is very high since the resistivity of the epitaxial layer would go out of control from impurity deposition and defective units would result.

### high purity metals

Metals used for contacts, interconnections, lifetime control, formation of *p-n* junctions, and thin film deposits are commonly Au, Al, Ga, As, In, Sb, Ta, Ti, Pd, etc. As in the case of the dopants the purity available may likely become the specification purity. Some very high purity material specifications have been established based on the limits of analytical sensitivities. Further specification of purity requirements would become academic. Typical percentage purities available are: Al 99.997$^+$; As 99.9999; Au 99.99$^+$; In 99.99$^+$; and Ta 99.95 (see Table 9-3).

It should be noted that some difficulties can be encountered when these materials are purchased in fabricated forms such as Al or Au pellets. It is

TABLE 9-3   Annealed Tantalum. Maximum impurities for 99.95% minimum
purity tantalum.

| Element | Maximum % | Element | Maximum % |
|---------|-----------|---------|-----------|
| O  | .01000 | Fe | .0025 |
| N  | .0050  | Mg | .0005 |
| H  | .0010  | Mn | .0005 |
| C  | .0040  | Mo | .0015 |
| Al | .0025  | Ni | .0015 |
| Ca | .0025  | Si | .0025 |
| Nb | .010   | Sn | .0005 |
| Co | .0005  | Ti | .0010 |
| Cr | .0005  | W  | .0060 |
| Cu | .0005  |    |       |

generally accepted that the above purities are guaranteed only before supplier's or manufacturer's fabrication. Cleaning may be necessary after fabrication to avoid serious contaminant problems, especially from metallic impurities introduced by handling. An example of serious difficulty can be illustrated by sodium contamination on high purity metals. Sodium can seriously affect any type of passivated device by migration through the passivated oxide layer.[8] It is also important to recognize that 1/10,000th of one monolayer of ionic impurity is capable of inverting the surface of 1 ohm-cm silicon.[9] With this apparent sensitivity any substance which can contribute contamination directly to the active semiconductor device should be a candidate for stringent controls on purity levels by restrictive specifications wherever economics will permit.

In practice, the purity of these metals is analyzed on a laboratory sampling basis when received from suppliers.

### plating solutions

Plating solutions can be classified as a process chemical of critical nature since they directly relate to the metal deposited from them. Au, Ag, Ni, Cu, Rh, and Sn are typical metals deposited from electroplating solutions for semiconductor and thin film applications.

Where these electroplated deposits are applied directly to the active parts of a semiconductor or thin film device the purity is of the same critical nature as explained under high purity metals and is controlled only by the capability of the electroplating system chosen for deposition. However, when a mechanical support, such as a lead frame, is being electroplated the deposit may be required to be pure enough to provide suitable bonding and solderability. In the case of gold on lead frames the purity is normally established at 99.9% minimum gold content in the deposit. Impurities which will harden the deposit, such as Ni, Fe, and Co, are limited to .03% each. A hardness measurement on periodic samples taken from the plating system provides the major mechanical check on the deposit.

### solvents and cleaning agents

Materials used in cleaning operations are typically trichloroethylene, methylene chloride, methyl alcohol, isopropyl alcohol, acetone and other ketones, trichlorotrifluoroethane, and water. Most of these solvents or cleaners are meant to remove surface contamination, prevent surface discoloration, or prevent immediate corrosion.

Specifications of solvents such as acetone and alcohols are based primarily on the ACS standards. These materials, described as ACS grade, are generally of sufficient purity for semiconductor use.

Trichlorotrifluoroethane, used as a water removal and cleaning agent, is one of the purest organic chemicals commercially available. It has a minimum assay requirement of 99.8% and a maximum limit of 0.2% on other chlorofluorocarbons. The nonvolatile residue is held to a maximum of 2 ppm by

**TABLE 9-4   Physical Properties of Trichlorotrifluoroethane. (Courtesy E. I. duPont deNemours and Company)**

| | |
|---|---|
| Chemical Formula | $CCl_2F$-$CClF_2$ |
| Molecular weight | 187.4 |
| Boiling point at one atmosphere, | 117.6°F |
| | 47.6°C |
| Freezing point | -31°F |
| | -35°C |
| Critical temperature | 417.4°F |
| | 214.1°C |
| Critical pressure | 495 psia |
| | 33.7 atm |
| Density at 77°F(25°C) | |
| Liquid | 13.06 lb/gal |
| | 97.69 lb/ft$^3$ |
| | 1.565 g/cm$^3$ |
| Saturated vapor at boiling point | 0.4619 lb/ft$^3$ |
| | 7.399 g/liter |
| Latent heat of vaporization at b.p. | 63.12 Btu/lb |
| | 35.07 cal/g |
| Specific heat at 70°F (21.1°C), Btu/lb · °F or cal/g · °C | |
| Liquid | 0.213 |
| Saturated Vapor (Cp) | 0.152 |
| Thermal conductivity at 70°F (21.1°C), Btu/hr · ft$^2$ (°F/ft) | |
| Liquid | 0.043 |
| Saturated Vapor | 0.00430 |
| Viscosity at 70°F (21.1°C), centipoises | |
| Liquid | 0.694 |
| Saturated Vapor | 0.0102 |
| Refractive index of liquid at 79.7°F (26.5°C) | 1.355 |
| Surface tension at 77°F (25°C), | 17.3 dyne/cm |
| Relative dielectric strength (nitrogen = 1) | 4.4 |
| Dielectric constant | |
| Liquid at 77°F (25°C), 100 Hz | 2.41 |
| Saturated Vapor (0.5 atm) at 79°F (26°C) | 1.010 |
| Solubility of water at 70°F (21.1°C), % by Wt. | 0.009 |
| Solubility in water at saturation pressure and 70°F(21.1°C) | 0.017% by wt |
| Diffusivity in air at 77°F (25°C) and 1 atm | 0.068 cm$^2$/sec |

Source: E. I. du Pont de Nemours and Company based on "FREON" TF Solvent.

weight.   Its physical properties are elaborately detailed as shown in Table 9-4.   This substance is certified by the manufacturer and analyzed as required by routine inspection techniques.   As a drying agent it has a distinct advantage of being nonflammable compared to the generally used flammable solvents.

## plastic encapsulants

These materials are usually proprietary formulations controlled by manufacturers' specifications.   It is important to know whether the impurities in these materials can affect the product being made. Silicone rubbers, silicone resins, epoxy resins, and molding compounds are typical of materials employed for plastic encapsulation.

Encapsulants, in most cases, being chemical in nature contain reactive substances. In the case of epoxy resins amine activation is needed to cure the resin. In controlling the effects from corrosive catalysts of this nature one should evaluate the resin with respect to the particular device to be encapsulated. If unreacted catalysts remain in the resin after cure, it is possible that electrical characteristics of the electron device may be degraded.[10]   Some silicone resins are cured by acetic acid which results from the reaction of acetic anhydride and moisture in the air. This catalyst is somewhat less reactive and may be more suitable. Molded plastics can be most useful since they are generally thermosetting, and the device must be able to withstand only the pressures and temperatures required during transfer molding operations.   Full understanding of the compound is needed to insure that contamination of the units does not result. Dielectric and physical stability is critical to the preservation of desirable device characteristics. The successful use of these materials, however, is often predicated upon a properly protected and passivated device junction.   Critical items such as flexural strength, compressive strength, dielectric strength, dissipation factor, and volume resistivity can be specified.   Ionic or gaseous contamination to the semiconductor device, however, remains to be tested and specified where possible.

### specialized products

Photoresist materials, solders, fluxes, ceramics, etc. are classified as specialized materials oriented to one specific operation in the semiconductor and thin film manufacturing process.   Most of these materials are of proprietary nature and consequently the specifications are controlled by the manufacturer.

Photoresist material is a substance whose purity, viscosity, and particle content must be carefully controlled since it is used to delineate critical device contact and interconnection geometry.   In some instances filtering of these substances immediately before use is imperative to guarantee the absence of particles in the pattern generation system. It is also important not to overlook the effects of residuals from the photoresist. These substances can cause mechanical defects in subsequent layers which are deposited.  Proper specification describing maximum particle size (less than 0.1 micron) can establish the needed material, and subsequent filtration will protect against inadvertent particulate contamination during transport or storage.

Solders and fluxes normally are easily specified in detail. A typical specification for Sn-Pb solder is Alloy Grade 60B per ASTM B-32-66T. Only the residuals can cause difficulties. Solders are generally physically controllable, and only gross negligence will result in serious contamination. Fluxes, however, present another problem.   Generally they are of a noncorrosive nature, i.e., they contain an organic material which, upon heating, decomposes into a very mild acid. This acid activates the surface to be soldered. There are also fluxes of the corrosive nature which will attack many metals and make the associated connections unsatisfactory. Primarily these are the $ZnCl_2$ fluxes which will, upon heating, decompose into HCl. The contaminant nature of this acid is well known and is to be avoided for this application.

Ceramics have become more important with the advent of thin film manufacture. Previously these materials had little control on their composition and physical properties other than those provided by the manufacturer. Control parameters have been established covering chemical, physical, and electrical properties, and where ceramic glazes are involved, chemical and physical properties have been established. Elaborate descriptions were formulated for the surface finish of glazed and unglazed ceramics. Table 9-5 shows the typical specification requirements for ceramics. In practice these requirements are measured on every lot of ceramics submitted since they are extremely critical to the production of satisfactory thin film devices.

## materials for containers and equipment

Steels, pure metals, plastics, glass, ceramics, rubber, and wood fall into this category. These materials contribute major-to-minute contamination to the processing liquids or gases during semiconductor and thin film processing. The sensitivity of these devices to contaminants has been elaborated upon previously. It is important to consider the construction material in light of these previous considerations. Since each of the materials is useful only with a certain number of chemicals each case should be examined on its own application. For example, polyvinyl chloride has wide application as a container and equipment material in acid and alkaline etching systems. It is inert to most of these materials at low temperatures. However, if solvents of the ketone family, such as acetone, were to contact polyvinyl chloride, softening and distortion would result.

Quartz is another example of a critical construction material.[11] This material is commonly used for diffusion boats in doping processes for silicon wafers. Purity and relative inertness are absolutely essential. At diffusion temperatures, out-gassing of contaminants from an impure boat can adversely affect the diffusion process. Out-gassing of dopant from boats which absorb the dopant can also be detrimental to successful diffusion.

It is apparent that knowledge of all the chemicals used in each specific process is imperative to proper selection of construction materials. The contribution of contamination from construction materials for containers and equipment is perhaps more significant than from any other area. In the case of the specific items mentioned in previous sections of this chapter, it is generally possible to control the chemical involved by adequate specification. If a contaminant is involved it is possible to identify the substance and treat accordingly. In the case of construction materials specification is limited, and the contaminants are insidious in nature. Close surveillance of each process step is required to ensure freedom from contamination by residues of container construction materials.

TABLE 9-5  Ceramic Substrates for Thin Film Devices

**Ceramic Body**
Chemical
The total alkali content (sodium, potassium, and lithium) of the ceramic body shall not exceed 0.3 percent.

The ceramic body shall not contain any material which will reduce at $1925°F$ ($1050°C$) in an atmosphere of dry hydrogen.

The conductivity of the aqueous extract of a section of substrate (either glazed or unglazed) having a surface area of 5 $in^2$ shall not exceed $1 \times 10^{-6}$ mhos/cm.
Physical
Coefficient of thermal expansion: $7.3 \times 10^{-6}$ to $8.1 \times 10^{-6}$ in/in/$°C$
Flexural strength: 40,000 psi min.
Grain size: 4 microns max. average
Warpage: 5 mils/ in max.
Electrical
Dielectric constant: 9.3–9.7
Loss Factor: .008 max.
Volume resistivity at R.T.: $1 \times 10^{-14}$ ohm-cm
**Glaze**
Chemical
The total alkali content (sodium, potassium, and lithium) of the glaze shall not exceed 4.0 percent
Glaze Fit
The glaze and substrate shall not show any cracks or crazing.
**Surface Finish**
Glazed Substrates
The glazed surface shall have a continuous smooth glossy appearance.

The surface roughness of the glazed surface shall not be more than 1 microinch, peak to valley, when measured over at least three 0.250-inch traces, one of which shall be 90° to the other.
Unglazed Substrates
The surface shall be smooth and uniform in texture and appearance.

The surface roughness of the smoother side of the unglazed substrates shall not be more than 10 microinches (CLA) when measured over at least three 0.250-inch traces, one of which shall be 90° to the others.
**Glaze Thickness**
The thickness of the glaze shall not exceed 0.004 inch including the meniscus area.
**Dimensions**
The substrates shall meet the dimensional requirements.
**Porosity**
There shall be no evidence of dye penetration after 5 minutes immersion at atmospheric pressure.

## control in the factory

Having established the chemical processing needs and the resultant specifications, one might conclude that sufficient purity control of process chemicals is ensured. Unfortunately, this is usually not the case. Materials do not always arrive at the factory as specified, or they may age or become contaminated before use. For these reasons inspection of incoming materials and proper planning of chemical storage, handling, and use are needed to maintain a continuing high quality of process chemicals.

### material inspection

Chemicals, like any other material, must be tested for compliance to the specification. The methods of test depend on the kind of substance being tested and the requirements involved. It is important that a definite understanding be made as to what methods are used to verify requirements. The methods of test must be clearly stated and understood by the suppliers. For example, the ACS reagent chemicals specification for hydrochloric acid[12] provides detailed information as to the procedures to be used for testing in accordance with the requirements of the specification. Further details regarding possible tests for chemicals are given in Chap. 3.

### chemical storage and handling

Storage of chemical raw materials is related to the maintenance of an inspected chemical once it has been accepted for manufacturing use. This is vital since serious process problems may result if the chemical deteriorates while being stored. It should be recognized that some materials will undergo chemical breakdown or reaction with time, resulting in a product different than that originally put on the shelf. The difference can be in concentration, composition, or physical form. For example, hydrogen peroxide can decompose into water and oxygen under some circumstances. If used in an etching process after decomposition takes place, the resultant etchant will react differently than anticipated. Adequate methods of marking should be employed to identify the age of the product. In addition, "first-in-first-out" inventory techniques should be employed to provide adequate guarantees against shelf deterioration.

In some extreme situations it may be necessary to provide a protective gas cover for the chemical being stored. Impurities from any atmosphere might cause difficulties at a later time; consequently, an inert gas cover can be provided to remove this potential danger.

There are two major storage methods employed for chemicals: 1) storage in the original container, and 2) bulk storage and subsequent repackaging to a functional container. The use of the original container has the advantage of eliminating the risk of contamination from transfer operations. This is the most desirable method of storage from the viewpoint of controlling contaminants. However, economics may dictate other methods of storage. If chemicals are purchased and stored in bulk quantities, considerable cost advantage can be gained in purchase price and reduced storeroom space requirements. In the case of high volume solvent degreasing agents such as trichloroethylene, a large "on-site" tank can be used to store relatively pure solvent for dispensing by appropriate means into clean transport containers. Large volume bulk storage has the advantage of reduced material cost and improved product line supply protection with inventory of approved chemical raw materials. It should be apparent that economic studies are required to determine the real benefits of bulk storage since labor and storage container costs affect the economics significantly.

The primary factors to be evaluated when considering bulk storage of process chemicals are: 1) risk of contamination during transfers; 2) maintenance of large chemical raw material inventories; 3) handling problems associated with repackaging; and 4) associated labor and material costs. Careful evaluation should be given to the methods of transfer, types of containers, piping, etc. to be used to avoid contamination, protection of employees during transfer, and satisfactory labeling.

### point-of-use control

Once chemical raw materials have reached their point of use, it may be necessary to treat the materials even further. The major techniques used to refine chemicals at the point of use are filtration, activated carbon adsorption, distillation, drying, and combinations of these systems. These methods have been summarized in other chapters.

## summary

This chapter has reviewed the necessity for adequate chemical specifications and factory inspection routines for microelectronic manufacture. Each type of chemical material is controlled on the basis of their overall effects on the production process. With this information it should be possible to anticipate where problems could arise, to plan for their elimination in advance, and to protect from inadvertent recontamination once materials have entered the manufacturing process.

## references

1. H. G. Verner and W. B. Haynes, "The Use of Chemicals in Solid State Device Fabrication," *Semiconductor Products and Solid State Technology,* July 1964, Vol. 7, No. 7, pages 13–17.
2. H. C. Theuerer, "Purification of SiCl$_4$ by Adsorption Techniques," *Journal of the Electrochemical Society,* January 1960, Volume 107, No. 1, pages 29–32.
3. V. S. Sotnikov and A. S. Belanovskii, "Adsorption of some metal ions from electrolytes when etching Ge, Si, or Quartz in them," *Dokl. Akad. Navk. SSSR,* 162 (5) 1105–8 (1965) (Russ).

4. V. S. Sotnikov, L. L. Dyner, and A. S. Belanovskii, "Adsorption of $Co^{60}$ from its solutions in NaOH on the surfaces of Ge, Si, and Quartz," *Radiokhimiya,* 9 (2) 253-6, (1967) (Russ).

5. D. Keiler, "Purity of Chemicals with respect to the Production of Semiconductor Elements," (VEB Werk fur Fernsehelektronik, Berlin) *Solid State Electronics,* Vol. 6, pp. 605-610 (1963), Pergamon Press.

6. W. Kern, RCA Laboratories, Princeton, New Jersey, "Deposition of Trace Impurities on Silicon and Silica," *RCA Review,* Vol. 31, No. 2, pages 234-264, June, 1970.

7. W. Kern, RCA Laboratories, Princeton, New Jersey, "Adsorption of Reagent Compounds," *RCA Review,* Vol. 31, No. 2, pages 207-233, June 1970.

8. B. Yurash and B. E. Deal, "A Method of Determining Sodium Content of Semiconductor Processing Materials," *Journal of the Electrochemical Society,* Solid State Science, pages 1191-96, November 1968.

9. M. M. Atalla, E. Tannenbaum, and E. J. Scheibner, "Stabilization of Silicon Surfaces by Thermally Grown Oxides," *Bell System Technical Journal,* Volume 38, p. 749, May 1959.

10. S. Schwartz, A. T. Tweedie, et al., "Final Sealing and Encapsulation," *Integrated Circuit Technology,* McGraw-Hill, New York, 1967, pp. 132-157.

11. R. E. Tucker, "Diffusion Boats," *SCP and Solid State Technology,* July 1964, pp. 30-32.

12. *Reagent Chemicals,* American Chemical Society Specifications, 4th Edition, 1968.

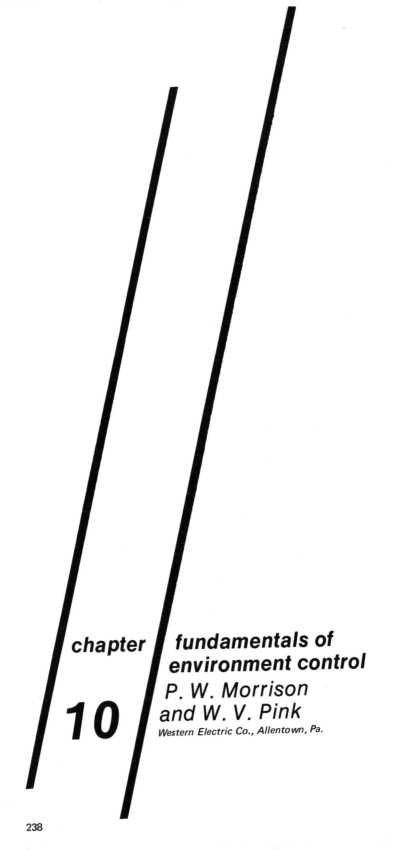

chapter

**fundamentals of
environment control**

*P. W. Morrison
and W. V. Pink*
Western Electric Co., Allentown, Pa.

**10**

Control of the manufacturing environment has for its primary purpose the protection of products of processes from undesirable substances or energy influences. This represents a formidable task in view of the complex mixture of solids, liquids, gases, and energy forms present in the atmosphere. Too often environmental contamination control design has been relegated to empirical solutions because the interaction of competing force systems is thought to defy theoretical analysis. The indeterminate nature of the mechanics of control can be made less formidable, and this chapter will lay a foundation on which practical control systems can be built.

The science of aerosols will be used in this chapter as a backdrop to describe the fundamentals of environmental contamination control which proceed from the nature of the contaminants in the air through the philosophy of control to the basic control element—the filter. This sequence sets the stage for Chap. 11, where practical designs are discussed in detail.

## airborne contaminants

In manufacturing activities where product contamination is to be controlled, the complete characterization of any environment would require a knowledge of the total physical, chemical, and biological composition and the concentrations of gaseous, liquid, and solid matter constituents found in the atmosphere. In addition, the energy levels present would be characterized: these include mechanical and acoustical vibrations; magnetic fields; electrostatic potentials; and heat, light, and other forms of electromagnetic radiations. Finally, the variations with time and location within the space would be evaluated for each of these potential contaminants. It is economically fortunate that such a complete contaminant survey is not often required in the manufacture of microelectronic products. Nonetheless, it is possible that one, several, or all might require some form of control for a specific product or process.

Of this list of potential contaminants the suspended solids (with their physical and chemical characteristics), temperature, and moisture content dominate the present design consideration of environmental control facilities. Their effects on product are obvious. The remaining factors may cause product variations either too subtle for observance or too specific to a particular process to warrant general concern. Further discussion on airborne contaminants will be limited to the physical and chemical nature of suspended solids and trace gases present in a factory environment. The latter are included since their chemical effects on product cannot be separated from those of solid deposits. The reader is referred to Chap. 14 for a discussion on energy effects.

### airborne contamination from external sources

A significant source of airborne material harmful to electron devices is the urban environment containing various forms of gaseous and particulate pol-

TABLE 10-1    Composition of the Atmosphere

| Gas | Composition by volume, ppm |
|---|---|
| $N_2$ | 780,900 |
| $O_2$ | 209,500 |
| A | 9,300 |
| $CO_2$ | 300 |
| Ne | 18 |
| He | 5.2 |
| $CH_4$ | 2.2 |
| Kr | 1 |
| $N_2O$ | 1 |
| $H_2$ | 0.5 |
| Xe | 0.08 |

lutants. The type and quantity of pollutants will vary with time and location, but all experts agree that the air pollution problem is worldwide and becoming more severe every day. Table 10-1 gives the average composition of the earth's natural atmosphere. In any one locale man-made additions in gaseous and particulate form may significantly compound the problem of controlling a specific manufacturing environment.

Man's chemical contributions to the earth's atmosphere are diverse in nature. Table 10-2 indicates the ranges of gaseous pollutants in Los Angeles in the 1950's which have now become typical of most industrialized areas. Tables 10-3 and 10-4 indicate typical inorganic and organic constituents of

TABLE 10-2    Typical Ranges of Air Contaminant Levels in Los Angeles Smoggy and Nonsmoggy Days[a]

| Contaminant | Typical contaminant range, ppm | | Record maximum value, ppm |
|---|---|---|---|
| | Smoggy day[b] | Nonsmoggy day[c] | |
| Aldehydes | 0.05–0.60 | 0.05–0.60 | 1.87 |
| Carbon monoxide | 8.00–60.00 | 5.00–50.00 | 72 |
| Hydrocarbons | 0.20–2.00 | 0.10–2.00 | 4.66 |
| Oxides of nitrogen[d] | 0.25–2.00 | 0.05–1.30 | 2.65 (3.93)[e] |
| Oxidant | 0.20–0.65 | 0.10–0.35 | 0.75 |
| Ozone | 0.20–0.65 | 0.05–0.30 | 0.90 |
| Sulfur dioxide | 0.15–0.70 | 0.15–0.70 | 2.49 |

[a]"Technical Progress Report, Air Quality of Los Angeles County," Vol. II, Los Angeles County Air Pollution Control District, 1961 (data through 1959).
[b]Defined as a day with severe eye irritation in central Los Angeles.
[c]Defined as a day with no eye irritation in central Los Angeles.
[d]$NO_x = (NO + NO_2)$
[e]Highest observation on January 13, 1961.

## TABLE 10-3    Mean and Maximum Concentrations of Selected Particulate Contaminants in U.S. Atmospheres[a] 1957-64

| Pollutant | Pollutant concentration $(\mu g/m^3)$ | |
| --- | --- | --- |
| | Geo. mean | Max. |
| Suspended particulates | 98. | 1706. |
| Benzene-soluble organics | 7.4[b] | 128.3 |
| Nitrates | 1.68 | 24.8 |
| Sulfates | 9.35 | 95.3 |
| Antimony | c | — |
| Bismuth | c | — |
| Cadmium | c | — |
| Chromium | 0.020 | 0.710 |
| Cobalt | c | — |
| Copper | 0.063 | 10.00 |
| Iron | 1.99 | 74.00 |
| Lead | 0.54 | 17.00 |
| Manganese | 0.064 | 4.70 |
| Molybdenum | c | — |
| Nickel | 0.028 | 0.830 |
| Tin | 0.024 | 1.00 |
| Titanium | 0.042 | 1.14 |
| Vanadium | c | — |
| Zinc | 0.09 | 58.00 |
| Radioactivity (in $\mu c/m^3$) | 4.7[d] | 5435.00 |

[a]Based on samples collected during a 24-hour period
[b]1957-1963 Mean, 1964 samples have been composited by quarters.
[c]Concentrations in most samples are below minimum detectable quantity.
[d]Arithmetic average of national monthly averages.

atmospheric particles. The problems of air pollution are beyond the scope of this text; however, it suffices to say that the varieties and quantities of chemicals present in a polluted atmosphere can have an adverse effect on chemically active product surfaces.

The major physical property affecting the occurrence and control of atmospheric particulate matter is particle size. It is obvious that particle dimensions have a direct bearing on filtration efficiency. Figure 1-3 indicates a particle size spectrum from $6 \times 10^{-4}$ micron to 1000 microns. Particles larger than 20-25 microns settle out rapidly and thus are few in number in a sample of airborne particles. Particles smaller than 0.1 micron will coagulate rapidly, and their concentration tends to decrease with time in the atmosphere.[1]    Figure 10-1 illustrates the typical size distribution for atmospheric particles from combined light microscope, electron microscope, and sedimentation measurements.    Note the different frequency distributions which occur from the same sample depending on particle parameter utilized. In the microelectronics field size distributions are usually taken on a numerical

**TABLE 10-4  Concentrations of Large Organic Compounds
in the Average American Urban
Atmosphere**

| Compound | Airborne particulate (μg/gm) | $\dfrac{\mu g}{100 \ m^3 \ air}$ |
|---|---|---|
| Benzo($f$)quinoline | 2 | 0.2 |
| Benzo($h$)quinoline | 3 | 0.3 |
| Benzo($a$)acridine | 2 | 0.2 |
| Benzo($c$)acridine | 4 | 0.6 |
| 11 H-Indeno(1,2-b)quinoline | 1 | 0.1 |
| Dibenz($a$, $h$)acridine | 0.6 | 0.08 |
| Dibenz($a$, $j$)acridine | 0.3 | 0.04 |
| Benz(a)anthracene | ~30. | ~4. |
| Fluoranthene | ~30. | ~4. |
| Pyrene | 42. | 5. |
| Benzo($a$)pyrene | 46. | 5.7 |
| Benzo($e$)pyrene | 42. | 5. |
| Perylene | 5.5 | 0.7 |
| Benzo($ghi$)perylene | 63. | 8. |
| Anthanthrene | 2.3 | 0.26 |
| Coronene | 15. | 2. |
| $n$-Heptadecane | 20. | 2.5 |
| $n$-Octadecane | 110. | 14. |
| $n$-Nonadecane | 160. | 20. |
| $n$-Eicosane | 180. | 23. |
| $n$-Heneicosane | 320. | 40. |
| $n$-Docosane | 480. | 60. |
| $n$-Tricosane | 620. | 77. |
| $n$-Tetracosane | 480. | 60. |
| $n$-Tentacosane | 480. | 60. |
| $n$-Hexacosane | 85. | 11. |
| $n$-Heptacosane | 260. | 32. |
| $n$-Octacosane | 340. | 43. |
| Total | 3800. | 480. |

From E. Sawicki, S. P. McPherson, T. W. Stanley, J. Mecker, and
W. C. Elbert, *Intern. J. Air Water Pollution*, **9**, 515 (1965).

basis using a light scattering particle counter. In cases of chemical contamination by airborne particles the weight or mass distribution is more significant.

Several conclusions can be drawn from the previous data. The nearly linear plots of Fig. 10-1 suggest a log-normal distribution relationship for airborne particles of polydisperse nature.[2] This tendency is used in defining the cleanliness classes of Federal Standard 209[3], which will be discussed later on in this chapter. Note the numerical predominance of airborne particles which are under 0.5 micron in size and, conversely, the weight predominance for those particles over 1 micron in size. Thus, a particle-free environment can only be attained with filters designed to collect minute particles without becoming clogged by the mass of larger particles also present in the atmosphere.

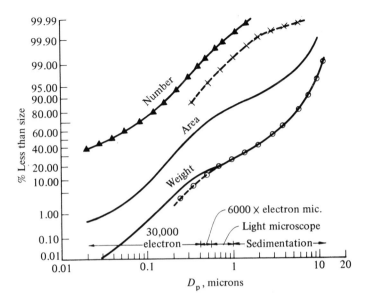

Figure 10-1 Typical airborne size distributions from combined light and electron microscope and sedimentation data. The triangles represent measurement by electron microscope; crosses represent measurements by light microscope; and circles show measurement by sedimentation. (After Transactions, ASHRAE, Vol. 64, p. 135, 1958, New York.)

## the factory environment

While the community environment contains an assortment of gaseous and particulate pollutants, how much of these materials enter the factory environment? A typical microelectronics factory will condition its air, i.e., the temperature will be controlled in the 70–80°F range, relative humidity will be kept below 50%, and low to intermediate efficiency filters will be included in the system to reduce the particulate contributions from outside. Gaseous pollutants, other than water vapor, will enter the factory in their original forms while the larger particle pollutants are generally excluded. The processes, facilities, and people within the factory add their contribution, and the resultant mixture of outside and inside sources of contamination is the factory environment.

From a chemical analysis viewpoint we would expect to find all the chemical constituents of air pollution vapors, plus whatever process vapors which escape into the factory air. Table 10-5 illustrates the major organic vapor contaminants identified at one microelectronics factory[4]; they were found to be the organic solvents most utilized in microelectronic cleaning operations.

Regarding the chemical nature of factory airborne particles, one can expect to find data similar to the air pollution analyses of Tables 10-3 and

TABLE 10-5 Vapor Contaminants (ppm) for a Typical Electronics Factory

| | Chemical | High | Average |
|---|---|---|---|
| Main Floor | Acetone | 1.8 | 0.7 |
| | Trichloroethylene | 2.3 | 0.6 |
| | Xylene | 0.4 | Trace |
| Lower Floor | Acetone | 2.7 | 0.7 |
| | Trichloroethylene | 1.9 | 0.4 |
| | Xylene | Trace | — |
| | Oil (as octane) | 3.1 | 0.4 |

TABLE 10-6 Inorganic Nature of Airborne Particles

| | Concentration (uglm³) | | | | | |
|---|---|---|---|---|---|---|
| | K | Ca | Fe | Cu | Zn | S |
| Factory | 0.97 | 0.55 | 1.99 | 1.88 | 0.71 | 1.52 |
| Outside Air | 1.01 | 6.40 | 3.48 | * | 1.29 | 1.11 |

*— None detected in samples taken

10-4. This assumption is supported by Table 10-6, showing the comparative inorganic content of particles collected from the factory and outside environments using x-ray fluorescence as the detection technique.[5]

The number and size distribution of airborne particles within a microelectronics factory do not appear to differ significantly from outside conditions. Figure 10-2 compares light scattering particle size distribution of both

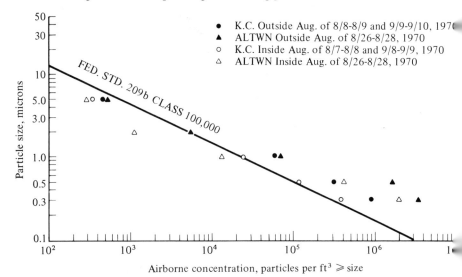

Figure 10-2 Typical average airborne dust concentration in general manufacturing area.

the inside and outside particle concentrations at two Western Electric plant locations. The curves approximate a log-normal distribution with the factory concentrations being essentially equal to or somewhat less than the concentration level of the community environment. The Allentown plant is located in an industrial area which has a high concentration of small particles in the environment. The Kansas City plant is located in a suburban-agricultural area. While the two locations show comparable concentrations of large particles, the Allentown environment is noticeably dirtier in the submicron size range. The higher proportion of large airborne particles in the Kansas City plant is caused by the greater number of dirt generating processes peculiar to specific products being manufactured.

In addition to the variations which can occur in different parts of the country and for different manufacturing situations, factory particle concentrations will vary greatly with time. Figure 10-3 shows a typical time variation

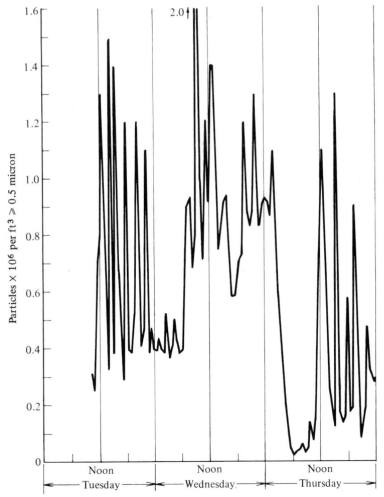

*Figure 10-3  Particle concentrations in general plant areas.*

of particle concentrations in the Allentown plant caused by the complex interaction of personnel, machinery activity, and climatic conditions. Generally speaking, a factory with a comfort air conditioning design can be classified as an uncontrolled environment from a small particle concentration viewpoint.

## classes of clean environment

The uncontrolled environments of the community and the factory form the background from which the various categories of special environments can be examined. Clean environment has come to mean particle control in some form. Practically, this is logical since the need for particle control is more obvious than for vapor control in most manufacturing situations. The technology of vapor control has not been developed to the point where its cleanliness classification can be defined.

Federal Standard 209, "Clean Room and Work Station Requirements, Controlled Environment,"[3] was established in 1963 (with 209b as the latest revision) as a basis of comparing degrees of particle control regardless of the method of clean environment design. Classes of environment were defined in a form which recognized the log-normal relationship of environmental size distributions, and yet they acknowledge the current limitations of the particle counting and sizing technology. One very important point about these environment classes which is sometimes overlooked is that they are intended as measurements of operating performance rather than as construction specifications. Federal Standard 209 states, "Particle counts are to be taken during work activity periods and at a location which will yield the particle count of the air as it approaches the work location." Insofar as is practical the environment classes are intended to measure the concentration of airborne particles in the immediate vicinity of a sensitive product surface when particle deposition is likely to occur.

The environment classes are defined in Federal Standard 209 by the maximum number of airborne particles/cu. ft of air in the diameter ranges of $\geqslant 0.5$ micron and $\geqslant 5.0$ microns. Bell System practice has recognized the usefulness of the classes shown in Fig. 10-4.

These definitions are plotted on a log-log basis to establish a particle distribution reference. Other classes may be designated using the particle distribution slope and the particle counts at the $\geqslant 0.5$ micron and $\geqslant 5.0$ micron sizes. The extrapolation of these curves is acceptable for particles sizes down to 0.3 micron and up to about 10 to 20 microns since the ratios defined by the distribution slope have been confirmed in many clean room environments.[6] However, further extrapolation should be avoided. Light scattering counting techniques are not commonly reliable in size ranges below 0.3 micron while particles over 20 microns settle rapidly and do not remain airborne for significant periods of time.

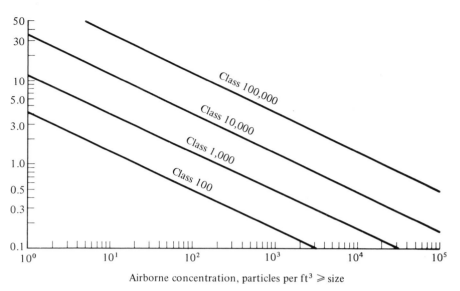

Airborne concentration, particles per ft$^3$ ≥ size

*Figure 10-4 Classes of clean environment.*

## deposition mechanisms

Airborne particles must deposit onto sensitive surfaces to cause damage to microelectronic products. An understanding of the mechanisms causing particle transport and deposition is fundamental to environmental particle control. The major deposition mechanism of airborne particles in a factory environment is sedimentation, but the circumstances are complicated by the effects of Brownian diffusion, coagulation, inertia, turbulent flow, convection, aerosol electrical properties, and artificial agitation. Thermal and other secondary forces are referenced briefly to give a complete picture of particle motion and its complexities.

### sedimentation

Gravity is probably the most commonly recognized mechanism for particle movement and deposition. Particles in any liquid or gas of a lower density will tend to settle out, reaching a terminal velocity determined by the balance between gravitational forces and the viscous drag of the fluid. This relationship is described by the Stokes and Stokes–Cunningham equations:

$$V_t = \frac{2}{9} \frac{r^2 g}{\mu} (\rho_1 - \rho_2) \qquad \text{Stokes} \quad (10\text{-}1)$$

$$V_t = \frac{2}{9} \frac{r^2 g}{\mu} (\rho_1 - \rho_2) \left(1 + \frac{A\lambda}{r}\right) \qquad \text{Stokes–Cunningham} \quad (10\text{-}2)$$

where $V_t$ = terminal settling velocity, $r$ = particle radius, $g$ = acceleration of gravity, $\mu$ = fluid viscosity, $\rho_1$, $\rho_2$ = densities of particles and fluid, $A$ = constant (0.86–0.90), $\lambda$ = mean free path of fluid molecules.

The Stokes equation is valid for cases where particle size is large compared to the mean free path of fluid molecules and settling is hydrodynamic in character. Green and Lane[7] cite accuracy within 1 per cent by Stokes' law for particle sizes from approximately 10 to 30 microns in diameter settling in air at 20°C and 1 atm pressure. The Stokes–Cunningham equation is the more accurate for smaller sizes where particle size becomes comparable to or smaller than the mean free path of the fluid molecules and slip between fluid molecules occurs. For larger particles (where inertial forces due to fluid displace-

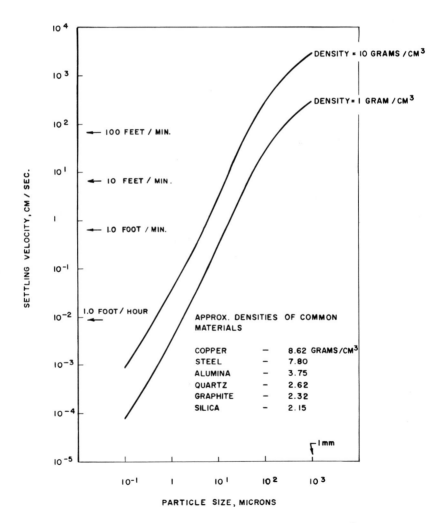

Figure 10-5 *Gravitational settling velocity for rigid spheres in air at 20°C and 1 atm.*

ment become large) the terminal velocity calculation is made by reference to drag coefficients determined from Reynolds' number correlations.[7]

Figure 10-5 shows calculated settling velocities vs. particle sizes for particles of 1 and 10 g/cm³ densities. Also shown are approximate densities for several common materials. Actual airborne particles consist of cubical, flake, fibrous, and chain shapes in addition to spheres, causing deviations from calculated terminal velocities to occur. From numerous experiments cited by Green and Lane[8] and Cadle,[9] it would appear that an equivalent Stokes diameter (diameter of sphere of same density as actual particle for the same terminal velocity) of $1/1.3$ to $1/2.05$ of the actual mean projected diameter would allow reasonable correlation. For practical purposes deductions from Fig. 10-5 are reasonably valid.

### Brownian diffusion

As particle diameters become smaller than 1.0 micron diffusion via Brownian motion becomes significant. As defined in many texts Brownian motion is the irregular motion of particles suspended in a fluid as a result of bombardment by atoms and molecules. Figure 1-3 shows the diffusion coefficients for particles in air while Table 10-7 compares the root mean square of

TABLE 10-7  Comparison Between Root Mean Square Brownian Displacement per Second and Terminal Velocity of Spheres of Unit Density in Air at 760 mm Hg Pressure and 20°C

| Diameter μ | Displacement cm | Terminal velocity cm/sec |
|---|---|---|
| 0.1 | $3.70 \times 10^{-3}$ | $8.71 \times 10^{-5}$ |
| 0.2 | $2.01 \times 10^{-3}$ | $2.27 \times 10^{-4}$ |
| 0.4 | $1.30 \times 10^{-3}$ | $6.85 \times 10^{-4}$ |
| 1.0 | $7.43 \times 10^{-4}$ | $3.49 \times 10^{-3}$ |
| 2.0 | $5.06 \times 10^{-4}$ | $1.29 \times 10^{-2}$ |

From H. L. Green and W. R. Lane, "Particulate Clouds: Dusts, Smokes and Mists," 2nd Ed, p. 74, Van Nostrand Reinhold, New York, 1964.

Brownian displacement per second and terminal velocities for spherical particles under stated conditions. As can be seen, Brownian diffusion is very slow. In gross movements of particles in air its influence is negligible. Its effects become significant in the coagulation of small particles and in the capture of these same particles in high efficiency filters.

### coagulation

Airborne particles will adhere to each other if they come in contact, and this process of coagulation is continuous in nature. The mechanics of coagulation, which are caused in part by Brownian motion, were investigated by Smoluchowski.[10] The theory of coagulation for a monodisperse aerosol caused by Brownian motion is expressed in the following equations:

$$\frac{1}{N_t} - \frac{1}{N_0} = Kt \qquad (10\text{-}3)$$

$$K = \frac{4\,RT}{3\,\mu N}\left(1 + A\,\frac{\lambda}{r}\right) \qquad (10\text{-}4)$$

where $N_0$ = initial aerosol concentration, $N_t$ = aerosol concentration at time $t$, $R$ = gas constant, $T$ = absolute temperature, and $N$ = Avogadro's number.

The coagulation rate is influenced by temperature, viscosity of the gas, and the mean free path (pressure). Other factors besides Brownian motion are influential. Green and Lane[11] show that aerosol heterogeneity, small particle size, high aerosol concentrations, turbulent motion, electrical charge, and acoustic fields will increase the rate of particle coagulation. For these reasons Whitby[1] is able to state that an aged aerosol rarely has a significant portion of its mass below 0.1 micron in size and the number concentrations of such particles in the aerosol will approach an equilibrium level of $10^5$. It is fortuitous that nature provides the self-cleansing mechanism of particle coagulation for the environment, otherwise we would soon choke to death from the smokes, mists, and fumes generated by industrial and natural processes.

### particle inertia

The dynamics of particle transport and deposition are not limited to sedimentation or Brownian diffusion. Moving particles of sufficient mass will possess momentum causing them to deviate from gaseous flow lines around obstacles. The resultant particle path may cause deposition either by impaction, which is the adherence of a particle being projected even across gas flow lines onto a surface, or entrapment, where the physical size of the particle precludes its passing through a porous obstacle. The mass dependence of inertial forces precludes any significant influence on particles in the submicron size range. In the case of large particles in motion they will be simultaneously acted upon by gravitational and inertial forces during the course of the trajectories. Both impaction and entrapment will be discussed further in the section on filters.

### turbulent flow

The main concern of this text is particle motion in a factory environment which is a turbulent flow condition even in the case of so-called "laminar flow" facilities. Unfortunately, theoretical and experimental study of turbulent flow is difficult, and according to Fuchs,[12] little is known about the motion of particles suspended in a turbulent fluid. The following example by Fuchs illustrates the typical known behavior of particles in turbulent flow.

Turbulent flow is characterized as an average flow velocity with a spectrum of transients superimposed. The root mean square eddy velocity transverse to flow is approximately 0.03 to 0.1 $\overline{U}$, $\overline{U}$ being the mean flow velocity. Given an aerosol passing through a horizontal tube with $\overline{U}$ equal to several m/sec, the vertical component of the mean eddy velocity (at least 3 to 10 cm/sec) would be considerably higher than the particle settling rate for particles sized 20 microns and under, and hence should establish a fairly uniform mixture of particles in the tube. The number of particles deposited in unit time on a unit length of tube of radius $R$ is $2RNV_t$, where $N$ equals the particle concentration in the gas flow. When a unit length of aerosol traverses a distance $dx$, in a time $dx/\overline{U}$, $2RNV_t$ $(dx/\overline{U})$ particles fall out of it. Since the volume of this length is $\pi R^2$,

$$-\frac{dN}{dx} = \frac{2\,RN\,V_t}{\pi R^2\,\overline{U}} = \frac{2N\,V_t}{\pi R\,\overline{U}} \tag{10-5}$$

which integrates into

$$N = N_0 \, \exp\left(\frac{-2\,V_t x}{\pi R\overline{U}}\right) \tag{10-6}$$

Equation 10-6 shows how the concentration of particles of a given size will vary with the linear distance $x$ traversed by the aerosol. Being proportional to $N$, the deposition rate of the aerosol decays exponentially in the direction of turbulent flow.[13]

While this example is just one special case in the general subject of turbulent air flow, it does provide some insight into a subject where scientific knowledge is limited. Particles traveling in turbulent flow will tend to drop out of the flow stream as they are being transported because of the eddy currents present, and the deposition rate from the aerosol is an exponential decay function.

### convection

Closely related to the concepts of turbulent flow are the convection patterns established in any factory environment. If there is a convective movement of gas containing suspended particles, there will be an interaction of forces between the gas medium and the particles tending to transport the particles on a bulk basis. Atmospheric clouds are examples of the convection phenomenum where droplets of water are transported great distances as a unit because of the temperature and concentration gradients between the cloud and the surrounding atmosphere. The magnitude of convection currents is illustrated by the following example given by Fuchs:

"Experiment shows that if a gas is in contact with a wall which is warmer by an amount $\Delta T$, a vertical flow occurs with a maximum flow velocity at a height $z$ from the base of the wall given by

$$U = 0.55 \sqrt{gz\alpha\Delta T} \tag{10-7}$$

where $\alpha$ is the coefficient of thermal expansion of the gas, equal to $1/T$. Thus, with $\Delta T = 0.01°C$ in a chamber one meter high the convective flow velocity reaches 1 cm/sec, which is the rate at which coarse particles, about 10 microns in radius, settle under gravity."[14]

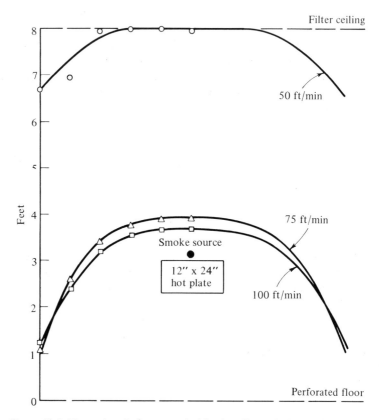

*Figure 10-6 Thermal updrafts vs. vertical laminar flow velocities. Hot plate surface temperature = 200° C.*

While in most cases convection is disordered and unpredictable, this controlled example shows the significance of convection to particle transport and deposition.

Convection currents have influenced the design of clean environment facilities. Figure 10-6 shows the dispersion envelope of cigarette smoke in a vertical laminar flow room resulting from convection currents created by an ordinary hot plate. The dispersion envelope is determined using a light scattering particle counter.[15] When the parallel air flow rate is 50 ft/min the smoke will rise completely to the ceiling as shown. The smoke envelope is much smaller at the higher flow rates in the vertical laminar flow room.[16] Thus, where high temperature gradients exist in a factory environment the influence of convective air flow must be a design consideration.

### electrical and other properties

Electrostatic charge can influence particle transport and deposition depending on the degree of charge carried by the particle and the intensity of the

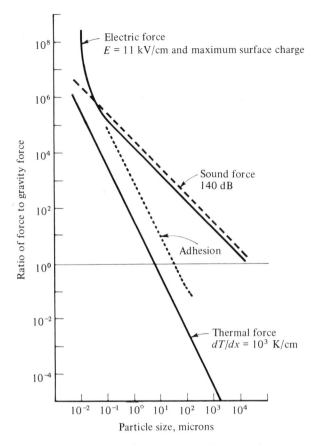

Figure 10-7 Electrical and other forces that can be exerted on aerosol particles. (Source: From K. T. Whitby and B. Y. H. Liu, "The Electrical Behavior of Aerosols," p. 60, in Aerosol Science, ed., C. N. Davies, Academic Press, New York, 1966.)

electric field present. Figure 10-7 provides a comparison between electrical and other forces which can be exerted on an aerosol relative to gravity. This example is for a moderately charged particle in a strong electrical field. As mentioned earlier the electrical properties of aerosols are a major cause of fine particle coagulation, especially since all atmospheric aerosols are charged to some extent.[17] The electrical properties of particles also have been used as a basis for filtration and particle size measurement.

The aerosol scientists have been successful in identifying other forces causing particle motion which are usually masked in the turbulent air flows of factory environment. Thermophoresis, photophoresis, diffusiophoresis, and sonic fields are described by Fuchs in considerable detail.[18]

## artificial agitation

Considerable particle transport and deposition are caused by movement of personnel, process machinery motion, and the air distribution system of the factory. The complex interaction between these force systems and the already mentioned mechanisms of settlement, diffusion, and convection make it impossible to establish quantitative guidelines for the effects of artificial agitation. Logic indicates that they amount to a series of random occurrences which are not self-cancelling. Thus, even the most elaborate calculations of environmental particle deposition must be considered only a rough approximation for any specific situation unless data are collected under carefully controlled circumstances.

## factors of environment control

The discussion on aerosol transport and deposition has shown the complexity of interacting forces which makes a rigorous examination of environment control based on first principles an impossible task at this time. Hence, an approximate model of the environment will be developed to allow predictive design in the control of airborne vapors and particles.

Let us designate the mechanisms available to limit within specified tolerances the approach of vapor or particle contaminants to a product surface in any given space. One technique is *exclusion* of the contaminant from entering the environment, either by its corollaries of *filtration* of the incoming atmosphere or *isolation* of the product surface from the harmful environment. The principle of *dilution* can be used. Contaminant-free atmosphere can be passed through the given space in sufficient quantities to remove the contaminants generated within the space. An important corollary to control by dilution is *air distribution* design whereby recirculating eddy currents can be minimized. A third technique is *reduction of contaminant generation* within the space through control of processes, facilities, and personnel. Finally, since contaminant deposition from the environment is time dependent, the *reduction of exposure time* of any contaminant-sensitive surface to the environment is an important form of control. All other things being equal, a process of short duration can often tolerate a less clean environment than a process of longer duration.

### mathematical model

The relationships between these mechanisms can be made more graphic by constructing a mathematical model of the transfer of contaminant mass within a space for a given time period. Referring to Fig. 10-8, let us assume a space with fixed dimensions where the atmospheric flow is sufficiently turbulent to approximate a uniform mixture of the contaminants throughout the space. The model is equally applicable to a room environment, an individual work position, or a special enclosed atmosphere system if the basic assumptions can be considered valid. The model also can be used for either a vapor or particle contaminant.

The differential equation for the impurity mass balance as a function of time may be written as:

$$V\,dC_s = Q_m C_m\,(1 - E_m)\,dt + Q_r\,C_s\,(1 - E_r)\,dt + GV\,dt - Q_r C_s\,dt$$

$$- DV\,dt - Q_{(e+p)}\,C_s\,dt \qquad (10\text{-}8)$$

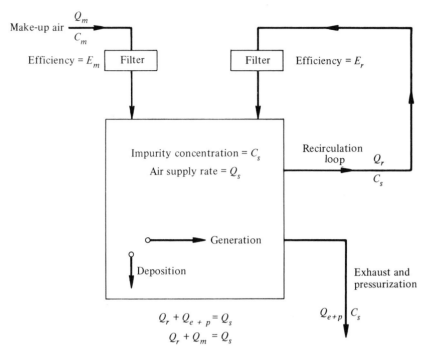

*Figure 10-8 Model of contaminant mass transfer.*

where

$V$ = space volume

$Q_m$ = rate of make-up air flow to space (volumes/time)

$Q_r$ = rate of recirculated air flow returned to space (volumes/time)

$Q_{(e+p)}$ = rate of positive exhaust air flow out of space and leakage air flow out of space (volumes/time)

$C_s$ = impurity concentration in space (parts/volume)

$C_m$ = impurity concentration in make-up air (parts/volume)

$E_m$ = overall fractional efficiency (mass basis) of make-up air filtration

$E_r$ = overall fractional efficiency (mass basis) of recirculated air filtration

$G$ = rate of impurity generation within space, averaged throughout the space (parts/volume-time)

$D$ = rate of impurity deposition from air to surfaces within space, averaged throughout the space (parts/volume-time)

$t$ = time

In Eq. (10-8), $V$, $Q_m$, $Q_r$, $Q_{(e+p)}$, $E_m$, and $E_r$ are determined by the construction, and are therefore constants. The internal impurity generation rate $G$ is a function of activity and may vary with time; however, for the present analysis it is also considered a constant. The deposition rate $D$ is a function of the force potentials causing the various modes of deposition and is also a direct function of the space impurity concentration. Thus, $D = AC_s$, where $A =$ (average deposition velocity) $\times \dfrac{(\text{exposed surface area})}{\text{unit volume}}$, will be considered a constant. The concentration of the make-up air $C_m$ will also be held constant for the purposes of this analysis. Then simplifying the terms for the rate of air exchange:

$$Q_s = \text{rate of total air supply to space}$$
$$= Q_m + Q_r$$
$$= Q_{(e+p)} + Q_r$$
$$m = Q_m/Q_s$$
$$r = Q_r/Q_s \text{ where } (m + r) = 1$$

And, rearranging Eq. (10-8)

$$\frac{dC_s}{C_s - X/Y} = -\frac{YQ_s}{V}\, dt \tag{10-9}$$

where

$$X = m\, C_m\, (1 - E_m) + \frac{G}{Q_s/V}$$

$$Y = r\, E_r + m + \frac{A}{Q_s/V}$$

Integrating between limits of $C_{s_0}$ at $t = 0$ and $C_{s_t}$ at $t = t$ and rearranging terms:

$$C_{s_t} = \frac{X}{Y}\,(1 - e^{-YQ_s t/V}) + C_{s_0}\, e^{-YQ_s t/V}$$

or

$$C_{s_t} = \frac{m(1 - E_m)}{Y}\, C_m\,(1 - e^{-YQ_s t/V}) + \frac{G}{Q_s Y/V}\,(1 - e^{-YQ_s t/V}) + C_{s_0}\, e^{-YQ_s t/V}$$

Now letting

$$K_1 = \frac{m(1 - E_m)}{Y} = \frac{m(1 - E_m)}{m + r\, E_r + \dfrac{A}{Q_s/V}}$$

$$K_2 = \frac{YQ_s}{V} = \frac{Q_s}{V}\left(m + r\, E_r + \frac{A}{Q_s/V}\right)$$

Then

$$C_{s_t} = K_1\, C_m(1 - e^{-K_2 t}) + \frac{G}{K_2}\,(1 - e^{-K_2 t}) + C_{s_0}\, e^{-K_2 t} \tag{10-10}$$

Equation (10-10) is in a form which allows investigation of transient conditions such as the recovery rate from a single high level burst of internal impurity concentration.

For steady state conditions,

$$C_{s_t} = C_{s_0} = C_s$$

And Eq. (10-10) reduced to

$$C_s = K_1 C_m + \frac{G}{K_2} \qquad (10\text{-}11)$$

Equation (10-11) also represents the equilibrium condition of $C_{s_t}$ in Eq. (10-10) after a long period of time has elapsed from a disturbed condition of $C_{s_0}$.

Equations (10-10) and (10-11) provide a qualitative insight into the parameters which are basic to environment control. If we combine the interpretations of these equations with practical experience we can avoid debate on the assumptions necessary to construct this simplified model of a complex aerodynamic situation.  Both equations illustrate that the equilibrium impurity concentration level within a space being studied will increase with the impurity generation rate within the space and the concentration of the incoming atmosphere.

Similarly, both equations indicate that the space impurity concentration will decrease as the air exchange rate of the space is increased ($K_2 \propto Q_s/V$). This hypothesis is in fact the basis for the high rate of flow of laminar flow clean rooms and the policy of purging gas lines with high flow rates of pure gas.  Equation (10-10) enables transient phenomena such as the recovery time from an impurity generation disturbance back to initial conditions to be rationalized.  Equation (10-11) allows some additional interpretations regarding the relationships of contaminant generation, dilution, and exclusion under equilibrium conditions:

1. Let $G = 0$ (no impurity generation within the space); then $C_s = K_1 Cm$. Now if $Q_m$ and $Q_r$ are assumed to be first mixed and then passed through a single final filter such that $E_m = E_r$, then the impurity concentration in the mixed stream before filtration must be $(m\, C_m + r\, C_s)$, and it can be shown that

$$C_s = \frac{1 - E}{1 + \dfrac{A}{Q_s/V}} \; (m\, C_m + r\, C_s) \qquad (10\text{-}12)$$

   In the absence of deposition on surfaces, $\dfrac{A}{Q_s/V} \approx 0$; then the space impurity concentration can never be less than that in the air supply after final filtration.

2. In clean rooms at rest and where $G = 0$, the equilibrium impurity concentration will be zero for all impurities for which filtration

efficiency $E$ is 1.0. As will be shown later, particle removal efficiency of properly tested and installed final filters will be virtually 100% for particle sizes of 0.5 micron and larger. Thus, under at rest conditions all such rooms should eventually come to a zero equilibrium particle count.

3. Let $\frac{Q_s}{V} \to 0$; then $K_2 \to 0$ and from Eq. (10-11):

$$C_s = \left(K_1 C_m + \frac{G}{K_2}\right) \to \infty \text{ whenever internal generation exists.}$$

This confirms the adverse effects of reduced air change rates in clean rooms.

4. Let $E = 1$; then $C_s = G/K_2$. Even when the filtration capability is essentially 100%, as is the case for most clean environment designs and some gaseous filters, one can see that the generation rate within the space must be offset by an adequate dilution rate.

The mathematical model as presented should be looked upon as a guide to the logic behind environmental control design. Further interpretation causes the assumptions to break down since the ability to measure many of these factors is very difficult in practice.

To reiterate, environmental control hinges on the proper application of contaminant exclusion by filtration or isolation, contaminant dilution via adequate air distribution and flow rate, and process limitation on contaminant generation rates and exposure time.

## filtration

In environmental control design there will come a point where some means of contaminant removal from a stream of air or other gas is required. Filtration is a basic building block on which the total concept of environment control is built. An understanding of filtration principles, filter grades and capacities, and methods of comparing performance is desirable to assist in the consideration of facility selection and design for specific purposes.

Particle filtration must, of necessity, be discussed separately from vapor filtration since both the principles and designs are different. Also, particle filtration is more widely practiced and provides a wealth of theory and experience on which to draw. Conversely, the experience with vapor filtration or collection from the environment is much more limited.

### particle filters

Particle filters for contamination-control application in manufacturing environments are predominately of the fibrous type. Other types of air cleaners such as air washers, scrubbers, or electrostatic precipitators have not been widely applied to manufacturing environment control due either to lower efficiency, higher initial cost or to problems of maintenance.

The theoretical analysis of filtration is a special case of the mathematical treatment of aerosol transport and deposition discussed earlier in this chapter. As applied to fibrous filters the mechanisms of deposition to be considered are inertia, interception, diffusion, and electrostatic forces, with sedimentation being of secondary importance. Under a model of a fibrous filter Fuchs demonstrates mathematically that:

"Inertial deposition increases with the size and density of the particles and with the flow velocity; sedimentation also increases with the size and density of the particles but decreases with the flow velocity; deposition due to the interception effect increases with particle size but does not depend on their density or on the flow velocity. Diffusive deposition varies inversely with the particle size and flow velocity and does not depend on the density of the particles. All types of deposition increase rapidly as the distance between fibers becomes smaller. The dependence on fiber thickness at constant separation is more complicated, but at constant filter porosity the deposition increases rapidly with decrease in fiber thickness. All these results are independent of the filter model taken and of the assumptions as to the nature of the flow: they are, therefore, of general significance."[19]

Both particles and filter fibers can carry electrostatic charge which influences particle deposition, but the charged conditions are usually not stable and will decrease with time.

One might expect that the effects of the various mechanisms could be summed to calculate a total efficiency for a given filter. Unfortunately, the net efficiency is always less than the calculated figure. While the mathematical function describing the interrelationships is unknown, experiments have established that filter efficiency will tend toward unity for very large and very small particles and for very large and very small flow velocities.[20] An economic balance must be selected between the particle size range to be collected, the particle collection velocity, the pressure drop across the filter, and the net collection efficiency desired.

In addition to efficiency, the relative performance of a filter is influenced by its ability to retain particles after deposition and its dust holding capacity. The sloughing of particles is a question of reliability which must be weighed in an economic evaluation. Likewise, rapid clogging of a filter, even though highly efficient, would be unprofitable. The mechanisms which affect filter sloughing and clogging are the aerosol velocity at the filter fibers, the particle size distribution and concentration of the aerosol, the physical phase of the aerosol, and the characteristics of the filter media. The filter characteristics affecting retention and clogging performance are the type of fiber (monofilament vs. high nap), the fiber diameter (greater surface area for deposition for a given filter porosity), the assembly of the fibers (a coherent cloth filter clogs more rapidly than a flocculent felt material), the porosity of the fiber assembly (compare a household furnace filter against a high efficiency industrial filter), and the depth of the filter media (increasing depth improves the probability for deposition by inertia, diffusion, interception, etc.)[21,22]

## types of fibrous filters

There are two generic classes of fibrous filters—viscous impingement and the dry type—which are commonly used to provide control of factory environments. Both types are "true filters" in that the pressure drop increases with dust loading to an operating point at which they are replaced or sufficiently cleaned for re-use.

Impingement filters are generally constructed of coarse fibers arranged in a porous bed to operate at high velocities (250–700 fpm) through the media at a low pressure drop. Filter face velocities of 300–500 fpm are common. Particle deposition occurs mainly by impaction or interception, which is aided by the application of a viscous substance to the media. The media may be glass, hair, synthetic fiber, plastic foams, or metal mesh. They are often used as prefilters in a multi-filter system, being of low cost, low pressure drop, and more effective in the capture of larger particles and lint. They are also used in building air conditioning systems where they may be mounted in a panel or used as a moving curtain. In the latter case the media is in roll form and is intermittently advanced across the air stream as the exposed section becomes loaded.

Dry type filters differ from viscous impingement type in the use of smaller size fibers which are more densely spaced in the blanket media. Also, the viscous coating is not applied to the fibers. The media mats or paper-like blankets may be composed of fine glass fibers, cellulose, wool felts, asbestos, or many synthetic fibers. The filament diameters range between 10 microns and submicron dimensions. Approach velocities to the filter of 250–625 fpm are commonly used, but the velocities through the media range from 2–90 fpm because of the extended filtration surface area created by pleats or deep folds. The low media filtration velocities allow diffusion to become effective as a deposition mechanism in addition to impaction, interception, and entrapment resulting from the closely interwoven, small-diameter fibers. Thus, dry type filter construction is commonly used for intermediate and high efficiency systems where the capture of small particles is sufficiently important to counteract their higher cost, lower dust holding capacity, and higher operating pressure losses.[23] The highest quality dry type design—the HEPA* filter—is constructed to be "fail-safe" as far as sloughing is concerned. The HEPA filter, as the heart of any clean environment design, will be discussed later in more detail.

---

*High efficiency particulate air: A throw-away extended media dry-type filter in a rigid frame having minimum particle collection efficiency of 99.97% for 0.3 micron thermally-generated dioctyl-phthalate (DOP) particles, and maximum clean-filter pressure drop of 1.0″ water gage, when tested at rated air flow.

## filter tests and standards

The complexities of filtration principles and the many varieties of aerosol filters can present a perplexing choice to the designer of a practical system. There is no single simple answer for comparing filters under all application conditions or for the proper filter for a specific job. Rating standards for general ventilation or air conditioning grade filters are covered in ASHRAE Standard 52-68.*[24]    For clean room grade filtration, performance rating is covered by Mil Std 282[25] and the AACC Standard CS-1T.[26]

The rating of a filter is based on its resistance to air flow, its aerosol removal capabilities and its dust holding capacity. ASHRAE Standard 52-68 provides two measures of removal capability: 1) "Arrestance," a weight percentage removal rating based on tests with a synthetic dust, and 2) "Atmospheric Dust Spot Efficiency," a percentage removal rating based on staining tests with normal atmospheric dust. Mil Std 282 and AACC Std CS-1T provide procedures for determining removal efficiency based on penetration tests with artificially produced monodisperse aerosols of submicron size. Dust holding capacity is determined in accordance with ASHRAE Standard 52-68 procedures using synthetic dust.

These rating tests are not necessarily relevant to actual particle-laden environments with particle size distributions different from those used for the tests. The ASHRAE tests, being destructive, are made on a sample basis; this assumes that the filter being tested is a representative sample of the product—which is actually susceptible to variations in manufacturing quality control. The Mil Std 282 or AACC Std CS-1T procedures are nondestructive but, while made on a 100 percent inspection basis for HEPA filters, are related to monodisperse aerosol behavior only.

Ratings may be expressed for clean conditions or as averages over the dust holding capacity life of the filter. Thus, in interpreting filter test data it is important to realize that they provide a comparison of different filter types and that actual performance in any one application may differ significantly from the test results.

**Arrestance (Weight Efficiency) Test.** Per the ASHRAE Standard this test is based on an artificial dust mixture consisting of "72 percent standardized air cleaner test dust, fine; 23 percent Molocco Black; and 5 percent No. 7 cotton linters ground in a Wiley mill with a 4 mm screen (percentages by weight)."[24] A known amount of dust is introduced up-stream of the filter under test, and the dust penetrating the filter is collected and weighed to give a measure of efficiency. Referring to Fig. 10-1, a small percentage by number of large particles represents a large percentage of the total weight of normally occurring particles in the atmosphere. The remaining weight of small particles represents

*American Society of Heating, Refrigerating & Air Conditioning Engineers Standard 52-68 contains in a single standard rating methods previously developed by the Air Filter Institute and the National Bureau of Standards.

a very large number. Thus, a high arrestance percentage by this technique does not indicate a high removal capability for the smaller particles of concern to microelectronics applications.

**Atmospheric Dust Spot Efficiency Test.** This standard test will reflect the effect of fine particles, with the challenge aerosol being the naturally occurring dust in the outdoor environment. The optical densities of filter paper samples upstream and downstream of the filter under evaluation are compared to produce a rating of the dust's staining abilities after filtration. Since smaller particles more readily cause staining, this technique is more indicative of small particle removal and less sensitive to large particle penetration.

**Dust Holding Capacity Test.** Usually performed concurrently with arrestance and dust spot efficiency tests, this procedure is an accelerated life test based on synthetic dust. Dust holding capacity is defined as the amount of dust fed to the filter times its average arrestance during test to the point of maximum rated resistance (or to the point where arrestance falls to 85 percent or 75 percent of maximum).

**DOP (Dioctyl-Phthalate) Penetration Test.** This test was originally established to evaluate the ability of HEPA filters to provide particle-free environments. By definition a HEPA filter must demonstrate a minimum collection efficiency of 99.97 percent for 0.3 micron thermally generated DOP smoke and a maximum clean filter pressure drop of 1.0 in w.g. when tested at the manufacturer's rated air flow capacity. The 0.3 micron challenge aerosol was selected since both theory[27] and experiment indicate that a particle of this size is the most difficult size to capture by a fibrous filter. The DOP penetration test is intended as a measure of a filter's ultimate ability to capture particles of any size at a practical operating pressure.

The actual test compares the upstream and downstream aerosol concentration using a light-scattering photometer to establish the filter efficiency. In practice all HEPA filters are individually tested to guard against variations in manufacturing quality control.

This test has also been applied to other than HEPA filters and is preferable to the ASHRAE procedures for all filters having ASHRAE method efficiencies greater than approximately 98 percent.

### comparison of efficiency standards

It is readily apparent that each of the filter test standards are biased in one form or another, and extrapolation beyond their intended range of applicability is questionable. Claims of 99 percent efficiency for filter products should be challenged as to the test method as can be seen in Fig. 10-9. This illustration provides an approximate comparison of the three different filter efficiency rating methods based upon published tests of a number of commercial filters. For example, a Cambridge Aerosolve 85 filter (normally considered in

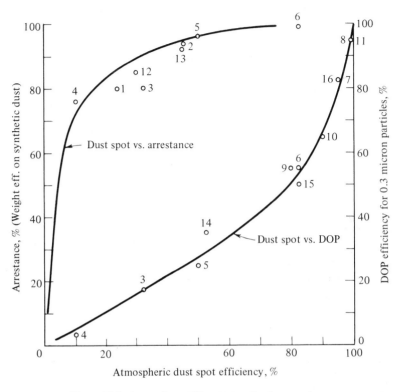

Figure 10-9 Approximate filter test method comparisons.

| Point No. | Manufacturers' Ratings Identification |
|---|---|
| 1 | Continental Dycon CA-24 |
| 2 | Continental Condflo 45 |
| 3 | Cambridge Hi-Cap |
| 4 | Cambridge 2″ Throwaway |
| 5 | Cambridge Aerosolve 45 |
| 6 | Cambridge Aerosolve 85 |
| 7 | Cambridge Aerosolve 95 |
| 8 | Cambridge Micretain |
| 9 | MSA Dustfoe M |
| 10 | MSA Dustfoe B |
| 11 | MSA Hospital |
| 12 | AAF PL24-NO. 5 |
| 13 | AAF PL24-No. 12 |
| 14 | AAF Dri Pak 60 |
| 15 | AAF Dri Pak 90 |
| 16 | AAF Dri Pak 100 |

Note: HEPA filters, 99.97% minimum DOP eff.

the intermediate to high efficiency class of filter) has a 99 percent weight efficiency, approximately 83 percent dust spot efficiency, and approximately 55 percent DOP penetration efficiency. Figure 10-9 permits estimation of the DOP penetration efficiency for those filters not yet evaluated according to the most stringent standard.

### the HEPA filter

As mentioned earlier the HEPA filter is the basic filter element for any clean environment design. It is desirable to have a more detailed understanding of its construction and performance characteristics. Figure 10-10 shows a new HEPA filter and a section of one that has been loaded past normal end of life conditions. One can see the paper-like media folded into deep pleats, the corrugated paper of aluminum separators, and the frame into which the media separator "pack" is cemented using rubber base, epoxy, or foamed sealants. Typically, a 24 in. × 24 in. × 11½ in. filter will have approximately 400 sq. ft of media surface which reduces the rated approach velocity of 275 fpm for this particular unit to 2.75 fpm through the media. Note that at end of life the downstream side of the media is not stained, giving subjective proof of the filter's fail-safe nature. While HEPA filters have been manufactured with various media, separator, and frame materials, the preferred construction is a filter with glass fiber media having less than 5 percent combustible binder, aluminum separators, and a frame made of plywood (treated for fire resistance), particle board, or metal. This is based on a balance of efficiency, life, stability, and fire resistance considerations.

The HEPA filter's ability to approach 100 percent removal of all particles regardless of size or composition from any gas has converted the concept of a particle-free work environment from wishful thinking into an attainable goal. However, the present DOP penetration test does not indicate whether the remaining 0.03 percent maximum permissible penetration* is uniformly distributed across the filter face or localized in one or more pinhole leaks. In the initial application of HEPA filters such pinholes were of no consequence. With the advent of laminar flow facilities with their parallel flow of output air the location of pinholes became a matter of serious concern.

Two additional performance criteria have been established in the AACC Standard CS-1 "HEPA Filters" to improve the reliability of HEPA filters. A characteristic of pinholed filters is an increase in percentage penetration with decreasing flow. This is relatively independent of particle size. The first criteria established under the above standard is to perform the thermally generated DOP penetration test at 20 percent as well as at 100 percent of the rated flow rate. Providing that penetration at 20 percent flow is not more than .04 percent as compared to .03 percent at 100 percent flow, this test eliminates most of the pinholes that are possibly existent. For those cases where even one pinhole is considered detrimental, the CS-1T Standard

*See footnote, page 260.

(a)

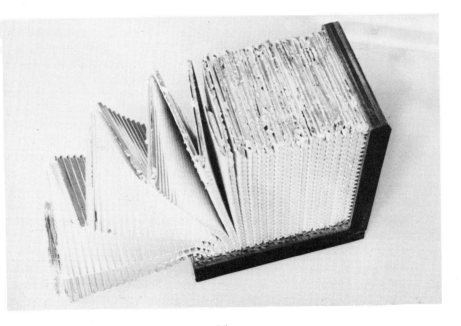

(b)

Figure 10-10  HEPA filter.

specifies the techniques to scan the downstream face and every inch of the edge sealant for pinholes using a light scattering aerosol photometer. Such filters are the type now recommended for use in laminar flow facilities to ensure that particle-free air is supplied to the clean work or clean work position within the bounds of present measurement ability experience.

## vapor collection

While particle control has been blessed with the potential of approaching perfection the same cannot be said for control of vaporous contaminants. Particle control is dependent on the contaminant's physical characteristics. Vapor control, on the other hand, involves the contaminant's chemical nature with all the possible variations implied in the periodic table of elements. In vapor control there is no counterpart to the HEPA filter.

There are several vapor removal techniques available to the design engineer in environmental systems: ventilation with pure air, air scrubbing, refrigeration, physical adsorption, chemical reaction, or the elimination of vapor sources. When interest is restricted to the range of vapor concentrations normally expected in a microelectronics factory environment (see Table 10-5), the choice of vapor removal methods is more limited.

Ventilation created by process exhaust facilities is the traditional technique of vapor or fume control where contaminant releases are high and the air pollution concentrations from the community are considered inconsequential. Air scrubbing also is given consideration in solving effluent problems of high vapor concentration. Neither ventilation, which is a dilution mechanism for the factory environment, nor air scrubbing, which acts to control the factory effluent, are suitable for influencing the character of the factory air supply. It is more appropriate to discuss these methods in Chap. 15 on effluent control.

Trace vapor concentrations can be destroyed by oxidation or catalytic combustion, but such techniques are usually found in combination with adsorption facilities. Another alternative, the removal of a vapor source, is a commendable goal but is not often possible in the common situations where there are sources of vapor contaminants in the factory environment. For the reduction of vapor concentrations in the factory environment to levels below that of the air supply (see Eq. (10-10)) the discussion narrows down to refrigeration and the forms of sorption as the most commonly-used techniques.

### refrigeration

Air conditioning via refrigeration is the common technique for controlling the presence of water vapor in a factory environment. As a water vapor control technique, refrigeration is practical for the control of relative humidities down to a level of about 40 percent, assuming a typical summer design condition of $95°F$ dry bulb and $78°F$ wet bulb. Attempts at extracting additional moisture from the air are often frustrated by frost build up on the cooling coils.

Unfortunately, refrigeration as commonly applied for air conditioning will not provide effective control for other vapor contaminants. Chemical vapors in the very low concentrations normally common to the factory and outdoor environments (see Tables 10-2 and 10-5) have dew points lower than the operating temperatures of air conditioning refrigeration systems and cannot be removed by this technique. For further information in air conditioning and refrigeration design considerations the reader is referred to the latest edition of the *ASHRAE Handbook of Fundamentals.*[28]

## sorption techniques

The phenomena of sorption have been described in Chap. 6 as causing adherence of environmental materials to a surface. Absorption occurs when absorbed gas forms a homogeneous solution with a liquid or a new chemical compound with a solid. Physical adsorption occurs when a gas is taken on only at the surface or in the porous areas of a solid to form a surface condensate. Chemisorption occurs when the adsorbed gases react with the surface material to cause chemical bonding between their respective atoms. Chemisorption does not play an important role in vapor control; therefore, the term adsorption will be used to describe physical adsorption only.

As might be expected in any interacting system of materials both absorbents and adsorbents have preferences in the types of gases each will collect. An absorbant will be influenced by the degree of solubility of the gas and the partial pressure differentials. In the case of solid absorbents the property of hydration is used as a dehumidification agent. Sorbent materials often exercise material preference because of their polar nature and retain water by this mechanism in preference to less polar materials.

The desiccants used in environmental air dryers include such absorbents as lithium chloride, sulfuric acid, ethylene glycol, and calcium chloride and such adsorbents as activated alumina and silica gel. In contrast, activated carbon adsorbs a broad spectrum of gases and vapors (especially organic vapors) in preference to water. For these reasons trace vapor control is further reduced to those absorbents and adsorbents which collect water vapor and activated carbon which is used in most of the other forms of vapor control.

**Moisture Control Systems.** Refrigeration is used in air conditioning design to maintain the normal manufacturing relative humidity requirement of $45 \pm 5$ percent. However, there are certain microelectronic processes (e.g., photolithography) which require dry environments, and in some cases moisture control as low as 10 percent R H has been specified. In these circumstances desiccant drying is combined with refrigeration to provide the dry conditioned air required for the process environment. The design of desiccators used in environmental control will vary with the types of sorbent. Figures 10-11, 10-12, and 10-13 illustrate respectively the liquid absorption, the solid absorption, and the solid adsorption systems. In each case there is a use period when air is dried as it passes through the system, and a reactivation period

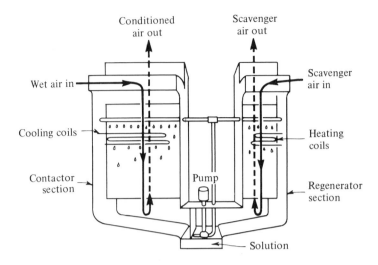

*Figure 10-11  Flow diagram for typical liquid-absorbent dehumidifier.  (Courtesy American, Society of Heating, Refrigeration, and Air Conditioning Engineers, 1967 ASHRAE Guide and Data Book, New York.)*

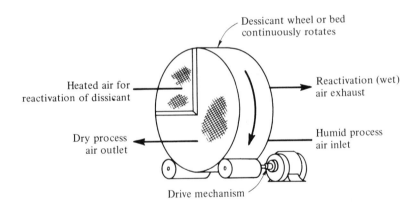

*Figure 10-12  Typical rotary dehumidification unit. (Courtesy 1967 ASHRAE Guide and Data Book, New York.)*

when the sorbent is exposed to a reverse flow of heated air or other heating media to drive off the collected moisture from the desiccant.

The liquid absorption system commonly uses lithium chloride as its desiccant. As shown in Fig. 10-11, the absorbent solution is sprayed over the contactor coils to bring about an intimate mixture with the incoming wet air. The moisture content of the air is regulated by the temperature and concentration of the liquid absorbent. This establishes the partial pressure differential between the air and the hydroscopic solution. The absorbent concentration is maintained by continually cycling a portion of the liquid through the

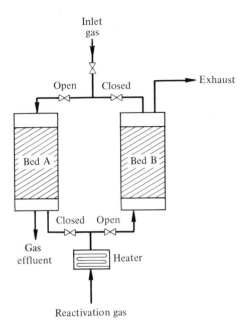

*Figure 10-13. Typical dual-bed adsorption unit. (Courtesy 1967 AHRAE Guide and Data Book, New York.)*

regeneration section where heat is applied and a scavenger air stream removes the moisture.

The solid absorbent system shown in Fig. 10-12 utilizes a rotating porous drum composed of pleated asbestos paper impregnated with lithium chloride to achieve the intimate contact with the incoming humid air. Approximately 25 percent of the rotating drum is devoted to reactivation which is accomplished by passing heated air through the quadrant in a reverse direction. This type of dehumidification system provides a high degree of absorption capability in a compact, easy-to-maintain unit and can withstand variations in air velocity without loss of desiccant to the supply air stream. It is well suited for low humidity, clean environment applications.

The solid adsorption system, as diagrammed in Fig. 10-13, is commonly composed of two beds filled with silica gel, activated alumina, or some other adsorbing desiccant. A time cycle is established to alternate each bed from use to reactivation, consistent with the design variables in the bed itself, the characteristics of the input air, and reactivation heating system. Solid adsorption units are manufactured in many different sizes for practically any application where dehumidification is required.[29]

**Trace Vapor Adsorption by Activated Carbon.** For general purpose adsorption of multicomponent systems activated carbon in fixed bed adsorbers is best suited. Activated carbon is considered nonpolar and will generally adsorb organic organic vapors in preference to water vapors. This is due to the nonpolar nature of the surface as well as the fact that larger organic molecules tend to displace physically the smaller water molecule. Actually the existence of polar surface impurities renders the surface slightly polar. These surface impurities are the reason that activated carbon will chemisorb reactive gases such as sulfur dioxide to a small extent.

Vapor phase activated carbons have an extremely large surface area, generally in excess of 1000 $m^2/g$. Granular size is not very significant with respect to surface area because the porous structure of the carbon—especially the micropores—is responsible for the large surface area. Physical adsorption into the porous structure is to a large extent reversible. Although a hysteresis effect is noticed upon desorption, it is possible to strip and recover collected material from fixed bed adsorbers. Hence, fixed bed operations often employ two columns, one of which is adsorbing while the other is being regenerated for recovery of valuable materials. This is usually done with high temperature steam (see Fig. 10-13). In normal air conditioning applications activated carbon filters for odor control are usually replaced when exhausted. Under certain conditions exhausted filters may be returned to the supplier for regeneration.

Fixed bed adsorption may be examined from both a qualitative and mathematical standpoint. The process of removing impurities from an air stream is dynamic in nature. Air which contains contaminants enters the bed at one end, is stripped of impurities in transit through the bed, and leaves the other end purified. This purification is accomplished by means of mass transfer of impurity molecules from the gas phase to the solid adsorbent phase. Once in contact with the adsorbent surface forces of adsorption hold molecules to the surface. As air flows through the adsorptive bed the adsorbent material becomes saturated with adsorbate molecules in a regular manner. A stable wavefront (Fig. 10-14) is produced which travels across the adsorptive bed and eventually results in a uniform exhaustion of the adsorbent material. This wavefront is essentially a concentration history of the adsorptive bed. In other words, for any section in the adsorbent bed the S curve of the wavefront describes the manner in which the adsorbent becomes exhausted.

The existence of a mobile wavefront of adsorbate is the key to the adsorption process. Much work has been done to characterize this wavefront, or breakthrough front, in terms of physical parameters of the system. Among the physical parameters thought to affect this concentration history are column volume, volumetric and mass flow rate, contact time, linear and superficial velocities, bed depths, fluid volume passing over the column, concentration in the influent, temperature, pressure, mass transfer rates, and the adsorbent particle diameter.

Results[30-34] indicate that the time for the breakthrough front to reach the end of the bed is almost linear with respect to the bed depth. It has also been shown that bed depth and contact time are linear[31,33] and that the shape of the breakthrough wave or the breakthrough zone length is independent of

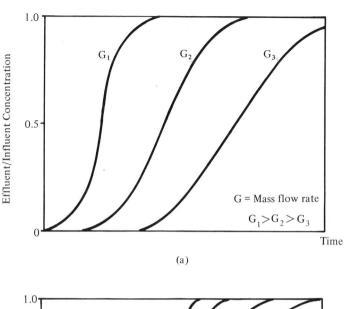

(a)

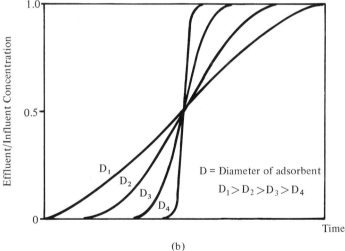

(b)

*Figure 10-14 Breakthrough wavefront activated carbon bed. (a) Effect of mass flow rate for constant adsorbent size; (b) effect of adsorbent size for constant mass flow.*

bed depth.[34-37] Increased flow or velocity tends to sharpen the breakthrough front;[31-37] however, better saturation has been reported at low flow rates.[37] Increased influent concentration also tends to steepen the wavefront,[30,31] while increased adsorbate mass flow rate has been shown to sharpen the wavefront.[38] Temperature and pressure do not seem to play a significant role;[31,33] however, quantitative adsorption can only be accomplished at temperatures below the boiling point of the adsorbate being collected.

Results with regard to particle size indicate improved bed performance with decreased adsorbent particle size, the only restriction being the increased pressure drop that accompanies decreased particle size. However, with decreased particle diameter, the wavefront more closely approaches a square wave,[39] the mass transfer rate increases,[37,40] efficiency of collection improves,[36,37] and the time before breakthrough increases.[32,39]

For environmental control mathematical analyses are not generally performed because one deals with numerous impurities that may fluctuate in concentration over the lifetime of the bed. The analytic approach to vapor control using activated carbon for a single impurity can be useful in guiding engineering judgment, but actual design solutions must still rely upon practical engineering experience. The design criteria for both environmental particle and vapor control will be discussed in the next chapter.

## summary

The nature of airborne contaminants and their occurrence in both particulate and gaseous forms in the atmosphere and the factory have been discussed as an introduction to clean environment control. After defining classes of cleanliness, the mechanisms affecting environment control have been examined and related to a mathematical model to illustrate basic design criteria. Descriptions of particle and vapor filtration methods and equipment have been given with emphasis placed on HEPA and carbon filters. The details of environment control design and application will be presented in the next chapter based on these fundamentals.

## references

1. K. T. Whitby, "The Starting Point for Contamination Control," American Association for Contamination Control, Boston, 1963, p. 5.
2. Richard D. Cadle, *Particle Size, Theory, and Industrial Applications,* Reinhold, New York, 1965, p. 33.
3. Federal Standard 209, "Clean Room and Work Station Requirements, Controlled Environment," General Services Administration, Washington, D.C. (latest issue).
4. C. L. Fraust, Unpublished Data from Western Electric Co., Allentown, Pennsylvania.
5. R. W. DeMott, Unpublished Data from Western Electric Co., Allentown, Pa.
6. Philip R. Austin and Steward W. Timmerman, *Design and Operation of Clean Rooms,* Business News Publishing Co., Detroit, 1965, pp. 21–30.
7. H. L. Green and W. R. Lane, *Particulate Clouds: Dusts, Smokes, and Mists,* 2nd Ed., D. Van Nostrand, Princeton, New Jersey, p. 68.
8. Ibid, p. 72.
9. Richard D. Cadle, op. cit., p. 84.
10. M. von Smoluchowski, *Physik,* 1916, Volume 17, p. 557.
11. H. L. Green and W. R. Lane, op. cit., pp. 140–176.
12. N. A. Fuchs, *The Mechanics of Aerosols,* Pergamon Press, Oxford, 1964, p. 257.
13. Ibid, p. 264.
14. Ibid, p. 250.

15. G. L. Schadler, "Evaluating Characteristic Air Patterns in Clean Rooms and Clean Air Devices," Journal of American Association for Contamination Control, March 1969.
16. A. J. Agacan, Unpublished Data from Western Electric Co., New York, New York.
17. K. T. Whitby and B. Y. H. Liu, "The Electric Behavior of Aerosols," in *Aerosol Science*, ed. C. N. Davies, Academic Press, New York, 1966, p. 62.
18. N. A. Fuchs, op. cit., pp. 56–69, 83–89.
19. Ibid, p. 217.
20. Ibid.
21. J. Pich, "Theory of Aerosol Filtration," in *Aerosol Science*, ed. C. N. Davies, Academic Press, New York, 1966, p. 271.
22. H. L. Green and W. R. Lane, op. cit., p. 332.
23. *1969 ASHRAE Guide and Data Book-Equipment*, ASHRAE, New York, 1969, pp. 111–113.
24. ASHRAE Standard 52-68, "Method of Testing Air Cleaning Devices Used in General Ventilation for Removing Particulate Matter," ASHRAE, New York, 1968.
25. MIL-STD-282, "Filter Units, Protective Clothing, Gas-Mask Components, and Related Products: Performance Test Methods," U.S. Government Printing Office, 1956.
26. AACC Std, CS-1T, "Tentative Standard for HEPA Filters," American Association for Contamination Control, Boston, 1968.
27. G. Langmuir and K. B. Blodgett, "Smokes and Filters," OSRD Report 3460, 1944.
28. *ASHRAE Handbook of Fundamentals*, ASHRAE, New York, 1967.
29. *ASHRAE Guide and Data Book-Equipment*, ASHRAE, New York, 1969, pp. 87–93.
30. G. S. Bohart and E. Q. Adams, "Some Aspects of the Behavior of Charcoal with Respect to Chlorine," *J. Am. Chem. Soc,*, 1920, Volume 42, p. 523.
31. M. Dale and T. M. Klotz, "Adsorption of Chloropicrin and Phosgene on Charcoal from a Flowing Gas Stream," *Ind. Eng. Chem.*, 1946, Volume 38, p. 1289.
32. T. M. Klotz, *Chem. Reviews*, 1946, Volume 39, pp. 241–268.
33. S. H. Jury and W. Licht, Jr., "Drying of Gases, Adsorption Wave in Desiccant Beds," *Chem. Eng. Progr.*, 1952, Volume 48, Number 2, p. 102.
34. J. M. Cambell, F. E. Ashford, R. B. Needham, and L. S. Reid, "More Insight into Adsorption Design," *Hydrocarbon Process Petr. Refiner*, 1963, Volume 42, Number 12, p. 89.
35. L. C. Eagleton and H. Bliss, "Drying of Air in Fixed Beds," *Chem. Eng. Progr.*, 1953, Volume 49, Number 10, p. 543.
36. G. H. Dale, et al., "Dynamic Adsorption of Isobutane and Isopentane on Silica Gel," *Chem. Engr. Progr.*, 1961, Series 57, Number 34, p. 42.
37. D. E. Marks, et al., "Dynamic Behavior of Fixed Bed Adsorbers," *J. Petrol. Technol.*, 1963, Volume 15, Number 4, pp. 443–459.
38. C. L. Fraust and E. R. Herman, "The Adsorption of Aliphatic Acetate Vapors onto Activated Carbon," *Am. Ind. Hyg. Assoc. J.*, 1969, Volume 30, p. 494.
39. *Basic Concepts of Adsorption on Activated Carbon*, Pittsburgh Activated Carbon Company, Pittsburgh.
40. T. Vermeulen, *Advances in Chemical Engineering*, Academic Press, 1968, New York.

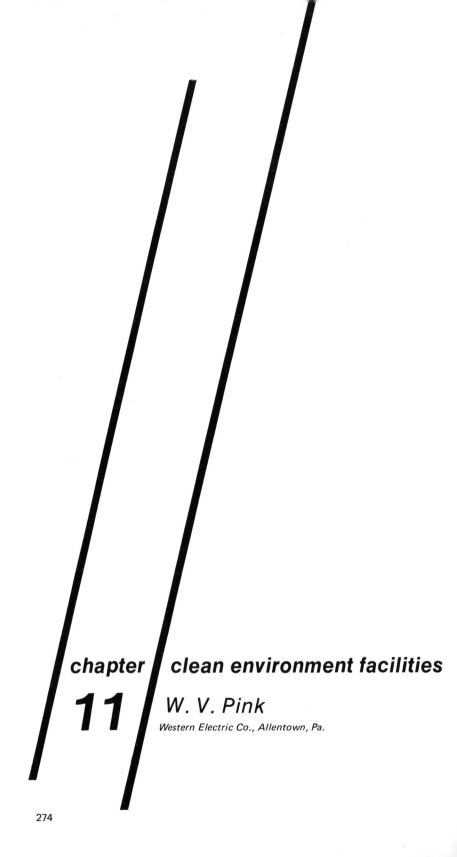

chapter

11

clean environment facilities

W. V. Pink

*Western Electric Co., Allentown, Pa.*

The factory atmosphere is the starting point for control of a processing environment aimed at minimizing product contamination. Within the factory a series of successively cleaner working environments can be provided by increasing the amount, quality, and distribution control of air-conditioned air with proportionately greater initial investment and operating cost. Depending on the character of the factory environment and the type, activity level, and extent of the work to be performed, the clean facility may be a type of local work station or an area control system (clean room). Local work stations are utilized to protect an individual process and may be categorized as dust control hoods, laminar flow work stations, or artificial atmosphere systems. Area control systems for the protection of several process steps or an entire production line may be of the mixed air flow, vertical laminar flow, or horizontal flow type. Each facility type will be discussed in turn in relation to capabilities, design parameters, and cost considerations.

### clean environment evolution

Some degree of particulate control has actually existed for a long period of time using electrostatic filters or the best available fibrous filters. The development of the HEPA type filter, originally for radioactive particle containment in air conditioning systems, started the design evolution of clean rooms for reliable control of all particle sizes. Figure 11-1 shows an early typical clean room design approach. The rooms were conventional air conditioned spaces with additional emphasis directed toward "clean" conditions: HEPA

*Figure 11-1 Traditional clean room.*

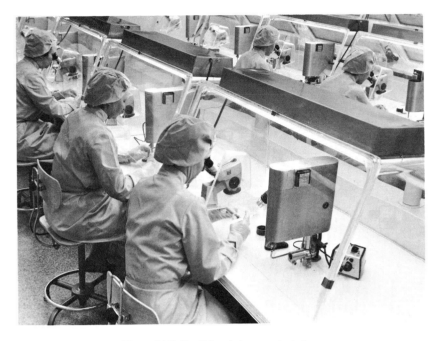

*Figure 11-2 Traditional clean work station.*

filtration of the air entering the space; air return usually at or near floor level; specification of smooth, washable, nonshedding surface finishes; elimination of sharp corners or dirt catching protrusions; and dependency on complex housekeeping and personnel control procedures. Assessment of the cleanliness level was accomplished by impingement or settlement sampling with microscopic counting in manners similar to those prescribed in ASTM F 24-65 or ASTM F 25-68. Little or no consideration was given to gaseous contaminant control. Local work stations were of the dust control hood type design of Fig. 11-2 for particle control or of the artificial atmosphere design type of Fig. 11-3 for elimination of harmful gaseous contaminants.

The introduction of the automatic light scattering particle counter and the laminar flow concepts[1] of clean room air handling stimulated the further development of clean environment techniques to their current state. HEPA filter advancements, facility design innovations, and the establishment of performance standards[2-6] have kept pace to such an extent that particulate clean environment engineering is now reasonably well developed. Control of gaseous contaminants in clean environments has not been as well understood and much less frequently applied. However, the need, particularly in microelectronics activities, is being more seriously recognized.

## performance criteria

Based on microelectronic manufacturing experience, performance criteria for evaluation of clean environment facility designs have been derived. These

*Figure 11-3 Illustration of enclosed environment system.*

are: (1) temperature and humidity control; (2) control of average particle and vapor concentrations; (3) control of contaminant dispersion within clean space; (4) rate of recovery from transient bursts of high contaminant concentrations; (5) control of noise and vibration caused by the air handling equipment; (6) general engineering considerations of construction quality, appearance, safety, etc.

Until recently quantitative evaluation was generally limited to average particle concentration, temperature, and humidity control. With one exception the remaining criteria are now incorporated into facility standards of the AACC.[5,6] In the case of vapor concentrations no mandatory or recommended control levels nor any standard test procedures exist to evaluate a facility's control of vapor contaminants.

It is apparent from these criteria that clean environment design is mainly an air conditioning problem. The interrelationships of contaminant concentrations and recovery rates and their dependency on air handling design may be appreciated by reviewing Eq. (10-10) and (10-11). The capability to control particle dispersion is a function of the air distribution control, i.e., the degree to which air movement through the space is unidirectional without major eddying effects. The control of noise and vibration of a clean facility is influenced by the mechanical design required to satisfy air conditioning requirements, while the general engineering criteria are primarily related to architectural and mechanical design.

## *clean room systems*

Clean rooms may be classified as either mixed flow or laminar flow depending on the manner by which the air flow within the work zone of the room is controlled. Clean rooms may also be classified according to the particle cleanliness level they are capable of sustaining in actual operation. This assumes compatibility of production equipment, materials, housekeeping procedures, and personnel behavior with the cleanliness classification. Whatever the design type or cleanliness class level, a clean room will usually consist of the elements shown in Fig. 11-4. Recirculating blowers provide the main air movement into and through the room, prefilters, and final filters. Air handling units provide the cooling and/or heating requirements. Make-up blowers provide air to pressurize the clean room and accommodate the leakage and process exhaust flowing out of the room (sometimes they are incorporated with air handling units). Final filters will be HEPA type for high order clean rooms or one of the high grade intermediate efficiency types for low order clean rooms. Prefilters will be of a low or intermediate efficiency to prolong the life of the more expensive final filters. Activated carbon filters may be provided where the presence of gaseous contamination is recognized and the need to control it is acknowledged.

The discussions to follow on specific types of room systems will be confined to particle cleanness considerations. The question of gaseous contamination control will be discussed separately.

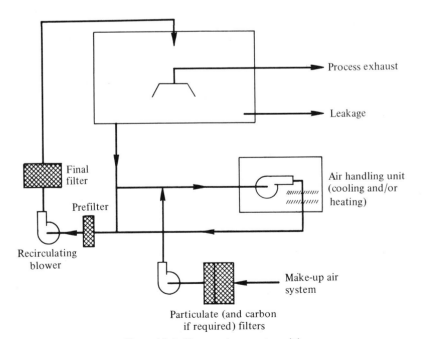

*Figure 11-4 Clean environment model.*

## mixed flow clean rooms

The mixed flow clean room is the traditional design concept where the air is introduced into the clean space at the ceiling, usually by some means of air supply units intermittently spaced, and is removed from the room usually by return air grilles intermittently or continuously spaced along the side walls at floor level (see Fig. 11-5a). As opposed to early clean room designs, modern facilities utilize air change rates considerably greater than those normally em-

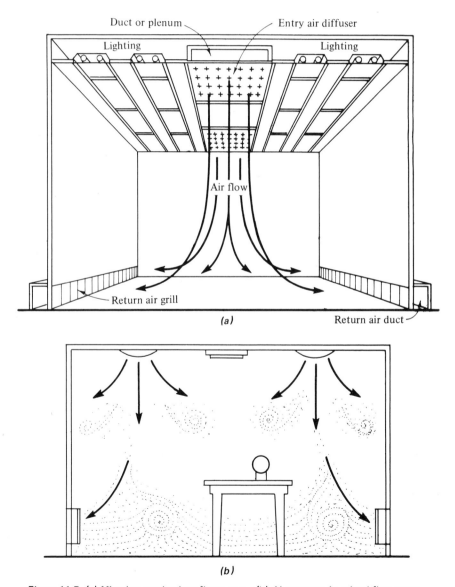

Figure 11-5  (a) Mixed or nonlaminar flow room. (b) Air patterns in mixed flow room.

ployed for comfort air conditioning. The term "mixed flow" stems from the air patterns to be expected in the "work zone" as shown in Fig. 11-5b. Since the air is introduced nonuniformly, secondary induction and mixing do occur, and recirculation within the room space is a natural occurrence. The air passing a work zone is not necessarily a "once-through" quantity as in laminar flow arrangements.

A mixed flow (or nonlaminar flow) clean room design is usually considered capable of sustaining working levels of cleanliness in the class 100,000 to class 10,000 ranges. It is not suited to handling large quantities of air without turbulence, and this handicaps its applications. The distance between walls where air returns are located should be limited to minimize this turbulence and the eddies, or major horizontal air movements above work surfaces. Wall-to-wall dimensions over 20-25 feet should be avoided for class 10,000 designs. On the other hand, where the air supply is extensively distributed over the ceiling and under optimum conditions (low occupancy, low internal generation, and low make-up air quantity) some mixed flow rooms have been observed to sustain class 1,000 levels. Under similar conditions narrow designs ($<$ 12 ft wide) have achieved class 100 conditions.

For class 100 to class 10,000 conditions final filters would be of the HEPA type. For class 100,000 conditions it is often possible to use less than HEPA quality filters, e.g., 85–95 percent atmospheric dust spot efficiency rated filters.

Supply air can be introduced into the room by several different designs of air diffusing elements (Figs. 11-6, 11-7, and 11-8). For class 10,000 room construction using a central blower–final filter system the following types have been proven acceptable: (1) slotted T-bar ceiling with a pressurized supply air plenum, the T-bars to be spaced on 4 ft centers (Fig. 11-6); (2) air supply lighting fixtures on similar spacing as slotted T-bars; (3) completely perforated suspended ceiling, with a 0.2–0.3 in. w.g. pressure drop being desirable for delivery uniformity; (4) intermittent spacing (maximum of 10–12 ft) of perforated ceiling panels in a T-bar suspension (Fig. 11-7).

Another variation places final filters or complete blower-filter modules (Fig. 11-8) in a suspended ceiling system on an intermittent spacing sufficient to guarantee the required air change rate. Standard air conditioning diffusers are not recommended for effective control in class 10,000 rooms since they are designed for intentional maximum secondary induction, which should be minimized as much as possible in clean room systems.

Air returns for class 10,000 rooms without raised floors should be placed at floor level in the side walls preferably continuously along the two opposing walls of the longer dimension of the room (shortest air path from center of room to return). Where room shape dictates it, all four walls may be provided with returns. Where returns cannot be continuous, spacing between outlets should probably not exceed 10 feet. To save space and simplify room design a double wall construction is often used for room partitions. The hollow space between the two walls can then be used as the return air chase.

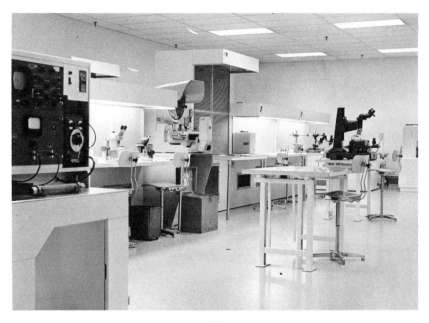

(a)

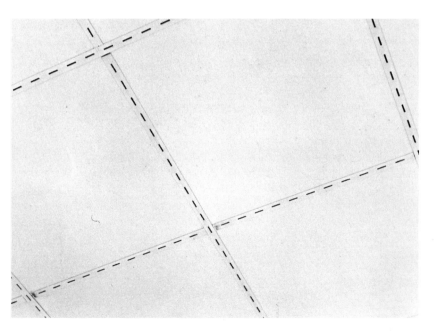

(b)

Figure 11-6 Mixed flow room showing slotted T-bar air supply.

*Figure 11-7 Mixed flow room with perforated ceiling panels.*

For class 100,000 conditions similar air supply units and returns are suitable, although limits on placement are less stringent. Likewise, room size limitations are not as restrictive.

**Air Quantity and Distribution.** Since air is not always supplied over the entire cross section of the mixed flow room, air change rate is used as a criterion rather than air velocity. Air volume changes of 40-75 per hour (with the higher ranges preferred) are common for class 10,000 rooms, and 20-25 changes per hour for class 100,000 rooms. The higher values correspond to apparent average velocities of 12.5 fpm and 4.2 fpm respectively for rooms of 10 foot high ceilings.

As the air supply is increased and extended over a greater portion of the room area the sustained cleanliness level will be enhanced. The recovery capability of the room from a burst of contamination will also be improved, and the distances to which contamination will disperse within the room will be reduced.

**Air Locks.** A mixed flow clean room is usually provided with an entrance air lock or vestibule to prevent influx of contaminant laden air from the outside. As the air change rate is reduced and as the air distribution becomes less

*Figure 11-8  Photograph of a class 100,000 clean room.*

unidirectional the value of the air lock increases.  The air lock also can serve as a smock or coverall storage and change area.

**Materials of Construction.**  Materials used in mixed flow clean room construction require some degree of special consideration to ensure their suitability for the purpose.  For example, floor finishing materials must have a high resistance to abrasion to minimize the particles scuffed loose and projected upward into the air.  Materials such as homogeneous roll vinyl flooring or the poured epoxy or chemically resistant composition floors have proved quite successful.

Early clean room walls almost universally were comprised of high gloss metal, melamine, or vinyl finishes with walls coved into floors for cleaning. The need for such precaution has generally been reduced in modern clean room design, and a nonshedding cleanable material for the walls is sufficient. Ceilings, of course, must not contribute particles that will fall into the clean room work zones.  Where T-bar ceilings are employed, ceiling panels are usually plastic covered or otherwise treated to be nonshedding.  Similar nonshedding or nonabrading characteristics are required for surface finishes in air supply plenums above ceilings and in duct work.  Production equipment and work benches must also be of a readily cleanable design and nonshedding.

Present clean room concepts generally recognize that the "hospital-white" and stainless steel materials once considered necessary are more useful psychologically to stimulate clean behavior of operators than to minimize particle generation.

**Operating Procedures and Controls.** The recirculation and mixing that occur in a mixed flow room can well necessitate the imposition of stringent requirements on operating procedures, personnel behavior, and housekeeping routines. To sustain the stipulated cleanliness level the amount of contaminant generation and entrainment into the air must be kept within the limits that the air change rates and the air distribution mode will tolerate (Eq. (10-11)). It is not uncommon to see mixed flow class 10,000 rooms equipped not only with air locks, but also with dressing rooms, shoe cleaners, and air showers (Fig. 11-9).

*Figure 11-9 Mixed flow clean room—entrance with air shower.*

People are frequently the greatest source of particulate contamination in a clean room.[7,8] Protective clothing suitable for the environment class and room design is required, and behavior patterns compatible with the environment must be established and enforced. While some relaxation of procedure has occurred in the evolution of room design, modern mixed flow rooms still require persistent and relatively elaborate housekeeping routines such as frequent floor vacuuming, floor washing, work surface wiping, etc.[9]

## laminar flow clean rooms

In contrast to the mixed flow clean room the laminar flow room design provides for a uniform parallel stream air flow through the work zone with a minimum of eddying and cross-mixing. The result, while if not truly "laminar" in the strict mathematical sense, is a "one-pass" flow of air from the filtered inlet point past the work point to the exit point. Laminar flow rooms may have horizontal air flow from wall to wall (Fig. 11-10), vertical air flow

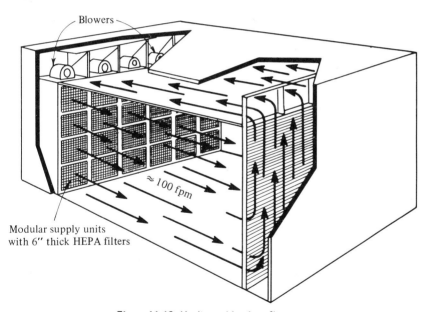

Figure 11-10  Horizontal laminar flow room.

from ceiling to floor, (Fig. 11-11), or, per the original concept by Whitfield,[1] the wall to floor air flow pattern shown in Fig. 11-12.

A laminar flow clean room of the vertical flow type is capable of sustaining at least a class 100 cleanliness level throughout its work zone. In contrast, a horizontal flow design will provide class 100 conditions at the first line of work positions with the cleanness level at successive work points deteriorating because of the release of generated particles. It is usually accepted that con-

*Figure 11-11   Class 100 clean room.*

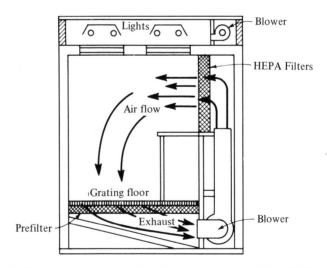

*Figure 11-12   Wall-to-floor airflow.   (As originally prepared by W. J. Whitfield, Sandia Corporation.)*

ditions will vary from class 100 at the first air exposure to about class 10,000 at the most distant work position from the air inlet.

In both vertical flow and horizontal configurations the supply air is arranged to be introduced as uniformly as possible over as much as possible of the flow cross-section.   Air inlet over at least 75 percent and preferably 80

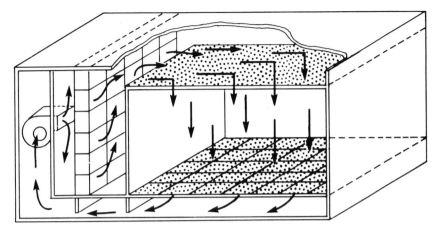

*Figure 11-13  Vertical laminar flow room with central filter bank.*

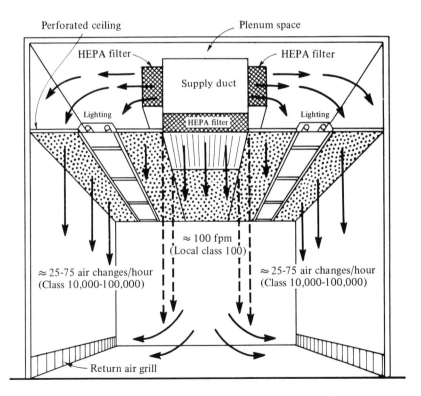

*Figure 11-14 Combination clean room design: class 100 locally and class 10,000–100,000 in room at large.*

percent of the wall or ceiling area is the general rule. HEPA filters are usually located at the air inlet face of the room, combining the functions of a filter and an air diffuser. Vertical flow facilities have also been designed with air diffusing ceilings of perforated panels and HEPA filters located remotely in an air handling plenum (Fig. 11-13). Laminar flow concepts have also been used in combination with mixed flow clean room design as shown in Fig. 11-14.

Several different design arrangements for HEPA filters located in the ceilings of vertical laminar flow rooms have been devised. Each is capable of providing comparable clean performance and has its own certain advantage for flow adjustment, installation ease and/or integrity, equipment space required per unit of clean space, or overall cost. Figure 11-15 shows an individual HEPA filter-plenum system where several plenum units are connected to a supply air duct. Figure 11-16 illustrates the single plenum design where HEPA filters are clamped into a ceiling grid system. Figures 11-11 and 11-17 show a scheme whereby HEPA filter-blower modules are suspended and sealed together to form a ceiling with the plenum above serving as a return air-blower supply plenum.

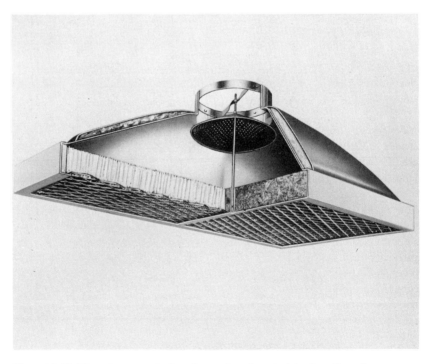

*Figure 11-15 Filter-plenum unit for air supply of vertical laminar flow-clean rooms. (Courtesy Weber Technical Products, Grand Rapids, Michigan.)*

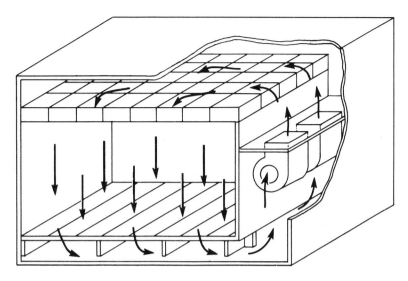

*Figure 11-16  Vertical laminar flow room filter ceiling, single plenum.*

Return air is usually taken through a perforated raised floor and under a floor chase back to the blower plenum. The filter-blower module system of Fig. 11-17 uses a perforated floor with return air towers spaced about in the clean area. In narrow rooms it is quite feasible to eliminate the raised floor and utilize low side wall returns.

**Air Quantity and Distribution.** Laminar air flow achieves the class 100 cleanliness level because of parallel air flow and the high air velocities which can be used without forming disrupting eddy currents. Uniform velocities have been specified from 50 to 100 fpm with a tolerance of 20 fpm around the selected average. Some attempts have been made to use even lower air flow rates in the interest of economy. Federal Standard No. 209, as originally issued, recommended an average air velocity of 90 fpm ± 20 fpm.

The high velocities and parallel flow patterns serve to prevent lateral dispersion of generated particles so that operations are protected from each other and any single position will recover rapidly from high level bursts of contamination. These capabilities are not automatically acquired. The laminar flow room requires careful adjustment and balancing to achieve its potential. In vertical laminar flow the lowest air velocity that can be tolerated without causing undesirable dispersion is a function of the physical design of ceiling and floor for air flow control, as well as of the temperature of heated surfaces which cause rising convection columns to overcome the downward air flow. It has been determined[10] that without heat columns 50

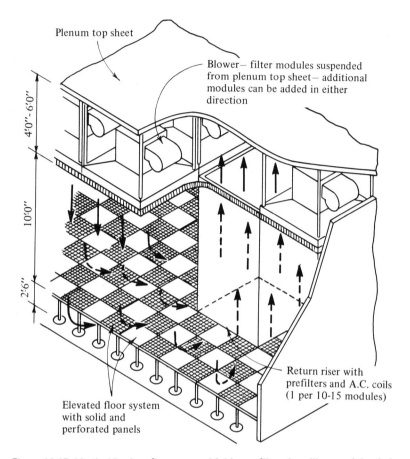

Plenum top sheet

Blower— filter modules suspended
from plenum top sheet— additional
modules can be added in either
direction

4'0"– 6'0"

10'0"

2'6"

Return riser with
prefilters and A.C. coils
(1 per 10-15 modules)

Elevated floor system
with solid and
perforated panels

*Figure 11-17  Vertical laminar flow room with blower-filters in ceiling, modular design.*

fpm velocities are suitable, and that in the presence of heated surfaces at 300°C velocities of at least 100 fpm are required. A design velocity of 75 fpm provides good control of convective forces for most of the typical types of heat sources connected with microelectronic manufacturing (Fig. 10-6). Velocities below about 45 fpm, even in the absence of heat columns, do not appear to provide the control considered necessary for class 100 designs.[11]

**Airlocks.** The nature of the air flow pattern as well as the air quantity used in vertical laminar flow rooms would make an air lock unnecessary for preventing the influx of contaminated air. In horizontal flow configurations the air lock might be required depending on the location of the entrance. Vestibules are normally constructed for laminar flow facilities to provide an area for smock storage, shoe cleaners, etc., and for preliminary materials handling prior to their introduction into the clean space.

**Materials of Construction.** The high air flow rates and parallel flow patterns make the materials of construction less a matter of concern in laminar flow than in mixed flow rooms as far as room airborne cleanliness level is concerned. Surface finishes must be durable for reasonable life and periodic cleaning requirements; however, except for nonshedding characteristics of materials in the clean air stream as it enters the room, abrasion resistance, etc., are of less importance. For example, unreasonably violent motions must be produced to cause entrainment of particles from the floor upward to working levels in a vertical flow room. Most construction materials could therefore be effectively utilized in laminar flow clean room construction. As with mixed flow rooms, however, the psychological benefit of clean appearing surfaces should not be overlooked.

Insofar as interior furnishings are concerned, laminar flow clean rooms do impose some limitations in the choice of materials and configurations as well as equipment and work position arrangements. In vertical laminar flow the desirability of perforated work surfaces has been demonstrated for maintaining the downward air flow pattern and prevention of the "bounceback" that occurs on a solid work surface.[11] The latter is the turbulent zone created in changing the direction of the vertical air flow striking a solid bench top. Excessively close spacing of equipment must be avoided to minimize upsetting of laminar flow and in horizontal flow systems the contamination of one process by another operation directly upstream.

**Operating Procedures and Controls.** Laminar flow facilities can permit some relaxation of the rigid rules of behavior necessary in high order mixed flow rooms. The influence, for example, of a person walking in an aisle of a vertical laminar flow room is only brief and limited at a work position (Fig. 11-18).

Some protective clothing is still necessary to prevent direct fallout on the work surface of large particles or fibers (skin, dandruff, hair, and clothing), and in horizontal flow configurations the generation of particles into the air stream at elevations above work surfaces will be evident farther down the air flow path. The degree of protection necessary is somewhat controversial. The author believes that head covers, smocks and shoe cleaners are a minimum requirement, while some feel that protective clothing is not necessary. Others prescribe even more protection. Documentation of need has not advanced as far as is desirable in this regard, and in most cases regulations are established by judgment related to production risk.

Regarding cosmetics, the possibility of workers' heads being directly over product must be considered. Face powders, etc., should generally be discouraged in favor of grease-type make-up which prevents skin drying and sloughing.

Housekeeping routines maintained on a regular basis are not as complex nor as frequent or costly as in mixed flow rooms. Once-daily floor mopping and work surface cleaning are recommended as a reasonable level. For further discussion see Chap. 13.

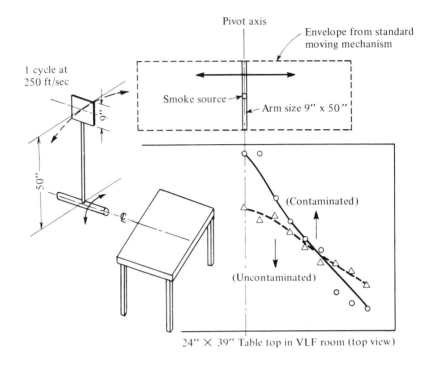

Figure 11-18 Dispersion envelope 2" above table top in vertical laminar flow room due to simulated walking movement.

## work stations

A clean work station is used for the protection of one or a few individual operations performed at the same place. Depending on its design type and its location, the particle level environment that it can sustain may be anything from worse than class 100,000 to better than class 1,000. Depending on design type it may or may not provide temperature, humidity, and gaseous contaminant control as well.

## dust control hoods

Dust control hoods typified by Fig. 11-2 are still used today for certain less critical situations where large particle control is required. The design intent is to provide a curtain of clean air at the work opening to prevent the intrusion of particles. Dust control hoods do not provide the degree of control possible with laminar flow units, but they are much less costly. With HEPA filtered designs high order cleanliness is possible immediately downstream of the filter before the influence of "back-streaming" turbulence is felt.

## laminar flow work stations

In addition to room design, laminar flow has been applied to individual work positions. They may be of the horizontal or vertical parallel flow configuration as shown in Figs. 11-19 and 11-20. Horizontal flow systems are suitable

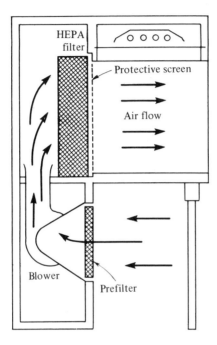

HEPA filter

Protective screen

Air flow

Blower

Prefilter

*Figure 11-19   Horizontal laminar flow clean work station.*

for assembly and other operations where toxic fumes are not encountered. Vertical flow units are applicable where chemical exhaust must be continual or where production equipment bulk precludes parallel horizontal flow across the work surfaces. Special air flow designs have been developed as shown in Figs. 11-21 through 11-23. These are not truly parallel flow systems but are ultra-clean and have evolved through efforts to apply the basic systems to specialized needs.

Laminar flow work stations, with no production equipment installed or operators present, can produce essentially zero particle level conditions for particle sizes $\geq 0.5$ micron regardless of the ambient particle level. The actual operating cleanliness level with an operator present will vary depending on the station design, the type of process, the design and placement of process equipment in the station, the operator's behavior pattern, and the location of the station. In an open manufacturing area of an air conditioned microelectronics plant, cleanliness levels from class 10,000 to class 1,000 are possible.[12] Where class 100 conditions are required, a laminar flow clean work station

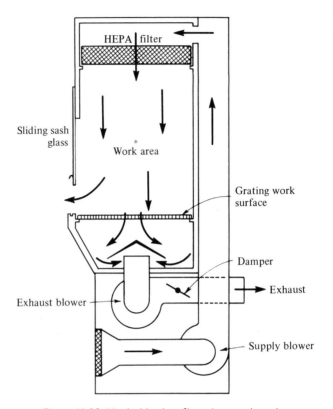

*Figure 11-20   Vertical laminar flow clean work station.*

must be installed in a clean room of at least class 100,000 and preferably of class 10,000 capability.

The AACC Standard CS-2T[5] covers in detail the basic minimum construction, performance standards, and performance verification test procedures for laminar flow work stations. A work station meeting these requirements and properly "proven-in"[13] can be expected to give satisfactory performance provided operating procedures are suitable.

While it is physically possible to equip some laminar flow clean work station designs with activated carbon filters,[14] very little has actually been done along this line on a commercial scale. Few laminar flow stations have any capability of controlling gaseous contamination. The control of moisture has been successfully accomplished commercially—as it has in clean rooms—through the use of refrigeration and/or desiccant drying in recirculating work station systems.

### special atmosphere systems

A special atmosphere system can be used where gaseous contamination other than moisture requires control (Fig. 11-3). The artificial atmosphere (nitro-

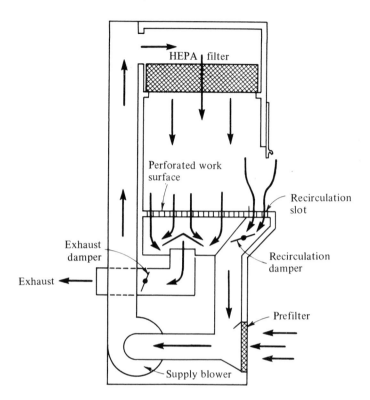

*Figure 11-21  Exhausted vertical laminar flow clean work station with recirculation.*

gen in most cases) provides protection from environmental oxygen, fumes, and moisture for a specific operation. Moisture levels < 10 ppm in nitrogen with < 200 ppm oxygen are typical. The design of such enclosures where access must be through glove parts to ensure the gaseous environment is not normally arranged for laminar flow. The cleanliness level sustained in ambient system enclosures is at least one order of magnitude less clean than in a laminar flow work station.

The atmosphere recirculating system usually consists of a closed loop gas piping system with a rotary type pressure blower, desiccant gas dryer, and HEPA type pipe line filter. The quantity of gas circulated is much lower than in laminar flow units due to the moisture load and level to be accommodated. While it might be possible to provide laminar flow in this type system, the equipment and operating costs would probably be extremely high.

### operating restrictions

Clean work stations of the open front design, laminar flow, or dust control hood type require the adherence to operating procedures that will not degrade the environment. Hand motions must not be so rapid as to overcome the

*Figure 11-22 Horizontal flow clean bench with recirculation. (Edgegard® Hood Courtesy the Baker Company, Biddeford, Maine.)*

outwardly flowing clean air and cause back flow of more heavily contaminated ambient air. For high order cleanliness levels in laminar flow stations within clean rooms the use of smocks and head covers is often a recommended practice. Materials introduced into the work station should be precleaned and must not be placed upstream of the work being prepared. The design of operating equipment installed in the clean work station should consider the effect of bulk on air flow pattern so that turbulence and eddies are minimized.

## *"particle clean" facility comparisons*

Tables 11-1 through 11-5 are presented as digests of the previous descriptive sections on particle control facilities. Table 11-1 compares the performance capabilities of the various clean facility design types. Desirable clean room design criteria are shown in Tables 11-2 through 11-5 for all the major performance items. The basic facility items are useful as acceptance limits in a

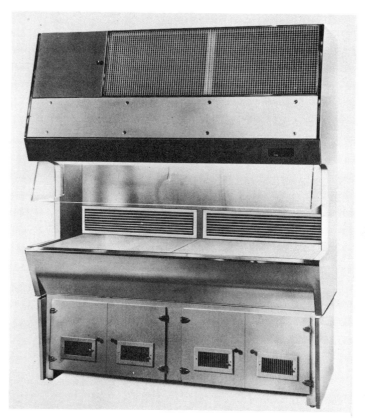

*Figure 11-23  Vertical flow chemical work station.  (Courtesy Air Control, Inc., Norristown, Pennsylvania.)*

purchase specification.  The "at rest" condition indicates the expected capability of a clean facility when all production equipment is installed and no operators are present.  The operating criteria show the combined effects of room capability and operator influences.

## gaseous contamination control

In clean environment manufacturing many freshly cleaned microelectronic surfaces cannot be exposed longer than 15 minutes in order to protect against deposition of organic vapors.  With the improvement of particle control and the increasing need for clean product surfaces the influence of gaseous contaminants is becoming more significant.  Where environmental gases or vapors other than moisture must be controlled, activated carbon filtration is the most suitable technique, as discussed in Chap. 10.

**TABLE 11-1. Comparative Capabilities of Clean Environment Control Facilities**

| Facility Design Type | Work Point Cleanliness Class | Ambient Cleanliness Class | Temperature and Humidity Control | Lighting Control | Gaseous Contamination Control |
|---|---|---|---|---|---|
| Laminar flow work station in mfg. area | 1,000–10,000 | To $10^6$ | Yes | Partial | No |
| Ambient atmosphere systems | 1,000–20,000 | To $10^6$ | Yes | Yes | Yes |
| Mixed flow clean rooms | 10,000–100,000 | 10,000–100,000 | Yes | Yes | Partial |
| Mixed flow clean room with laminar flow work stations | 100 | 10,000 | Yes | Yes | Partial |
| Horizontal laminar flow clean room | 100–10,000 | 100–10,000 | Yes | Yes | Partial |
| Vertical laminar flow clean room | 100 | 100 | Yes | Yes | Partial |

**TABLE 11-2. Desirable Performance Criteria for Class 100 Vertical Laminar Flow Room**

| Item | Basic Facility | Facility at Rest | Facility in Operation |
|---|---|---|---|
| AIRBORNE PARTICLE COUNT | $<10/\text{FT}^3 \geqslant 0.5\mu m$ & $0 \geqslant 1.0\mu m$ at any point from 3 ft. below ceiling to floor level. | $<10/\text{FT}^3 \geqslant 0.5\mu m$ & $0 \geqslant 1.0\mu m$ 12" upstream of work points and throughout room 30" above floor 24" and further away from prod. equipt. | $<$ class 100 conditions 12" upstream of work points and throughout room 30" above floor and 24" and further away from contaminant generators. |
| DISPERSION CONTROL | Not measurable beyond 24" from envelope of contaminant generator in zone 30" above floor to 36" below ceiling. | Same | Same |
| RECOVERY CONTROL | 1 minute from removal of contaminant generator. | Same | Same |

| | | | |
|---|---|---|---|
| AIR FLOW PATTERNS | Parallel downward from 3 ft. below ceiling to satisfy dispersion control. | Same to within 18″ horizontally from prod. equipment. | — |
| AIR FLOW UNIFORMITY | Velocity readings of 75–80 fpm avg.*, 80% within ±20 fpm of req'd avg. and remainder within ±30 fpm–in zone 30″ to 72″ above floor throughout room. | — | — |
| AIR DELIVERY SYSTEM | Provide 1% of total for pressurization; provide differential pressures between rooms. Permit adjustment locally without violating parallelism and dispersion control. Have reserve capacity for 1″ pressure drop increase thru HEPA filters. Provide emergency shutdown controls for fire and fume hazards. | Same | Same |
| TEMPERATURE & HUMIDITY CTL. | ±2°F and ±5% R.H. at any constant elevation throughout room. | Same to within 24″ of equipment and at 36″ elevation above floor. | Same as At Rest. |
| NOISE LEVEL | NC-55 maximum 30″ to 72″ above floor throughout room. | — | — |
| VIBRATION LEVEL | Peak to peak displacement at floor surface in any direction at or above specified frequency to be specified based on existing conditions before construction. | — | — |
| LIGHTING LEVEL | Minimum of 100 F.C. at 30″ above floor throughout room. | Same | Same |

*Currently recommended for microelectronic applications.

TABLE 11-3. Desirable Performance Criteria for Class 1,000 Vertical Laminar Flow Room

| Item | Basic Facility | Facility at Rest | Facility in Operation |
|---|---|---|---|
| AIRBORNE PARTICLE COUNT | $< 100/\text{ft}^3 \geq 0.5\,\mu m$ & $0 \geq 1.0\mu m$ at any point from 3 ft. below ceiling to floor level. | $< 100/\text{ft}^3 \geq .5\mu m$ & $0 \geq 1.0\mu m$ 12″ upstream of work points & throughout room 30″ above floor 24″ & further away from prod. equipt. | $<$ class 1,000 conditions 12″ upstream of work points & throughout room 30″ above floor 24″ & further away from contaminant generators. |
| DISPERSION CONTROL | Not measurable beyond 30″ from envelope of contaminant generator in zone 30″ above floor to 36″ below ceiling | Same to within 18″ horizontally from prod. equipment. | — |
| RECOVERY CONTROL | Same as for class 100. | Same | Same |
| AIR FLOW PATTERNS | Same as for class 100. | Same as for class 100. | — |
| AIR FLOW UNIFORMITY | Velocity readings at 50 fpm avg.,* 80% within ±20 fpm of avg. Remainder within ±30 fpm in zone 30″ to 70″ above floor throughout room. | — | — |
| AIR DELIVERY SYSTEM | Same as for class 100. | Same | Same |
| TEMPERATURE & HUMIDITY CONTROL | Same as for class 100. | Same as for class 100. | Same as for class 100. |
| NOISE LEVEL | Same as for class 100. | | |
| VIBRATION LEVEL | Same as for class 100. | | |
| LIGHTING CONTROL | Same as for class 100. | Same | Same |

*Currently recommended for microelectronic applications.

Activated carbon has long been applied for odor control in many air conditioning installations and for the recovery of valuable solvents from exhaust systems. Its use to provide "organic-vapor clean" environments in connection with particle clean facilities has been less advanced. An installed cost of $250-300 per 1000 cfm of treated air (based on 45 lb of activated carbon per 1000 cfm) is one major limiting factor. When the ranges of typical air loads, contaminant loadings, lifetimes, and estimated operating costs are better understood, such expenditures may be justifiable. A conceptual insight can be gained from Eq. (10-10), restated as Eq. (11-1), which gives the time relationship of contaminant concentration within a space:

$$C_{s_t} = K_1\, C_m\, (1 - e^{-K_2 t}) + \frac{G}{K_2}(1 - e^{-K_2 t}) + C_{s_0}\, e^{-K_2 t} \qquad (11\text{-}1)$$

$$K_1 = \frac{m\,(1 - E_m)}{m + rE_r + \dfrac{A}{Q_s/V}}$$

$$K_2 = \frac{Q_s}{V}\left(m + rE_r + \frac{A}{Q_s/V}\right)$$

Consider the case where no gaseous contaminant filtration is provided and surface deposition is negligible. Assuming generation inside the space is of burst character rather than continuous so that $G$ can be set equal to zero, and with $C_{s_0}$ equal to the initial average level due to the burst release, then

$$E_m = E_r = 0;\ \frac{A}{Q_s/V} = 0;\ K_1 = 1.0;\ K_2 = \frac{Q_s m}{V}$$

and

$$C_{s_t} = C_m + (C_{s_0} - C_m)e^{-(mQ_s t/V)} \qquad (11\text{-}2)$$

As time passes, $t \to \infty$, and $C_{s_t}$ will come to equilibrium at $C_m$. For practical considerations it should be noted that when $t = \dfrac{4V}{mQ_s}$ or $e^{-4} = 0.02$, then 98% of the burst concentration above the make-up air level will be removed. Consider a vertical laminar flow clean room at an air velocity of 100 fpm (100 cfm/sq ft of floor space) in a 10 ft high room where the total space volume (room + plenums + ducts, etc.) is twice that of the clean space alone. The time constant is then

$$\frac{mQ_s}{V} = \frac{m \times 100}{2 \times 10} = 5m\ \text{min}^{-1}$$

When $t = 4\left(\dfrac{1}{5m}\right)$ min the dilution by make-up air will remove 98% of the

TABLE 11-4. Desirable Performance Criteria for Horizontal Laminar Flow Room

| Item | Basic Facility | Facility at Rest | Facility in Operation |
|---|---|---|---|
| AIRBORNE PARTICLE COUNT | < 10/ft² ≥ .5μm & 0 ≥ 1.0μm at any point from 6" below ceiling to 6" below downstream filter bank to 12" upstream exit plane. | < 10/ft³ ≥ .5μm & 0 ≥ 1.0μm 6" below ceiling to 6" above floor 12" upstream first row equip.; 12" upstream equipt in direct air path from filters & 24" laterally from any equipt upstream; 6" below ceiling to 6" above floor in any direct line from filters. | < class 100 conditions 1'2" upstream of 1st row of equipt.; class 10,000 indirect air path from filters & 24" laterally from any equipt. at all succeeding rows. |
| DISPERSION CONTROL | Not measurable beyond 24" vert. or horizontally (laterally) from envelope of generator in undisturbed air path downstream from source. | Same | Same |
| RECOVERY CONTROL | Same as for class 100 VLF room. | Same | Same |
| AIR FLOW PATTERN | Parallel & horizontal to satisfy dispersion control from 6" above floor to 6" below ceiling & 6" downstream from filters to 12" upstream of exit plane. | Parallel and horizontal with 10% max. lateral or vertical component in undisturbed air path at least 24" laterally from any equipt., 6" downstream from filters to 12" upstream of exit plane. | — |

| | | | |
|---|---|---|---|
| AIR FLOW UNIFORMITY | Velocity readings of 100 fpm avg.,* 80% within ±20 fpm of req'd avg. & remainder within ±30 fpm— in any vertical plane 6" downstream from filters to 12" upstream of exit plane & 12" above floor to 12" below ceiling. | — | — |
| AIR DELIVERY SYSTEM | Same as for class 100 VLF room. | Same | Same |
| TEMPERATURE & HUMIDITY CTL. | ±2°F & ±5% R.H. from 12" above floor to 12" below ceiling in any transverse vertical plane 6" downstream of filters to 12" upstream of exit plane. | ±2°F & ±5% R.H. from 12" above floor to 12" below ceiling at any point in direct air path from filters & 24" or more laterally from any prod. equipt. | Same |
| NOISE LEVEL | Same as for class 100 VLF room. | | |
| VIBRATION LEVEL | Same as for class 100 VLF room. | | |
| LIGHTING CONTROL | Same as for class 100 VLF room. | Same | Same |

*Currently recommended for microelectronic applications.

**TABLE 11-5.  Desirable Performance Criteria for Mixed Flow Rooms**

| Item | Basic Facility | Facility at Rest | Facility in Operation |
|---|---|---|---|
| AIRBORNE PARTICLE COUNT | < 1/10 of class condition throughout room from 12" above floor to 72" above floor for class 10,000 and < 1/5 of class condition for class 100,000. | < 1/3 of class condition 12" upstream of all work positions and throughout room at 30" above floor & 24" or more from contaminant generator for class 10,000 and < 1/2 of class for class 100,000. | < Class condition 12" upstream of all work positions & throughout room at 30" above floor & 24" or more from contaminant generator. |
| DISPERSION AND RECOVERY CONTROL | Same as for class 1,000 VLF room except zone to be 30–72" above floor and recovery interval to be 2 minutes* | Same | Same |
| AIR FLOW PATTERNS | Distributed supply at ceiling level and distributed return at low sidewalls or floor | — | — |
| AIR FLOW UNIFORMITY | Velocity readings** at supply units 80% within ±10% of avg. and remainder within ±20%. | — | — |
| AIR DELIVERY SYSTEM* | Provide for 10% of supply for pressurization, otherwise same as for class 100 VLF room. | — | — |
| TEMPERATURE & HUMIDITY CONTROL | Same as for class 100 VLF room. | Same as for class 100 VLF room. | Same as for class 100 VLF room. |
| NOISE LEVEL | Same as for class 100 VLF room. | | |
| VIBRATION LEVEL | Same as for class 100 VLF room. | | |
| LIGHTING CONTROL | Same as for class 100 VLF room. | Same | Same |

*Desired air supply is uniformly distributed over the ceiling; otherwise this requirement must be waived burst increase. If $m = .08$ (make-up air in the amount of 8 cfm/sq ft of floor space, an amount quite common in microelectronic clean rooms), then $t = 10$ minutes—the recovery time of the room.

If activated carbon filtration is applied to the make-up air to provide removal of gaseous contaminants in that air stream, then:

$$E_m = 1; E_r = 0; \frac{A}{Q_s/V} = 0; K_1 = 0; K_2 = \frac{Q_s m}{V}$$

and again for burst contamination only ($G = 0$):

$$C_{s_t} = C_{s_0} e^{-(K_2 t)} = C_{s_0} e^{-\left(\frac{Q_s m t}{V}\right)} \tag{11-3}$$

The concentration will now decay exponentially to zero as $t \rightarrow \infty$, and for all practical purposes $C_{s_t} \approx 0$ at $t = 10$ minutes for the example cited above.

If one considers carbon filtration of all of the recirculated air with the make-up air untreated, then

$$E_m = 0; E_r = 1; \frac{A}{Q_s/V} = 0; K_1 = m; K_2 = \frac{Q_s}{V}$$

and with $G = 0$ for burst contamination:

$$C_{s_t} = m C_m + (C_{s_0} - m C_m) e^{-(Q_s t/V)} \tag{11-4}$$

As $t \rightarrow \infty$, $C_{s_t} \rightarrow m C_m$, which by comparison with the results of Eq. (11-3) shows that while the time constant may be much lower (0.8 minute vs. 10 minutes), the equilibrium level is higher. All this is at the greater expense of carbon filtering 100 cfm/sq ft of floor space rather than 8 cfm/sq ft of floor space.

It should be apparent that by carbon filtration of the make-up air to control gaseous contaminants is most essential when the vapor concentration in the make-up air is greater than that desired in the controlled space. In the absence of sufficient make-up air to provide a reasonable time constant the additional carbon filtration of a portion of the recirculation air will assist in reducing the time of recovery.

The economics of activated carbon filtration for gaseous contamination control may be appreciated from a consideration of lifetime vs. amount of air treated and inlet contaminant load. Activated carbon has an adsorption capacity of approximately 50 percent by weight[15] of high boiling point organic vapors such as trichlorethylene and others normally used in microelectronics. At 25°C and one atmosphere pressure the time required for a carbon bed to saturate may be expressed as follows:[16]

$$t = \left(\frac{136 \, W}{M.W.}\right)\left(\frac{1}{C}\right)$$

where $t$ = time in days, $W$ = weight of carbon/1000 cfm in lb, $M.W.$ = molec-

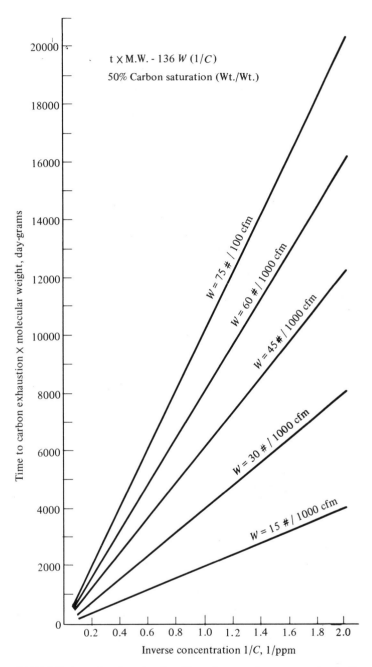

Figure 11-24 Lifetime of activated carbon filters for organic gaseous contamination control. In the figure, t = days to saturation for 100% absorbing efficiency; M.W. = molecular weight of adsorbate, grains; W = weight of activated carbon per 1000 cfm gain treated, lb; C = gaseous concentration at filter inlet, ppm.

ular weight of material adsorbed in grams, $C$ = vapor concentration in inlet air of material to be adsorbed in ppm.

This relation has been plotted in Fig. 11-24. It can be seen that for a typical condition of make-up air concentration of 1 ppm and a carbon weight of 45 lb/1000 cfm the lifetime under continuous use would be 6120/M.W. For trichlorethylene this amounts to 46½ days; for lower molecular weight materials such as acetone, the lifetime would be approximately double. For a clean room of 5000 sq ft of floor space with 5 cfm/sq ft of make-up air at 1 ppm vapor contaminant concentration, the 25,000 cfm of make-up air would require the installation of a filter unit of a size comparable to that shown in Fig. 11-25. Such a unit contains approximately 1100 lb of carbon in perhaps 30 or more separate trays. The total weight would be in the range of 2,500–3,000 lb, and the carbon replacement cycle would be in the range of 45–90 days. A maintenance cycle such as this might be difficult for a small maintenance staff, and the annual operating cost could approximate $12,000. The problems and costs of applying activated carbon filtration for gaseous contamination control are not insignificant.

*Figure 11-25 Activated carbon filter unit.*

# the economics of clean environment facilities

Obviously, the acquisition cost of a clean environment facility will increase with the order of particle cleanliness control, i.e., class 100 costs > class 10,000 costs > class 100,000 costs. This results from the increasing requirements of air circulation rates and degrees of air distribution control. Per the discussion on activated carbon filters, costs will increase with the amount of make-up air needed for the specified degree of gaseous contamination control.

The cost of a clean environment to a product in dollars per dollar of product should include: (1) the acquisition cost of the facility; (2) the rental or production value of the gross manufacturing space allocated to the clean facility; (3) the facility operating costs including depreciation.

Consideration should also be given to the comparative economics of different design approaches. Some designs require less gross floor space/clean space. Certain designs are adaptable to environment upgrading at a later date with a minimum of wasted dollars, while other approaches do not offer this flexibility. In other cases environmental designs may in themselves be more economical, but they require additional product handling with its attendant costs and risks in product contamination. All of these factors are a part of selecting the proper design approach with an understanding of the total cost implication of a clean facility to the expense of producing a given product.

## clean room acquisition costs

Too often confusion arises as to the costs of clean facilities because the basis of comparison is different. The total investment in a clean room is made up of the basic facility cost (without regard to occupancy) and the product related costs. The basic facility cost is a function solely of the particle cleanliness level and the selected design approach and would include the following factors:

### Basic Facility Costs

1. Room enclosure and lighting.
2. Ceiling structure.
3. Floor.
4. Air distribution system, including plenums, ducts, main recirculating blowers, prefilters, and final filters.
5. Air conditioning system, prorated for air moving, leakage, lighting, and transmission heat loads.
6. Make-up blowers with particle and activated carbon filters, prorated for pressurization air quantity.

The product related costs are those factors which remain constant for a particular case regardless of the environmental design approach. Different product requirements will cause cost variation for the identical clean room design. This is the major cause of confusion in clean facility cost comparisons.

*Product Related Costs*

1. Air conditioning system, prorated for the sensible and latent heat loads of people, process equipment, and process exhaust air replacement.
2. Make-up air blowers with particle and activated carbon filters, prorated for process exhaust.

When summing the basic facility and product related costs to arrive at the total incurred cost of a clean facility, one must allow for existing plant investment which will be utilized. For example, consider a clean room constructed in an air conditioned plant having ample central chilled water capacity and with the make-up air taken from within the plant. The total direct costs for acquisition would not include the air conditioning system costs for bringing the make-up air from outside conditions to plant conditions nor the refrigeration system costs for any of the heat load. While these are available in the plant, allocation to the specific job means that the plant capacity is accordingly reduced for other purposes. Consequently, the product should really be charged with the ownership costs of that capacity, not simply with the immediate acquisition costs. For other cases where refrigeration and total make-up air must be provided the total cost to the product is also the acquisition cost.

To illustrate, consider a typical clean room application in the microelectronic industry. First, Table 11-6 lists load ranges that are often encountered

**TABLE 11-6.   Typical Process Loads–Microelectronic Clean Rooms**

| | |
|---|---|
| People: | 5–20 People/1000 ft$^2$ |
| Process equipment: | 10–100 KW/1000 ft$^2$ |
| Chemical Exhaust: | 1,000–10,000 CFM/1000 ft$^2$ |
| Temperature: | $72-78°$F |
| Relative humidity: | $< 15-45\%$ RH. |

in this type of manufacturing. For a net clean space of 5,000 sq ft for the conditions listed, Table 11-7 gives an approximate cost comparison of different performance level facilities. Also, notice the total associated cost including the total make-up air and total refrigeration load, as opposed to the costs where central plant facilities are available.

## clean work station costs

In common with clean rooms, work stations also entail both basic facility costs and process affected costs. However, except for ambient atmosphere systems and those units specially equipped for close temperature and humidity control, most open work stations provide particle clean conditions only. Temperatures will be somewhat above ambient temperatures and humidity somewhat below due to the heat input from the air moving equipment and any lighting. This air conditioning load plus any process exhaust load is then

TABLE 11-7.  Typical Approximate Clean Room Costs for 5,000 FT² of Clean Space in Air Conditioned Plant

Plant conditions:     78°F and 45% RH
Room conditions:     75°F and 45% RH
Room load:           50 People, 500 KW,
    40,000 CFM Exhaust

|  | Costs per ft² of Clean Space | |
|---|---|---|
|  | Class 100 VLF Room at 80 FPM | Class 10,000 MF Room At 75 air chgs/hr |
| 1. Basic facility cost (See Note A) | 70–110 | 40–60 |
| 2. Product related costs (See Note B) | 10–20 | 10–20 |
|  | 80–130 | 50–80 |
| 3. Plant A.C. and chilled water systems— (See Note C) | 20–30 | 10–25 |
| Total associated cost | 100–160 | 60–105 |

*Note:* Cost ranges are given reflecting the variations caused by design details, labor and material cost changes with geography, and inflation.

*NOTES*

(A)  Basic facility costs as defined on page 308, except: (1) chilled water system (central plant) costs not included; (2) pressurization make-up air conditioned from plant to room conditions.

(B)  Product related costs as defined on page 309, except: (1) chilled water system (central plant) costs not included; (2) process make-up air conditioned from plant to room conditions.

(C)  Includes: (1) Costs of plant air conditioning system to condition total make-up air quantity from outside to plant conditions; (2) Cost of plant chilled water system for entire heat load including (1) above.

directly imposed on the general plant system. These hidden loads are as applicable to the total costs of the work station as they are in rooms.

For purposes of comparison the installed "basic facility" cost of a console-type laminar flow work station adapted for microelectronic processing averages about $1,600 apiece over and above the cost of standard factory benches, chemical hoods, etc. This includes the plant air conditioning system cost for the blower and lighting heat loads but does not include plant chilled water system costs. It is thus directly comparable with the basic facility costs for rooms in Table 11-7. It can be appreciated that below some installed density the cost of clean work stations will be less than a clean room of comparable cleanliness level. Above this point a room may be more economical.

For example, from Table 11-7 we see 5,000 sq ft of class 10,000 space may cost $50/sq ft. If 100 laminar flow work stations are placed in this room, the combined cost would be $82/sq ft and the resulting work point cleanliness would be class 100. From Table 11-7 note the cost range of a class 100 VLF rooms, and for the size quoted, the cost would be at the low end of the range. In this case a class 100 room would quite probably be preferable to the lesser room with clean benches—particularly since the cleanliness level would be class 100 throughout.

### floor space requirements

Floor space has value both from the depreciation and operating costs of the factory building and from the potential value of the product manufactured thereon. All other things being equal, clean facility designs are to be preferred which occupy less total floor space for the same amount of necessary clean operating space. Clean work stations only treat the work space area and not work aisles. Thus, particle control is limited to the critical points of need rather than all the associated space as occurs with clean rooms. Conversely, many clean rooms contain nonsensitive or even dirt producing operations because they cannot be separated physically from those processes requiring particle control. Some clean room designs also require greater support space than others of the same control level due to different equipment arrangements. In any event in comparing two design approaches, the amount of floor space required and its value should be included.

### operating costs

Operating costs due to process affected items will vary little regardless of the facility design type. However, operating costs for air handling, filter replacement, housekeeping, personnel control, and ownership will be different depending on the facility type. No average figure can even be presented here since such costs will vary from location to location and company to company.

### clean environment planning

Planning the procurement and use of clean environment facilities requires a systematic analytical approach. Emphasis must be placed on the compatibility of processes, equipment, operating procedures, and maintenance techniques to achieve economically the product manufacturing requirements. Thus, there is an order of procedure inherent in the selection of the type of facility to be provided:

1. Outline process steps and analyze cleanliness level needs.
2. Analyze present and future programs to establish the extent of process equipment and space required.
3. Compare environmental design appropriateness:
   a. Artificial vs. natural atmosphere systems.
   b. Local work stations vs. area control.
   c. Mixed vs. laminar flow systems.

4. Relate to product requirements:
   a. Product flow vs. exposure time to contaminants.
   b. Total costs vs. product risks.
5. Establish performance levels and design types and develop specifications for procurements.

The preparation of adequate facility specifications for procurement cannot be over-emphasized. Appendices A and B are included on equipment purchase specifications as indicators of the specific details which have proven to be important. Appendix A is a purchase specification for a vertical laminar flow room; Appendix B is for laminar flow work stations.

The clean facility, once erected or installed, deserves careful and extensive certification to ensure its meeting performance requirements. Of major importance here is the leak-free nature of the final filter installation and the air distribution characteristics. Other factors which require certification include lighting, noise, vibration, temperature, and humidity control. Verification procedures for vertical laminar flow rooms referenced to AACC Standard CS-6T. are included in the purchase specification of Appendix A. Certification of ultra-clean work stations should be performed in accordance with AACC Standard CS-2T. Standard test procedures do not exist for the other facility categories; however, verification by similar methods is advisable. For all facility types no standards exist to certify gaseous contamination control with activated carbon filtration. Here guidance by carbon filter manufacturers or experienced users should be sought.

Continued operation in keeping with the performance level provided is necessary. Proper personnel behavior and housekeeping is imperative (Chap. 13), periodic monitoring of the equipment performance will detect degradation of the clean environment (Chap. 16).

## summary

This chapter has examined the performance evaluation criteria and specifications of the major facility designs for clean environment control. The construction features and operational guidelines of laminar and mixed flow clean rooms have been outlined. The characteristics of dust control hoods, laminar flow work stations, and special atmosphere systems have been treated in a similar manner for environmental particle protection of individual work positions. Gaseous contamination control concepts have been presented through the use of dilution mathematics and the application of activated carbon filtration.

With these design parameters the economic factors of acquisition and operation can be evaluated as shown and a planning approach made for optimum selection, design, procurement, and verification of clean environment control facilities.

# references

1. W. J. Whitfield, "A New Approach To Clean Room Design," Sandia Corporation Report SC-4673 (RR), Office Of Technical Services, Dept. of Commerce, Washington, D.C., March, 1962.
2. Federal Standard No. 209, "Clean Room and Work Station Requirements, Controlled Environment," General Services Administration, Washington, D.C. (latest issue).
3. U.S. Air Force Technical Order 00-25-203, "Standards and Guidelines for the Design and Operation of Clean Rooms and Clean Work Stations," Office of Technical Services, Dept. of Commerce, Washington, D.C., July 1963.
4. AACC Standard CS-1T, "Tentative Standard for HEPA Filters," American Association for Contamination Control, Boston, May, 1968.
5. AACC Standard CS-2T, "Tentative Standard for Laminar Flow Clean Air Devices," American Association for Contamination Control, Boston, May, 1968.
6. AACC Standard CS-6T, "Tentative Standard for Testing and Certification of Particulate Clean Rooms," American Association for Contamination Control, Boston, 1970.
7. P. R. Austin, "Austin Contamination Index," Fourth Annual Technical Meeting and Exhibit, American Association for Contamination Control, Miami, Fla., May, 1965.
8. Contamination Control Handbook, NASA CR-61264, Clearing House for Federal Scientific and Technical Information, Springfield, Va., 1969.
9. P. R. Austin and S. W. Timmerman, Design and Operation of Clean Rooms, Business News Publishing Co., Detroit, 1965, Chap. 6.
10. A. J. Agacan, Unpublished Data from Western Electric Co., New York, N.Y.
11. B. J. Hollod, Unpublished Data from Western Electric Co., Allentown, Pa.
12. W. V. Pink, R. C. Whiteman, and F. J. Pordan, "Inspection, Maintenance, and Monitoring of Ultra-Clean Work Stations," The Western Electric Engineer, April 1967, Vol. XI, No. 2, pp. 39-45.
13. W. V. Pink, "Inspection, Maintenance, and Monitoring of Ultra-Clean Work Stations," Sixth Annual Technical Meeting and Exhibit, Washington, D.C., May, 1967.
14. J. C. Washburn and R. T. Sylvester, "Activated Charcoal Filtration, a Supplement to HEPA Filtration in Clean Rooms and Work Stations," Sandia Laboratories Report SC-DR-69-316, Sandia Corporation, Albuquerque, N.M., June 1969.
15. C. L. Fraust, "The Adsorption of Low Concentrations of Aliphatic Acetate Vapors onto Activated Carbon" (PhD Dissertation), Northwestern University, Evanston, Ill., 1969.
16. C. L. Fraust, Unpublished data from Western Electric Co., Allentown, Pa., 1970.

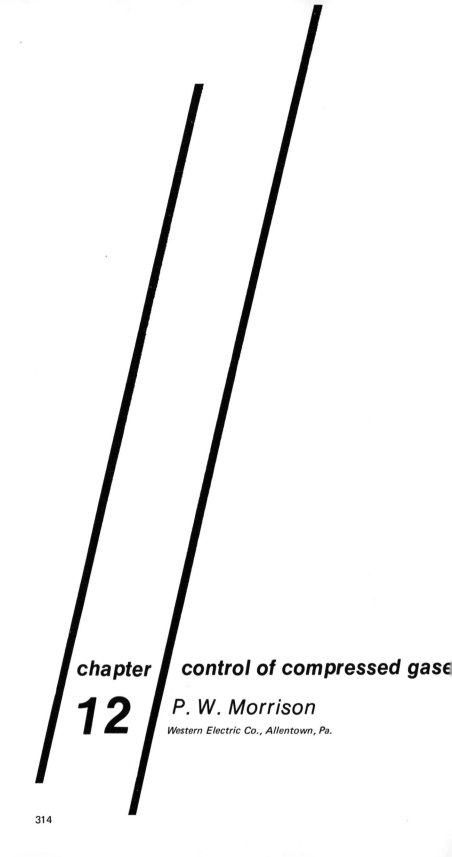

chapter

**12**

*control of compressed gases*

*P. W. Morrison*

Western Electric Co., Allentown, Pa.

One of the characteristics of microelectronic manufacture is the process use of many different compressed gases of high purity. The consumption of gases such as hydrogen, oxygen, nitrogen, and compressed air can reach gigantic levels. Natural gas and liquid nitrogen are also used in sizable quantities as sources of heat and coldness while gases such as argon, helium, silane, and diborane are used in more limited volumes.

This chapter will point out the need for high purity compressed gases, the likely forms of contamination, and the typical means of gas generation. The remainder of the chapter will emphasize the engineering design requirements to minimize contamination of the compressed gases during distribution and at the point-of-use.

## microelectronic usage

Compressed gases are usually required to provide a special processing environment, to act as a source of energy, or to serve as raw material ingredients in electron device fabrication. Interestingly enough, these process requirements also tend to parallel the consumption levels with environmental demands being the highest consumer while the raw material requirements would be analogous to spices in a food recipe. Although there are overlaps in these categories they will generally hold true.

It is easy to tie the high consumption of hydrogen, oxygen, and nitrogen to product environmental needs. All three gases are used for furnace environments for heat treatment of metal parts, with the choice of gas dependent on whether a reducing, oxidizing, or inert condition is required. Temperatures of $1100°C$ are often required for heat treatments of metal surfaces. At these temperatures impurities in the gas environments would tend to diffuse into the materials being processed. This causes requirements for gas purities of the highest practical level. High purity nitrogen is used in large quantities for both storage and processing environments where the presence of moisture and oxygen could cause tiny corrosion cells to form on production materials. In-process storage of clean silicon wafers and the encapsulation welding of metal-enclosed transistors are examples where the use of high purity nitrogen is a necessity. Figure 12-1 shows the typical history of process demand for high purity hidrogen, oxygen, and nitrogen at a microelectronics factory. The upward trend is indicative of both the growth of microelectronic production and the increasing demand for environmental protection during various process steps.

Compressed air is also used in copious quantities (1,000,000,000 cu ft/yr) both as an actuator of machinery and for environmental protection. The purity of the air is related only to the environmental requirements where particles, organic materials, and moisture content are of concern. Filtered air with less than 10 ppm of moisture is commonly used as an inexpensive alternative to nitrogen for silicon and thin film device storage environments.

Natural gas and liquid nitrogen are utilized in sizable quantities for process which require heat or coldness. Natural gas is commonly used   in

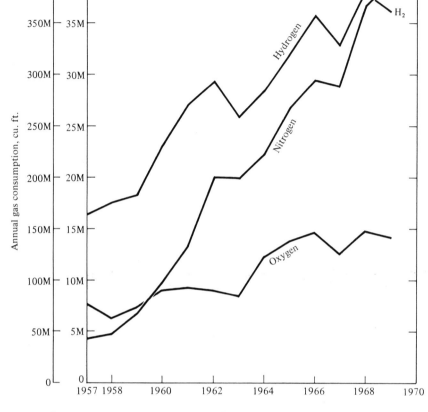

Figure 12-1  Typical process demand for nitrogen, hydrogen and oxygen.

the heat forming of glass and quartz parts for electron tube manufacture. Since the gas comes in direct contact with the glass where impurities can be adsorbed, it is important that the gas supply be of consistent quality.

There are situations when a gas as a source of energy does not come in direct contact with product parts. In these instances purity of the gas is unimportant. Liquid nitrogen fits into this category since it is chiefly used to cool sorption vacuum pumps. However, liquid nitrogen deserves mention in this text because of its common use in the microelectronics industry and the special design and safety problems caused by its cryogenic nature.

Compressed gases are also used at relatively small consumption levels as a raw material source for a process. For example, diborane $(B_2 H_2)$ is used as a source of boron to dope silicon into a p-type material. Silane $(SiH_4)$ is used as a source of high purity silicon for epitaxial deposition while hydrogen chloride gas is used in the same process as an etchant. Certain processes require environments which utilize the special characterisitcs of certain high purity gases. Argon is used as the bombardment material in vacuum sputtering of tantalum for thin film devices. Helium is used as the environment during the growing of single crystal silicon in the Czochralski process. In all cases from diborane to helium special care must be exercised to ensure that the desired purity is delivered to the production processes in a safe manner.

## types of contamination

The contamination present in compressed gases can be particles or chemical residues of a solid, liquid, or vapor form. These residues may result from the quality of gas generation and compression, or a high purity gas may be degraded by the pipe distribution system while in transit to the point of use. Both sources of contamination should be considered in planning the use of a high purity compressed gas system.

Gas purity is really a function of how much you are willing to pay to avoid contamination. Table 12-1 indicates the typical quality grades of

**TABLE 12-1.   Typical Grades of Nitrogen Gas**

| CGA Designation | E | G | H | L | N |
|---|---|---|---|---|---|
| *Nitrogen Minimum, % v/v | 99.5 | 99.95 | 99.99 | 99.998 | 99.9985 |
| Water, ppm | 26.3 | 26.3 | 11.4 | 3.5 | 1.5 |
| Dew Point, $^\circ$F | -63 | -63 | -75 | -90 | -100 |
| Total Hydrocarbons (as ppm methane) | 58.3 | | 5 | | 1 |
| Oxygen, % or ppm | 0.5 | 500 | 50 | 10 | 1 |
| Hydrogen, ppm | | | | | 1 |
| Argon, Neon, Helium, ppm | | | | | 5 |
| Carbon Monoxide, ppm | | | | | 1 |

*Unless shown otherwise % $N_2$ includes trace quantities of neon and helium and small amounts of argon.
Source: CGA Specification G-10.1, "Commodity Specification for Nitrogen," Compressed Gas Association.

nitrogen available from commercial suppliers. Note that these specifications concern the gaseous organic and inorganic residues which are controlled during generation and compression. While the cost for an individual gas cylinder generally increases with the grade of purity, the cost differentials will change when gases are purchased in bulk quantities. Usually, gas of the highest quality that can be purchased or generated in large quantities is specified by the microelectronics industry. Table 12-2 illustrates typical compressed gas specifications used by the microelectronics industry.

Once a source of high purity compressed gas is obtained it is often assumed that the gas arriving at the points of use is of the same quality. Unfortunately, this is not necessarily so. Large scale consumption implies a complicated maze of pipes feeding all parts of a factory. Such a pipe system is

TABLE 12-2.  Typical Specifications for High Purity Gases

| Impurities max. % v/v | $H_2$ | $O_2$ | $N_2$ | $He$ | $Ar$ |
|---|---|---|---|---|---|
| Argon | ⎫ | ⎫ | 0.01 | * | 99.95 min. |
| He | ⎬ 0.15 | ⎬ 0.45 | ⎫ | 99.98 min. | 0.03 |
| N2 | ⎪ | ⎪ | 99.985 min. ⎬ | | 0.02 |
| Inert Gases | ⎭ | ⎭ | * ⎭ | 0.01 | — |
| H2 | 99.85 min. | 0.04 | 0.002 | 0.0005 | 0.0005 |
| O2 | 0.001 | 99.5 min. | 0.002 | 0.0010 | 0.0005 |
| Water (dew point) | -85°F | -85°F | -85°F | -85°F | -85°F |
| Organics, CO2, CO (as CO2) | 0.0005 | 0.0005 | 0.0005 | 0.0005 | 0.0005 |

*Included in 0.01% limit.

frequently modified to provide gas services to new or altered work positions. There is great potential for contaminating a pipe distribution system by leaks at faulty joints and valves. Particles, pipe dope, and dirty pipe can be introduced during system modifications via poor pipefitting practices. Cross-connection of different gas services at process equipment can allow back diffusion of one gas into the other gas distribution system. The desiccants in gas dryers have been known to fracture from pressure shocks—generating fine particles which contaminate the process piping. Filters fail, depositing their collected load of particles or adsorbed gases back into the system. The reliability of an extensive gas distribution system is maintained only through good engineering specification of materials, pipe fitting procedures, and fail-safe contamination control designs plus good supervision of construction and periodic monitoring of the gas quality at the point of use.

## generation of high-purity gases

The generation of high purity compressed gases commonly used in the microelectronics industry is accomplished mainly by: (1) catalytic steam-hydrocarbon reforming plus refinements to produce hydrogen; (2) liquefaction of air to produce oxygen, nitrogen, and argon; and (3) liquefaction of certain natural gas supplies to produce helium. A brief summary of these processes will be given to illustrate the reliability of gas purity and the types of contaminants which have been removed.

### hydrogen

The generation of hydrogen by the catalytic steam-hydrocarbon reforming process involves the reaction of steam with hydrocarbon over a nickel catalyst at 1200-1800°F to produce carbon oxides and hydrogen. Either gaseous hydrocarbons (such as methane and ethane) or others with moderate vapor pressures (such as propane or butane) are commonly used. Purities of greater than 99.9 percent have been achieved by the process, but a 95 percent pure hydrogen is the usual output when further refinement is required.

The further purification of the hydrogen involves dehydration, cooling, and then scrubbing at $-180°C$ and 300 psig with liquid methane to remove $N_2$, CO, and Ar.[1-4] Residual hydrocarbons are removed by absorption in liquid propane followed by a sorption bed of activated carbon, silican gel, etc., to remove the trace impurities. Table 12-3 shows the purification levels achieved by each process and the quality of the final product.

TABLE 12-3   Properties of Process Streams

|  | Feed | Methane Absorber | Propane Absorber | Product |
|---|---|---|---|---|
| Pressure, psig | 300 | 298 | 297 | 295 |
| Temperature °C | -180 | -180 | -186 | -185 |
| Composition* |  |  |  |  |
| $H_2$ | 95.6 | 98.996 | 99.99 | 99.9999 |
| $N_2$ | 0.4 | 30 ppm | 30 ppm | |
| CO | 1.7 | 10 ppm | 10 ppm | < 1 ppm |
| $CH_4$ | 2.3 | 1.0 | 60 ppm | |
| $C_3H_8$ |  |  | neg. | |
| Total | 100.0 | 100.0 | 100.0 | 100.0 |

*In mole % unless otherwise indicated.
Adapted from "Hydrogen," Vol II of Kirk-Othmer Encyclopedia of Chemical Technology, Interscience, 1966.

An alternate refining technique utilizes the unique properties of palladium and hydrogen at elevated temperature. In the 575-750°F temperature range molecular hydrogen will disassociate into atomic hydrogen on a palladium surface, diffuse through the palladium, and reform into molecular hydrogen on the other side. Since hydrogen is the only molecule having this capability relative to palladium, the gas purity can be absolute if no mechanical defects exist in the palladium membrane. After removal of heavy hydrocarbons, hydrogen sulfide, or olefins, a feed gas containing 50 percent or more hydrogen is commonly refined into a highly purified hydrogen product.[5]

## oxygen, nitrogen, and argon

The liquefaction of air is used to manufacture pure oxygen and nitrogen— with argon, neon, krypton, and xenon being produced as by-products of the large commercial plants. Table 10-1 gives the typical constituents of the atmosphere. Most air separation plants are based upon: (1) air purification; (2) partial liquefaction of the air by regenerative heat exchange; and (3) separation into pure constituents by fractional distillation. Figure 12-2 shows the basic Linde cycle out of which has grown the cryogenic industry. Practical fractional distillation is more complicated than the figure shown since the generation of pure gas requires the discrimination between boiling points of

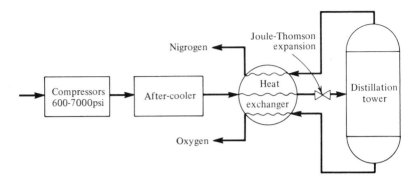

*Figure 12-2 Adapted from "Cryogenics," Vol. 6 of Kirk-Othmer Encyclopedia of Chemical Technology, Interscience, 1965.*

$-297.4°F$ for $O_2$, $-302.4°F$ for $Ar$, and $-320.4°F$ for $N_2$. Also, regenerative heat exchange involves the use of the cryogenic output product to cool several preceding processes, thereby achieving economical use of refrigeration energy. Water vapor and carbon dioxide are removed from the input air during the initial cryogenic steps while the hydrocarbon residues are removed in a sorption bed later on in the process chain. The resulting products typically can be 99.99 percent pure. The delivered purity level of these gases to a factory is more a function of transfer contamination from bulk containers, pipe lines, etc., rather than the purification process.

### helium

The manufacture of pure helium is achieved by liquefaction of natural gas containing at least 0.4 percent helium.[6] Certain natural gas wells may contain up to 8 percent helium. Like the liquefaction of air, natural gas is cooled to remove, in the descending order of their boiling points, water vapor, carbon dioxide, hydrogen sulfide, heavier hydrocarbons, methane, and finally nitrogen.

## high purity gas distribution

The distribution of gases like oxygen, hydrogen, nitrogen, and dry air in a microelectronics factory requires both a source of compression and an extensive grid network which are noncontaminating. The selection of the compressor is usually limited to the nonlubricated types to eliminate the possibility of oil entrainment into the gas stream. The concern for contamination can focus mainly on the piping grid. Since all of these gases are dry, there is a good possibility of moisture back diffusing into the gas stream through minute leaks in the pipe system. To ensure proper delivery of the high quality gas to the point of use, one possible choice would be to pipe directly from a tank supply to the use point with a welded pipe system which has no valves, joints, pressure regulators, etc., to become potential sources of leaks. Such a closed system would be inflexible under the circumstances of changing processes and equipment locations. Valves are needed to isolate sections of the pipe grid for either system repair or relocation of piping to equipment. Welded pipe is expensive compared to other means of joining pipe. Pressure regulators are essential to stabilize the gas delivery at a given pressure. A practical pipe grid must ensure both flexibility and reliability of design. As will be shown, this point affects the selection of construction materials and the installation techniques.

### construction materials

Pure nitrogen, hydrogen, oxygen, helium, and compressed dry air are non-corrosive and can be contained in any of the common metals. Oxygen does require the special precaution that all oil, grease, or other combustible material must be removed from pipe interiors because ignition temperature is lowered in the presence of pure oxygen. Black steel pipe with malleable iron threaded fittings can be considered as acceptable construction material from a strength, corrosion, and cost viewpoint. Silver soldered, hard copper pipe and fitting systems are preferred from a contamination control viewpoint to reduce the potential of joint leaks. (Hydrogen leaks also constitute a fire hazard.) Threaded joints are much more dependent upon the quality of construction labor for leak tightness than are soldered joints or special tubing connection systems. The latter have been used at final connection points and for piping headers behind bench rows to simplify equipment installation.

Special material selection also applies to such components as flowrators, pressure regulators, and valves. Diaphragm valves using teflon diaphragms minimize the leak and contamination potentials. Other valve types should have teflon valve seals and stem packing to reduce the generation of wear particles. Pressure regulators with stainless steel diaphragms are preferred over the more common neoprene diaphragm types.

In addition to the type of valve and regulator design the delivered state of cleanliness should be considered. A common requirement is for such equipment to be "cleaned for oxygen service," which is defined by the Compressed Gas Association by the following steps:

"1. Selection of a suitable cleaning agent.

  2. Removal of contaminants by a good cleaning method. These contaminants include oil, grease, threading compounds, flux, and other foreign material. Also removed are noncombustible metal particles and powders, such as weld metal, chips, and filings, because by impact and friction these particles may ignite nonmetallic parts and even stainless steel which are normally safe and are used in certain parts of oxygen equipment.

  3. Removal of all residual cleaning agent and the thorough drying of equipment with oil-free air or nitrogen.

  4. Inspection of the equipment for cleanliness.

5. Protection of cleaned parts and equipment from recontamination. Cleaned parts are stored and assembled in areas free from oil mist from other operations, dropping lubricants from cranes, and similar contaminant conditions. Gloves, tools, and slings free from grease are used for handling clean equipment. Parts temporarily stored on floor areas are wrapped or covered with clean paper."[7]

Supplementing this requirement, all valves, regulators, flowrators, etc., should be purchased with openings capped or plugged, and small items should be sealed in plastic bags to preserve the cleanliness of the assemblies.

### assembly techniques

Proper planning and piping material selection must be supported by special care in pipe assembly to ensure the purity of the delivered gas. Normal pipe-fitting practices ensure that the pipes are adequately assembled and that the gas distribution system will operate within the indicated pressure and temperature specifications. Deviations from good pipefitting practices, often tolerated in industrial and housing construction, simply are unacceptable where high purity compressed gases are involved. Perhaps the best way of illustrating what commonly occurs is to cite some examples.

The pipe as received from the mill has residues of dust, scale, drawing oils, oxides, etc. Most pipefitting specifications require some form of cleaning to remove this debris. A detergent or solvent wash plus swabbing is often required. Such regulations are not always adhered to. Sometimes after cleaning the pipe it is dropped on the floor where dust can redeposit on the interior walls. The end result is a dirty pipe installed to distribute a high purity gas to the point of use.

Because such situations are commonplace, it becomes necessary to plan accordingly. Copper tubing, valves, and fittings should be purchased already degreased for oxygen service and capped at both ends. A degreasing station should be made available to the pipefitters for the cleaning of pipe contaminated during cutting or assembly operations. Instructions on the proper techniques of pipe cleaning via degreasing and close supervision during assembly will reinforce the importance of clean pipefitting practices. Finally, a purge with highly pure nitrogen often is specified to attempt removal of any debris which is inadvertently introduced into the system. This flushing will also dehydrate the pipe system before introduction of the process gas.

The proper joining of pipes is another case where extra care during assembly is important. Cutting oil, particles, and pipe dope deposited inside the pipe at threaded joints and the greater potential for leaks in this type of joint make soldered joints more desirable. Note that contamination is possible even with soldered joints. The flux should be applied to the pipe—not the fittings. Overfluxing to the point where it runs to the end of the pipe should also be avoided. Nitrogen gas can be passed through the pipe during silver soldering to prevent formation of oxides inside the pipe. After assembly the system should be pressure-tested to 100 psig or 1½ times operating pressure, whichever is greater, to detect joint leaks. Rapping soldered joints a few times during the pressure test should cause flux seals to show.

If we assume that the previous examples have established a pipe grid of high integrity, we can appreciate the need for care when reopening the grid for modifications and additions. Detailed re-entry procedures should be prepared in a step-by-step format which the pipefitters can understand and put into practice. Training programs and adequate supervision of work will help motivate workers to appreciate their role in contamination control of the pipe system. These procedures should also be explicit on the ways to avoid exposure to safety, fire, and health hazards which are associated with compressed gas pipe systems.

### gases in cylinders

Many compressed gases are used in small volume and do not warrant a pipe distribution network. Cylinders of various types and sizes provide an adequate gas supply. Figure 12-3 shows the variety of shipping containers for

*Figure 12-3 Compressed gas cylinders. (Courtesy Pressed Steel Tank Co., Inc.)*

*Figure 12-4 Cryogenic containers. (Courtesy Union Carbide Corp.)*

compressed gases while Fig. 12-4 shows containers used for cryogenic materials.

While gas purity is an inherent result of most generation processes, subsequent transfers of bulk gas from storage reservoirs through compressors to tank trucks and gas cylinders cause the resultant contaminants in the gas delivered to a factory. Such problems can be minimized during cylinder or tank changes by making leak-tight connections and purging lines which have been opened to the atmosphere. These rules apply equally to the gas vendor when filling the cylinders and the industrial user at the point of consumption. The very simplicity of these rules to avoid contamination makes cylinder gases the most reliable form of high-purity gases. The fact that such obvious regulations must be constantly reiterated indicates that forgetfulness and carelessness are common human characteristics.

### hazard potential

The basic purpose of compression or liquefaction of gases is to concentrate the substance for later release or to store energy in the form of pressure or coldness. The effects of poisonous, flammable, or oxidizing gases thus become more severe in this concentrated state. The mishandling of the stored energy can likewise be extremely hazardous. The sudden release of 2000 psi cylinder pressures will cause an unsecured cylinder to thrash about like an errant rocket, knocking down anything or anyone in its path.  Similarly, accidental contact with cryogenic gas will cause a freezing injury similar to a burn.  These examples illustrate the hazard potential of compressed or liquefied gases and the importance of proper training of people who utilize such materials.  It is beyond the scope of the chapter to discuss the many safety precautions associated with the safe handling of cylinder and cryogenic gases.  The many pamphlets of the Compressed Gas Association listed in Table 12-3 are excellent references on the specific details of proper gas handling.

TABLE 12-4  Compressed Gas Association Pamphlets

| Pamphlet No. | Title |
| --- | --- |
| C-1 | Methods for Hydrostatic Testing of Compressed Gas Cylinders. |
| C-2 | Recommendations for the Disposition of Unserviceable Compressed Gas Cylinders. |
| C-3 | Standards for Welding and Brazing on Thin Walled Containers. |
| C-4 | American Standard Method of Marking Portable Compressed Gas Containers to Identify the Material Contained, Z48.1-1954. |
| C-5 | Cylinder Service Life—Seamless, High-Pressure Cylinder Specifications ICC-3, ICC-3A, ICC-3AA. |
| C-6 | Standards for Visual Inspection of Compressed Gas Cylinders. |
| C-7 | A Guide for the Preparation of Labels for Compressed Gas Containers. |
| C-8 | Standard for Requalification of ICC-3HT Cylinders. |
| G-1 | Acetylene. |
| G-1.2 | Recommendations for Chemical Acetylene Metering. |
| G-1.3 | Acetylene Transmission for Chemical Synthesis. |
| G-1.4 | Standard for Acetylene Cylinder Charging Plants. |
| G-2 | Anhydrous Ammonia. |
| G-2.1 | American Standard Safety Requirements for the Storage and Handling of Anhydrous Ammonia, K61.1. |
| G-3 | Sulfur Dioxide. |
| G-4 | Oxygen. |
| G-4.1 | Equipment Cleaned for Oxygen Service. |
| G-4.2 | Standard for Bulk Oxygen Systems at Consumer Sites. |
| G-4.3 | Commodity Specification for Oxygen. |
| G-5 | Hydrogen. |
| G-5.1 | Standard for Gaseous Hydrogen at Consumer Sites. |
| G-5.2 | Standard for Liquefied Hydrogen Systems at Consumer Sites. |
| G-5.3 | Commodity Specification for Hydrogen. |
| G-6 | Carbon Dioxide. |
| G-6.1T | Tentative Standard for Low Pressure Carbon Dioxide Systems at Consumer Sites. |
| G-7.0 | Compressed Air for Human Respiration. |

**TABLE 12-4 (continued)**

| Pamphlet No. | Title |
|---|---|
| G-7.1 | Commodity Specification for Air. |
| G-8.1 | Standard for the Installation of Nitrous Oxide Systems at Consumer Sites. |
| G-9.1 | Commodity Specification for Helium. |
| G-10.1 | Commodity Specification for Nitrogen. |
| G-11.1 | Commodity Specification for Argon. |
| OA-1 | Oxy-Acetylene Cutting. |
| OA-2 | Oxy-Acetylene Welding and Its Applications |
| OA-4 | Braze Welding of Iron and Steel by the Oxy-Acetylene Process. |
| OA-5 | Safe Practices for Installation and Operation of Oxy-Acetylene Welding & Cutting Equipment. |
| OA-6 | Flame-Hardening by the Oxy-Acetylene Process. |
| OA-7 | Hard-Facing by the Oxy-Acetylene Process. |
| OA-8 | Carbide Lime—Its Value and Its Uses. |
| P-1 | Safe Handling of Compressed Gases. |
| P-2 | Characteristics & Safe Handling of Medical Gases. |
| P-2.1 | Standard for Medical-Surgical Vacuum Systems in Hospitals. |
| P-2.3T | Standard for Hyperbaric Facilities Intended for Use in Medical Application. |
| P-3 | Standards for Solid Ammonium Nitrate (Nitrous Oxide Grade) |
| P-4 | Safe Handling of Cylinders by Emergency Rescue Squads. |
| P-5 | Suggestions for the Care of High-Pressure Air Cylinders for Underwater Breathing. |
| P-6 | Standard Density Data, Atmospheric Gases and Hydrogen. |
| S-1.1 | Safety Relief Device Standards—Cylinders for Compressed Gases. |
| S-1.2 | Safety Relief Device Standards—Cargo and Portable Tanks for Compressed Gases. |
| S-1.3 | Safety Relief Device Standards—Compressed Gas Storage Containers. |
| S-3 | Frangible Disc Safety Device Assembly. |
| S-4 | Recommended Practice for the Manufacture of Fusible Plugs. |
| V-1 | American-Canadian Standard Compressed Gas Cylinder Valve Outlet and Inlet Connections; ASA-B57.1; CSA-B96. |
| V-5 | Diameter-Index Safety System. |
| SB-1 | Hazards of Refilling Compressed Refrigerant (Halogenated Hydrocarbons) Gas Cylinders. |
| SB-2 | Oxygen Deficient Atmospheres. |

Source: Compressed Gas Association, Inc., New York, N.Y.

# point of use control

If the preceding design intent for high purity gas generation and distribution were carried out every time in minute detail, there would be no need for additional control at the point of process use. However, people and equipment do not always perform as expected. When gas purity must be guaranteed to protect electronic product from contamination, scavenging equipment to collect particulate or vapor contaminants is often required. The potential for generating particles in the gas distribution system is so high that particulate filters at the point of use are mandatory for most microelectronic applications. To a lesser degree gas dryers and purifiers ensure a constant gas supply of the highest quality.

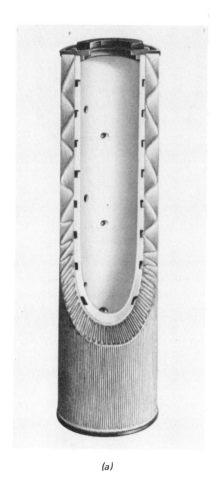
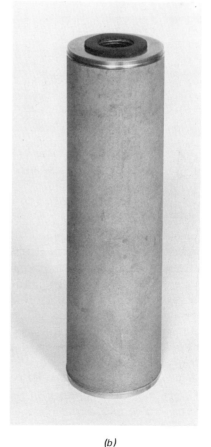

(a)                                   (b)

*Figure 12-5  Filters for compressed gases.  (Courtesy Pall Trinity Micro Corp.)*

*(a) Inert cartridge: Inorganic fibers in organic binder*
*(b) Sintered cartridge*

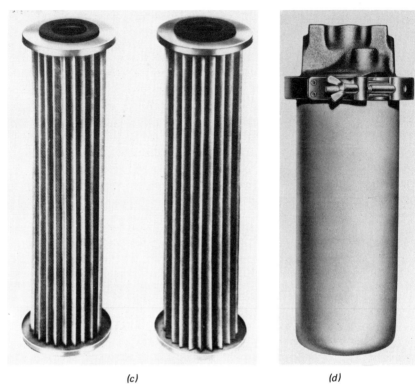

*(c)*                                    *(d)*

*Figure 12.5  continued*
*(c) Sinter woven wire mesh cartridge*
*(d) Filter housing.*

## particulate filters

Compressed gas filters conform to the principles of particle filtration discussed in Chap. 10 and can be classified as depth filters, screen filters, or combinations thereof (Figs. 12-5, 12-6, and 12-7). Depth filters usually consist of a random mat of fibers creating tortuous flow paths which ensnare particles as the gas flows through the media. A different approach is the sintered filter, whereby a porous solid is created by powder metallurgy techniques. In either case the pore size is not specifically established. By design intent depth filters have high dust holding capacity along with the potential for high collection efficiency.

Screen filters can be made with metal, cloth, or polymeric materials arranged in a given geometry to achieve a given pore size within specified manufacturing tolerances. The construction may vary from coarsely woven wire for an inexpensive strainer to precision etching of pores (down to 15 microns) using photolithographic techniques. Since the pore size is specified in a screen filter, an absolute relationship for particles greater or equal to that pore dimension is inherent: Generally screen filters are limited in dust holding capacities because of the surface loading effects.

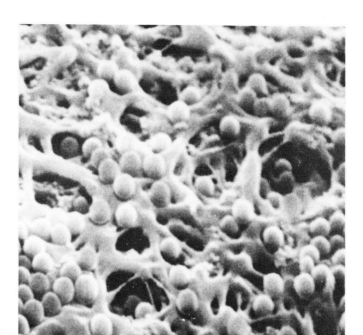

(a)

*Figure 12-6  Membrane filters. ((b) Courtesy Gelman Instrument Co.)*

*(a) Polystyrene latex spheres on membrane filter (10,000✕).*

The membrane filter is a special form of screen filter which is used extensively to filter gas and other fluid systems in the electronics industry. Made by the gelling and drying of colloidal solutions, membrane filters are porous (void volume of 80–85 percent), and their pore sizes range from 10 microns down to 0.01 micron with approximately 5 percent variation from mean pore size.[8,9]  Membrane filters are constructed for use in both flat and pleated forms as are most screen filters.

Gas line filters should be rated for particle retention efficiency just as is required for air conditioning type filters. Some filters (Fig. 12-7) warrant the HEPA filter designation since a 99.98 percent retention efficiency on 0.3 micron particles can be guaranteed by the DOP Test (see Chap. 10). Many screen filters can guarantee an absolute retention of particles above a specified size because the manufacturing process is precise in establishing pore size.

The bulk of compressed gas filters are rated according to some form of the bubble point test (ASTM Method D2499-66T[10]). In this method a filter is wetted completely with fluid. Then air pressure is gradually applied through the filter until bubbles begin to form on the filter surface. The air pressure value at that point is used to calculate the maximum pore size using capillary rise relationships. Additional application of pressure can be used to

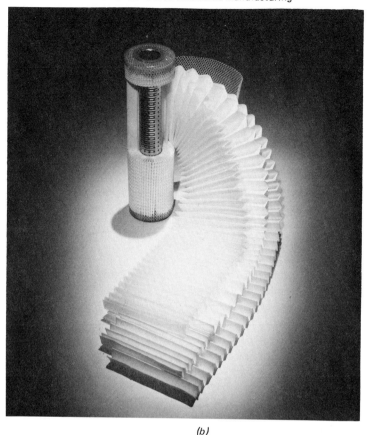

*(b)*

*Figure 12-6 continued*

*(b) Pleated filter.*

measure smaller pore sizes and hence, the pore size distribution. The maximum pore size establishes an absolute retention value as a means to guarantee operating performance. Any pore size less than the maximum would have a nominal efficiency which is less than 100 percent.

*Figure 12-7 Combined particle and vapor filter assembly. MSA filter showing sandwich of activated carbon unit between particulate filter cartridges. (Courtesy Mine Safety Appliances Co.)*

While the different tests measuring retention efficiencies are technically well founded there is little information on the comparison between methods. One manufacturer evaluated a 0.45 micron membrane filter by the DOP test at a 5.3 cm/sec face velocity and achieved a retention efficiency of 99.9995 percent.[11] This experiment only applies to one case among the many designs for screen and depth filters. Many filter manufacturers do not state what filtration efficiency technique has been used for performance certification nor is information readily available on the production sampling basis for claims of high retention efficiency. One experiment using light scattering measurement techniques[12] has shown that filters rated by the DOP test demonstrate excellent particle retention characteristics without the pressure drop limitations inherent in membrane and other screen filters.

### vapor scavengers

The need for further refinement of high purity gases at the point of processes may be necessary to: (1) provide a factor of safety against pipe leaks or other contaminating mechanisms common to gas distribution; (2) remove a trace impurity, such as residual hydrogen or oxygen in nitrogen, which might be acceptable to other bulk users of the process gas; and (3) recycle gases whose intrinsic value is high or where large volumes would otherwise be consumed. Because scavenging of vapors can be expensive the equipment designs generally are tailored to collect a specific contaminant residual in the process gas.

Scavenging methods usually depend upon either sorption techniques or chemical reactions in the presence of a catalyst. The sorbents, the catalytic process, and the complexity of the scavenging equipment will vary from a simple sorbent bed to a completely automatic system of purification, regeneration, and instrumentation. One technique sandwiches an activated carbon cartridge between two particle filter elements as an inexpensive option to particulate air filtration (Fig. 12-7).

It should be noted that the breakthrough limitations cited in Chap. 10 will apply to sorbent filters which do not have a regeneration or replacement cycle.

Moisture can be controlled at the point of use with activated alumina, silica gel, or molecular sieve desiccants in gas dryers with automatic reactivation provisions.

Where absolute control over a gas impurity is specified, by it hydrocarbons or any residual impurity in a process gas, catalytic chemical reactors are used as a preliminary step to convert the contaminant into a form whereby the resultant products can be readily absorbed. Various noble metal catalysts are used to oxidize organic vapors into $CO_2$ and $H_2O$ before absorption, and similarly, oxygen in hydrogen is converted into $H_2O$ before absorption in a desiccant column. The type of catalytic reactor will vary with the impurity to be removed. Figure 12-8 shows a small gas dryer/purifier and its various components.

*Figure 12-8  Point-of-use gas dryer/purifier.  US dynamics dryer with cover removed. (Courtesy United States Dynamics.)*

One precautionary note is that scavenging dryers and purifiers can become a mixed blessing to any process.  A properly installed unit can indeed provide excellent control over the common impurities as shown in Table 12-5; however, the addition of a scavenging dryer or purifier can also cause problems.  Failure of timers, improperly functioning solenoid valves, pipe leaks within the unit, and cross contamination of gas supplies due to back diffusion are examples of what can happen.  Most of these failures are due to improper installation.  The equipment itself is generally very reliable.  Nonetheless, it is important to recognize that gas dryers and purifiers can, under certain circumstances, have a contaminating rather than a beneficial influence on com-

TABLE 12-5  Typical Control by
Impurity Scavengers

| Impurity | |
| --- | --- |
| Moisture | < 1 ppm |
| Hydrocarbon vapors | < 0.1 ppm |
| Oxygen | < 1 ppm |
| Hydrogen | < 1 ppm |

pressed gas quality. The manufacturer's recommendations on proper installations should be taken seriously.

## monitoring

There is a saying made popular by the aerospace program which makes the following point: If something can fail, it will fail! Guaranteed performance of gas quality at various points of use requires a program of periodic measurements to detect unwanted particles, organic vapors, moisture, and other contaminants. The sample size and frequency of measurement should be based on the degree of risk of poor product yield or reliability. The actual measurement techniques are explained in Chap. 16 on production monitoring and in Chaps. 3, 4, and 5 on contamination analyses. The point here is to ensure that provisions for monitoring compressed gases are incorporated into installation design at the point of use.

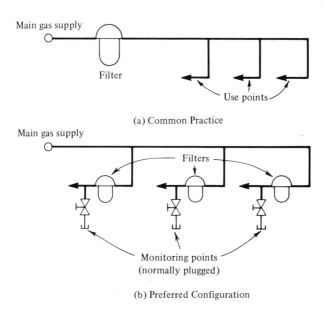

Figure 12-9 Point-of-use filter installation.

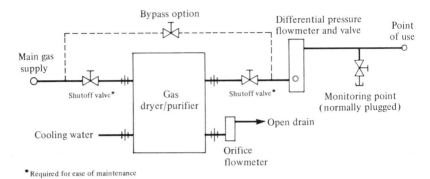

Bypass option

Differential pressure flowmeter and valve

Point of use

Main gas supply

Shutoff valve*

Gas dryer/purifier

Shutoff valve*

Monitoring point (normally plugged)

Open drain

Cooling water

Orifice flowmeter

*Required for ease of maintenance

Figure 12-10  Installation of a compressed gas dryer or purifier at point-of-use.

### point-of-use installations

Installation techniques of gas distribution systems are generally part of good engineering practice. Figures 12-9 and 12-10 illustrate important common sense points of designs at process stations which are frequently overlooked. When providing contaminant scavenging facilities they should be installed in a way such that: (1) they are as close as possible to the point of use; (2) they can be easily maintained or replaced; and (3) sample point plumbing is included to allow for monitoring of their performance.

Another precautionary note concerns the use of the sampling points for monitoring of particles. A light scattering particle counter is not designed normally to function at high gas pressures. If such instruments are used for monitoring, the sampling points must be installed such that a maximum of 2 to 3 psig is exerted on the instrument.

### summary

The use of many different compressed gases of high purity has become a distinguishing characteristic of microelectronics manufacturing, with volume requirements being highest for special processing environments. The means of generating high purity gases are well established, but the gases tend to be contaminated during distribution to the point of use by particulate and chemical contaminants introduced by pipe leaks, improper cleaning of piping, and non-adherence to good pipefitting practices. The selection of more expensive piping materials is recommended to simplify the installation routine and thereby reduce the potential for human error. To guarantee the delivery of high purity gases, point-of-use particulate filtration is essentially mandatory while vapor scavenging may also be required. Monitoring for contaminant in compressed gases at the point of use is desirable as the final insurance of gas purity.

# references

1. C. R. Baker and R. S. Paul, *Chem. Eng. Progr.*, **59** (8), 61–64 (1963).
2. U.S. Pat. 3,073,093 (1963), to Union Carbide Corp.
3. *Chem. Eng.*, **70** (10), 150–152 (1963).
4. *Chem. Eng. News*, **40** (20), 68–71 (1963).
5. "Hydrogen," *Kirk-Othmer Encyclopedia of Chemical Technology*, 1966, Vol. 11 Interscience, New York, pp. 366–368.
6. "Helium-Group Gases," *Kirk-Othmer Encyclopedia of Chemical Technology*, 1966, Vol. 10, Interscience, New York, p. 872.
7. Robert M. Neary, "Equipment Cleaned for Oxygen Service," Pamphlet G-4.1, Compressed Gas Assoc. New York, 1959, pp. 3, 4.
8. James L. Dwyer, *Contamination Analysis and Control*, Reinhold, New York, 1966, p. 263.
9. J. Pich, "Theory of Aerosol Filtration by Fibrous and Membrane Filters," in *Aerosol Science*, ed. C. N. Davies, Academic Press, New York, 1966, p. 273.
10. ASTM D2499-66T, "Pore Size Characteristics of Membrane Filters for Use with Aerospace Fluids," *1969 Book of ASTM Standards*, ASTM, Philadelphia, 1969, Vol. 18, pp. 656–662.
11. Gelman Instrument Co., Ann Arbor, Michigan 48106.
12. R. A. Yeich, Unpublished Data, Western Electric Co., Reading, Pa.

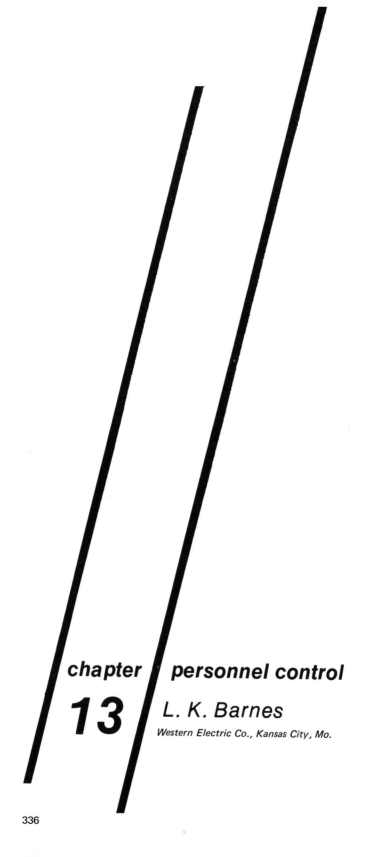

chapter **personnel control**

**13** *L. K. Barnes*
*Western Electric Co., Kansas City, Mo.*

Harmful effects on products may be the result of individual contributions or from large group activities related to factory routines. The distinction between individual and group generation of contaminants aids in the appropriate selection of personnel control techniques. The discussion on control techniques includes the general rules on personnel regulations, the logic behind the rules, and examples to illustrate their proper implementation.

It must be pointed out that much of this information represents a consensus of engineering judgment and established precedent rather than results from experimental data. Experimental data which can isolate and prove the benefits of personnel control are very difficult to collect. Many of the conclusions are derived from ideas which are logical to assume but difficult to prove. But this is also true of many other cases in the control of contaminants for product manufacture. Dependence on potential cost avoidance, estimated risk of failure, logic, or just plain common sense is a frequent basis for engineering judgment.

## contaminant sources

### the individual

It has been said that people are the dirtiest elements that are put in a factory. The human body is a constant source of contaminant emission. Contaminant types may be pieces of human hair and skin, droplets of moisture, lint and starch particles, fibers and frayed particles of clothing, cosmetics, fingerprints, and physiological problems.

The droplets of moisture can come from breathing, coughing, and perspiring; lint and starch particles can come from washable fabrics while fibers and frayed particles slough off wool, cashmere, or similar loose weave fabrics. Cosmetics include hair spray, aerosol deodorants, eye shadow, powders, etc.; and physiological contaminants are allergies to chemicals and other irritants in the work area, itching and scratching, illness, scabs from injuries, sunburn, and excessive nervousness. Other sources of personnel contaminants are personal belongings, food, food residues on clothes and body, and the various microbial organisms living in or on human beings.

There is general acceptance among clean environment designers that people contribute a wide variety of contaminants to the environment, but the quantity produced and its significance to product manufacture are sparsely documented. Austin[1] has developed the estimating guides shown in Table 13-1, and in a study concerning airborne bacteria Cown and Kethley[2] found that an average person in a laboratory can generate 8000 bacterial particles/min. These normal emissions can vary with the personal hygiene of individuals and the effects of off-time activities. Many other industrial investigations on individual emissions have been made on the benefits of protective clothing (smocks, hats, etc.) with the general result that most clean room users specify some form of clean room garments.

In a similar vein, the detrimental effect on product is logical to assume

TABLE 13-1.  Estimates of Personnel Particle Generation

| | Particles emitted/min ≥ 0.3 Micron |
|---|---|
| Standing or sitting-no movement | 100,000 |
| Sitting-light head, hand and forearm movement | 500,000 |
| Sitting-average body and arm movement, toe tapping | 1,000,000 |
| Changing position-sitting to standing | 2,500,000 |
| Slow walking     2.0 MPH | 5,000,000 |
| Average walking 3.5 MPH | 7,500,000 |
| Fast walking     5.0 MPH | 10,000,000 |
| Calisthentics | 15,000,000 to 30,000,000 |

Source: Philip R. Austin, "Austin Contamination Control Index," *Contamination Control*, Vol. V, No. 1, June 1966.

but difficult to prove. Western Electric has found chips of fingernail polish and eye shadow in products during failure mode analyses, but these direct correlations are only symptomatic of a larger, wide spread problem. Contamination from humans can be deposited onto product directly or indirectly via the water, solvents, acids, gases, and material handling equipment used in the manufacturing processes. The handling of containers will leave behind fingerprints, body oils, and moisture to contaminate the contents of the containers. Product contamination related to people can also result from process errors. Interestingly enough, it has been found that when employees do not perform their work in accordance with established procedures the product contamination level usually increases. Carrying particle sensitive product in uncovered containers in the factory environment or forgetting to turn on the clean bench blowers are common examples of improper job performance. Daily experience in practically any clean manufacturing area will provide other examples of the existence of these forms of personnel contaminants.

## mass activity

A study of personnel control and particle emission must take into account the mass activity of the factory routine. Documentation is useful here in exposing the contaminating effects of authorized personnel activity. The mass movements of people during shift changes, coffee or smoking breaks, lunch periods, etc., tend to increase the contamination level of the entire factory. This is illustrated in Fig. 10-3, where the time-dependent mass activity of the factory routine can be easily recognized.

This same problem is even more obvious in high grade clean rooms. Tables 13-2, 13-3, and 13-4[3] show the effect on particle count and time duration for a clean room capable of providing a class 1,000 environment (a combination clean room–clean bench facility). In this example the 1 minute particle counts measured during factory routine events have been defined as high activity levels; the other high counts due to unknown causes have been

TABLE 13-2.   High and Normal Activity Level Airborne Particles (Particles/ft³)

| Event | Time Duration | High Activity Count | Avg. | Normal Activity Count | Avg. | Class 1,000 Limits |
|---|---|---|---|---|---|---|
| μm | | ≥0.5   ≥5.0 | | ≥0.5   ≥5.0 | | ≥0.5   ≥5.0 |
| Shift 1 | | | | | | |
| Arrival | 7:00 to 7:19 AM | 700 | 90 | | | |
| Break (AM) | 9:07 to 9:25 AM | 730 | 140 | | | |
| Lunch | 11:05 to 12:05 AM | 650 | 260 | 150 | 17.0 | 1000 7.0 |
| Break (PM) | 1:49 to 2:13 PM | 1060 | 70 | | | |
| Departure | 3:11 to 3:31 PM | 1030 | 110 | | | |
| Shift 2 | | | | | | |
| Arrival | 3:33 to 3:55 PM | 950 | 190 | | | |
| Break 1 | 5:11 to 5:45 PM | 920 | 100 | | | |
| Lunch | 6:55 to 7:47 PM | 570 | 180 | 114 | 14.0 | 1000 7.0 |
| Break 2 | 9:51 to 10:09 AM | 630 | 190 | | | |
| Departure | 11:39 to 11:59 PM | 460 | 140 | | | |
| Shift 3 | | | | | | |
| Cleaning | 1:43 to 5:15 AM | 2700 | 190 | 54 | 13.0 | 1000 7.0 |

TABLE 13-3.   Time Duration of High Activity Routine

| Event | Authorized Time Minutes | High Activity Time Minutes | % Shift-High Activity Levels |
|---|---|---|---|
| Shift 1 | | | |
| Arrival | 5 | 19 | 27.6% |
| AM Break | 10 | 18 | |
| Lunch | 30 | 60 | |
| PM Break | 10 | 24 | |
| Departure | 5 | 20 | |
| TOTALS | 60 | 141 | |
| Shift 2 | | | |
| Arrival | 5 | 22 | 26.6% |
| 1 Break | 10 | 34 | |
| Lunch | 30 | 52 | |
| 2 Break | 10 | 18 | |
| Departure | 5 | 10 | |
| TOTALS | 60 | 136 | |

TABLE 13-4.   Frequency of High Activity Particle Counts

| Shift | High Particle Counts, # | | High Activity Particle Counts, # | | % High Activity Counts | |
|---|---|---|---|---|---|---|
| | ≥ 0.5 μm | ≥ 5.0 μm | ≥ 0.5 μm | ≥ 5.0 μm | ≥ 0.5 μm | ≥ 5.0 μm |
| 1 | 136 | 132 | 55 | 78 | 40.4% | 59.1% |
| 2 | 133 | 144 | 68 | 70 | 41.1% | 38.6% |

called random activity levels; the remaining particle counts have been defined as normal activity levels. Notice that the high activity routine increases the airborne particle level approximately 5 times over the normal activity levels. The duration of activity is more than 2 times greater than the authorized time periods, and the factory routine causes approximately 50 percent of all high particle counts witnessed in the clean room. Table 13-2 also shows the effect of housekeeping activity during the third shift. It is important to recognize that much of the personnel activity which contributes contamination is actually permitted or even mandated by established factory routines. Stringent control of mass activity may require extensive revisions of factory procedures using experimental data as the basis for judgment.

The effect of mass activity in a factory also is noticed in the performance of clean benches. It is common knowledge that while work is being performed in the clean bench air stream, airborne particles from the dirtier factory environment are inadvertently brought into the bench environment on the operator's hands and arms and are scattered by the worker's movements. Figure 13-1 shows the typical degradation of the clean bench environment as the average factory particle count level is increased. The "No Activity" curve shows the basic cleaning ability of a clean bench when no work operations are being performed. According to this study the average factory air condi-

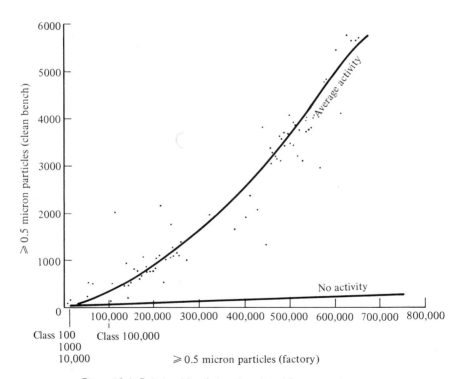

Figure 13-1 Relationship of clean bench and factory environments.

tions should be maintained at approximately 200,000 particles/cu ft to maintain a class 1,000 environment with a clean bench.[4]

## control techniques

Certain techniques have evolved to protect against personnel emissions, to regulate behavior that produces particles and other emissions, and to aid in the reduction of contamination from tools. For the imaginative and conscientious individual there is an endless list of personnel regulations and clean room rules that can be generated to ensure a required cleanliness. If all rules that could be listed were implemented and enforced, the clean room operation would be expensive, inconvenient, and time consuming, but probably very clean. There could even be an employee revolt against such restrictions.

The first consideration of personnel control techniques then centers around the questions: What equipment and procedures are necessary? How much is necessary? Of the seemingly necessary procedures, which are more important and which might be deleted so that speed, convenience, and cost savings might be obtained? Some general provisions will be discussed with their individual relationship with control. They are far from being a complete listing. Consideration should be given to necessity, economics, and employee relationships. Many techniques will not need to be fully implemented, depending upon the actual circumstances. They should be used as guidelines and adapted to each condition.

### control equipment

The variety and quantity of personnel contaminants have caused a demand for special equipment which is intended to protect product by providing some form of isolation. Of the many means to accomplish this goal special garments, product handling equipment, and clean room entrance equipment are almost universally specified for work situations where personnel control is warranted. Logically, such engineering planning recognizes their potential benefits, but it must be mentioned that proper care and utilization are also essential.

**Smocks and Caps.** Clean room garments—smocks, hats, and in some cases face masks and shoe covers—are intended as "people filters." Their effectiveness depends on their design, how well they are cleaned, and the manner in which people use them. Table 13-5 shows typical clean garment requirements as a minimum guide for their specification and use.

Clean garments are usually made from synthetic monofilament materials. These are less likely to fray or be a contributor of lint. The weave of the cloth is intended to be sufficiently tight to keep particles from sloughing off the individual, yet not tight enough to cause discomfort because the cloth fails to

**TABLE 13-5.   Typical Clean Garment Requirements**

*Clean Room Garment Design*
1. The fabric of smocks and caps should be of a synthetic type (dacron polyester or nylon), exhibit limited linting properties, and a very low electrostatic generating property.
2. Smocks should be of simple design with no pockets and as few seams as possible.
3. Seams should have no open ends of fabric which might become frayed.
4. Seams should be double-stitched with synthetic thread of the same fiber as the garment.
5. Smocks should have adjustable neck bands and cuffs which provide a snug fit when worn.
6. Snap fasteners should be of rust proof metal, and located at the neck and cuffs. There should be sufficient snap fasteners for snug fit adjustment.
7. The cap should be of the style worn in hospital operating rooms. It should fit snugly around the head, covering the hair so as not to allow hair to fall in the clean room area.
8. Hoods, coveralls, or foot coverings, if required, should conform to these same general requirements.

*Clean Garment Laundry*
   Clean room garments should be processed in a suitable environmentally controlled laundry using equipment reserved for clean room garments only. The laundering technique must be capable of producing a clean garment which can pass all tests for cleanliness.  Failure to meet these requirements will be cause to reject that particular lot of garments.  These rejected lots should be returned to be relaundered at no additional cost to the user.  Cleaned garments should be neatly folded, and individually packaged and sealed in clean plastic (polyethylene) bags.  The identification mark of each garment should be clearly visible without opening the bag.

*Garment Use and Storage*
1. Each operator should have three changes of garments: one in use, one in standby, and one in laundry.
2. Putting on and removing garments should be done in a way to avoid contamination such as touching the floor or other items that may contaminate it.
3. Clean room garments should never be worn beyond the clean room and clean room dressing area.
4. During inactive periods garments should be stored consistent with the integrity of the garment.
5. The number of garment changes should be consistent with the cleanliness of the area in which it is worn.  Soiled garments should be changed immediately.

breathe.  There are many commercial designs of clean room smocks and caps allowing a latitude of choice.

   The design of smocks and hats is only part of the solution to proper clean room garments.  Any clean room clothing will become dirty during wearing, thus losing its value as a means of isolating personnel contaminants from the product.  Normal clean room practice allows about 24 hours use before laundering is required.  Clean garment specifications should include provisions for cleaning garments in an environmentally controlled laundry which will pass cleanliness test requirements. ASTM F 51-68[5] is the current-standard used by many industries. Some companies have developed their own techniques utilizing a light-scattering particle counter as the test instrument.

**Product Handling Techniques.** The hands of the clean room operator are a common source of product contamination and so gloves, finger cots, special soaps and skin lotions, and special tools often are provided as a means of isolation. Whether any of these techniques other than special tooling are entirely satisfactory is a matter of opinion. For instance, a person wearing clean gloves or finger cots may scratch his head or touch a greasy portion of a machine during working operations and then continue to handle product sensitive to such contaminants. The question arises whether a dirty glove or finger cot is any better than a dirty hand. The same argument may be used for special soap and skin lotions. The answer to this question depends largely on the sensitivity of the product being handled and the particular conditions of the work operation. The contaminants transferred to the product by bare hands may be of a different character than the materials picked up on gloves and finger cots.

Tweezers, vacuum pickup tools, and other special handling equipment are desirable alternatives to provide isolation of the operator's hands from the product. If used, the tools must be kept clean to the same level as specified for the product. This may mean special cleaning of the tools, a special rack or storage space for them, and instructions as to their proper use.

**Clean Room Entrance Equipment.** The entrance to any clean room is the logical point to create a buffer zone against the dirtier environment of the factory and the contaminants which people can drag with them into the clean room. For this reason most clean rooms have vestibules which are maintained at an intermediate air pressure between the high level of the clean room and the base level of the general factory. Door interlocks may be provided to minimize any disruption of the air pressure balance. A vestibule, as the only entrance to the clean room, gives the supervisor-in-charge an opportunity to minimize unnecessary traffic into the clean room. Here both incoming people and materials can be properly prepared for entrance into the clean environment. In the case of people this involves the installation of such equipment as lockers for personal belongings, smock storage, shoe cleaners, carpets, tacky mats, sinks, and air showers.

Lockers for personal belongings are provided usually near the factory entrance for storage of coats, umbrellas, etc., but there is a need for small lockers in the vestibules of clean rooms to hold women's purses while employees are in the clean room. No items used in the clean room should be stored in the lockers with the possible exception of personal safety equipment.

Shoe cleaners often are installed near the entrance to reduce the tracking of dirt into the clean room. They should be equipped with a HEPA filter on

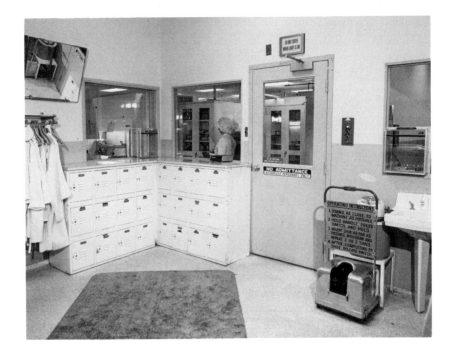

*Figure 13-2  Typical clean room vestibule.*

the exhaust of a portable vacuum cleaner or be attached to a central vacuum system to prevent redispersion of the collected dust.  As an added insurance against dirt from shoes, treated carpets or tacky mats can be installed at the entrance.  Carpeting at all the entryways of the factory has proven to be a useful housekeeping technique to reduce the tracking in of dirt from outdoors.

Other equipment commonly associated with clean room vestibules are smock storage racks and desks for supervision while sinks and air dryers (for employees to clean their hands before beginning work) and air showers (to blow loose dirt off of personnel) are options sometimes included by clean room designers.  Figure 13-2 shows a typical vestibule installed for a class 10,000 clean room.

### work and housekeeping rules

If personnel contaminants are of sufficient importance to require special equipment for personnel control, engineering planning also should regulate personnel behavior and housekeeping routines.  Conceptually, the behavioral rules are intended to minimize the generation of human-related contaminants, while the housekeeping procedures should collect the dirt generated by mass activity of people and production processes before migration to sensitive product surfaces.  Tables 13-6, 13-7, and 13-8 show typical restrictions on

**TABLE 13-6.    Typical Requirements for Clean Room Personnel Access and Conduct**

1. Only authorized personnel assigned to the clean room should be permitted to enter the area. Visitors to the clean room should have prior approval from the supervisor-in-charge.
2. Each person in a clean room should be properly attired in a smock and head covering.
3. Smocks should be worn over street clothes and fully closed with snug fitting closures at neck and wrists. Head coverings are to be worn in such a manner so as to cover all hair.
4. All personnel should remove their clean room garments and hang them in the change room upon leaving the work area.
5. Activity of all personnel should be kept in a minimum consistent with production requirements and operating procedures.
6. Smoking is prohibited in all clean rooms.
7. Contamination producing materials should not be taken into clean rooms. Examples are: tobacco, food, matches, cosmetics, chewing gum, cardboard, and paper.
8. No pencils, pens, or other writing instruments should be allowed in the clean room with the exception of ball point pens.
9. The following rules are to be enforced to assist in the successful operation of the clean room:
    a. Wash hands often.
    b. Wear gloves or finger cots when required.
    c. Keep finger nails clean.
    d. Never comb your hair in a clean room.
    e. Do not wear finger nail polish.
    f. Always wear the proper clothing.
    g. Never wear or apply cosmetics in a clean room.
    h. Personal items should not be carried into the clean room.
    i. Avoid wearing jewelry in the clean room.
    j. Work on clean surfaces.
    k. Nervous type mannerisms such as scratching head, rubbing hands, or similar type actions are to be avoided.
10. Personnel with skin or upper respiratory diseases should not be allowed to work in clean rooms. Some examples of problems that are detrimental to clean rooms are:
    a. Allergies to synthetic fabric.
    b. Profuse nasal discharge.
    c. Skin conditions which result in above average skin shedding, dandruff, or skin flaking.
    d. Severe nervous conditions.

personnel behavior and housekeeping requirements. When compared to the concepts on which they are based, such regulations become self-explanatory.

Notice that Table 13-6 places certain restrictions on personal habits. Other restrictions such as recommended bathing times and frequency, hair washing times and frequency, cleanliness and types of undergarments and outer clothes, and the hobbies and duties outside of working hours, shaving, etc., can be an extremely touchy subject. The screening of employees before working in a clean room area is perhaps easier and better than rigid rules on

**TABLE 13-7. Typical Clean Room Housekeeping Requirements**

| Class | Waste Baskets | Window Glass | Work Surfaces | Production Facilities[a] | | Interior Surfaces | Walls | Ceiling and Ceiling Fixtures |
| | | | | Exterior Surfaces | | | | |
|---|---|---|---|---|---|---|---|---|
| 100 | None permitted | Clean daily with urethane foam wiper (clean room grade) & Windex* | Damp wipe daily.[b] Wipe as required with detergent solution. | Damp wipe weekly[b] Wipe monthly with detergent solution Wipe stainless steel weekly with appropriate polish | | Wipe weekly[b] with detergent solution | Wipe monthly[b] with detergent solution. | Vacuum montuly |
| 1,000 | Empty daily | Clean weekly with urethane foam wiper (clean room grade) & Windex. | Damp wipe daily. Wipe as required with detergent solution. | Damp wipe daily. Wipe monthly with detergent solution. Wipe stainless steel monthly with appropriate polish | | Wipe weekly with detergent solution. | Wipe monthly with detergent solution. | Vacuum monthly |
| 10,000 | Empty daily | Clean weekly with urethane foam wiper (clean room grade) & Windex | Damp wipe daily. Wipe as required with detergent solution. | Damp wipe daily Wipe monthly with detergent solution. Wipe stainless steel monthly, etc. | | Wipe monthly with detergent solution. | Wipe monthly with detergent solution. | Vacuum monthly |
| 100,000 | Empty daily | Clean weekly with urethane foam wiper (clean room grade) & Windex. | Damp wipe daily. Wipe as required with detergent solution. | Damp wipe weekly. Wipe monthly with detergent solution. Wipe stainless steel monthly with appropriate polish. | | Wipe monthly with detergent solution. | Wipe bimonthly with detergent solution. | Vacuum bi-monthly |

[a]Production facilities include such items as benches, sinks, hoods, pass-throughs, racks, ovens, lockers, machines, shoe cleaners, and service pipes.
[b]Detergent solution consists of 4 oz. of liquid cleaner per gallon of city water (deionized water for class 100) mixed in a stainless steel bucket. After wiping down with cleaner, remove excess with water and dry with a dry urethane foam wiper or sponge of clean room grade.
*Registered trademark of Drackett Products Co., Cincinnati, Ohio.

TABLE 13-8.    Typical Clean Room Floor Maintenance Frequency

| Class | Vacuum | Damp Mop |
|---|---|---|
| 100 | End of each shift | Daily<br>Mop weekly with detergent solution & dry using wet vacuum equipment. |
| 1,000 | Daily | Daily<br>Mop monthly with detergent solution & dry using wet vacuum equipment. |
| 10,000 | Daily | Weekly<br>Mop monthly with detergent solution & dry using wet vacuum equipment. |
| 100,000 | Daily | Weekly<br>Mop monthly with detergent solution & dry using wet vacuum equipment. |

Note: The use of wax for floor preservation is controversial; the better appearance and floor preservation advantages of waxing must be weighed against the extrainment of abraided wax particles into the clean process air stream.

personal habits. Mention of these items can be made and controlled insofar as is practical.

The concept of minimum contaminant generation within the clean space should be extended beyond personnel control to include restrictions on manufacturing, maintenance, and housekeeping procedures. If possible, there should be no maintenance during periods of productive activity. Grinding, soldering, welding, and other processes that generate large amounts of particulate matter should be enclosed and properly exhausted. Sweeping should be prohibited (vacuum cleaning only) except for picking up glass, metal, and objects too large for the vacuum cleaning system. The housekeeping routines shown in Tables 13-7 and 13-8 should also be performed by properly garbed, trained, and supervised personnel. Like other facilities in the clean room, the cleaning equipment should be selected to minimize the generation of contaminants.

Since there are several types of clean environment designs no set of detailed rules or controls can solve all situations. It would be uneconomical to specify the same controls for a clean room of class 10,000 as are typically specified for a class 100 work area. The use of the facility (such as electronic manufacturing, pharmaceutical manufacturing, research) also would call for a variation of the personnel control and housekeeping techniques.

## training program

A great number of contamination control problems can be the result of poorly trained personnel and personnel that do not comply with good clean room practices. A training program is essential to effective personnel control.

The elements of good employee participation in any training program, regardless of whether it is a clean room operation or any task, include: (1) the creation of a receptive and cooperative attitude in the minds of employees; (2) an understanding of the necessity of the program; and (3) a knowledge of what must be done and how it can best be accomplished. It is obvious to any individual not accustomed to it that a clean room is a strange and unusually strict place to work. Each individual working in a clean room must become interested in the products being manufactured and alerted to the importance of controlling contaminants. Even with sophisticated equipment, procedures, and regulations, the attitude of operating employees and the involvement of higher management are the prime ingredient of a personnel control program.

A training program should be developed around the concept of putting contamination control into practice. Training should be given to the employee so that he or she has some basic knowledge of the engineering that goes into the facility, the parameters of the contamination control program that will be enforced by the supervisor, and the reasons for the maintenance and housekeeping routines as well as the employee's own work regulations. A training program should therefore include: (1) background information on contamination; (2) a definition of the problem with examples of contamination repeatedly illustrated until the worker becomes aware of the problem; (3) as always—safety precautions; (4) proper clothing; (5) biological precautions as needed; (6) a printed list of regulations for conduct and work in the clean area should be passed to each worker. Each item should be fully explained and discussed.

Success of any training program is measured only by the results accomplished. Instead of an examination, results from contamination control training should be measured by the employee's performance.

These instructions should be given by a specialist. Some companies, due to organizational structure or personnel policies, may be unable to do this. However, this training is like any other instruction—only one who completely and adequately knows his subject should teach in any type of training program.

## summary

Personnel control is essential to achieve a prescribed level of cleanliness in a clean environment area. It has the capability of making the success or failure of a contamination control program. People must be motivated and trained to achieve clean work habits so that high quality environments can be maintained. Equipment, housekeeping procedures, and personnel regulations are available to help achieve personnel control, but the attitude of the operating employees and their management is the prime ingredient of a personnel control program to obtain the special cleanliness required for the manufacture of microelectronic products.

## references

1. Philip R. Austin, "Austin Contamination Control Index," *Contamination Control,* Vol. V, No. 1, June 1966.

2. William B. Cown and Thomas W. Kethley, "Dispersion of Airborne Bacteria in Clean Rooms," *Contamination Control,* Vol. VI, No. 6, June 1967.
3. R. M. Seip, "Computerization of Airborne Particle Counts," Presentation at the Ninth Annual Technical Meeting, April 19–22, 1970, AACC (Boston).
4. M. K. Brumbach, "Factory Cleanliness for Manufacture of Microelectronic Devices," Presentation at the Plant Engineering and Maintenance Conference, March 16–18, 1970, ASME (New York).
5. ASTM F 51-68, "Standard Method for Sizing and Counting Particulate Contaminant in or on Clean Room Garments," *1969 Book of ASTM Standards,* ASTM, Philadelphia, 1969, Vol. 8, pp. 636–641.

## *bibliography*

1. *Contamination Control Handbook,* NASA CR-61264, Clearing House for Federal Scientific and Technical Information, Springfield, Va., 1969.
2. Philip R. Austin, *Design and Operation of Clean Rooms,* Business News Publishing Co., Detroit, 1970.
3. "A State of the Art Report," Garments & Laundry Committee, AACC, May 1970, Boston.

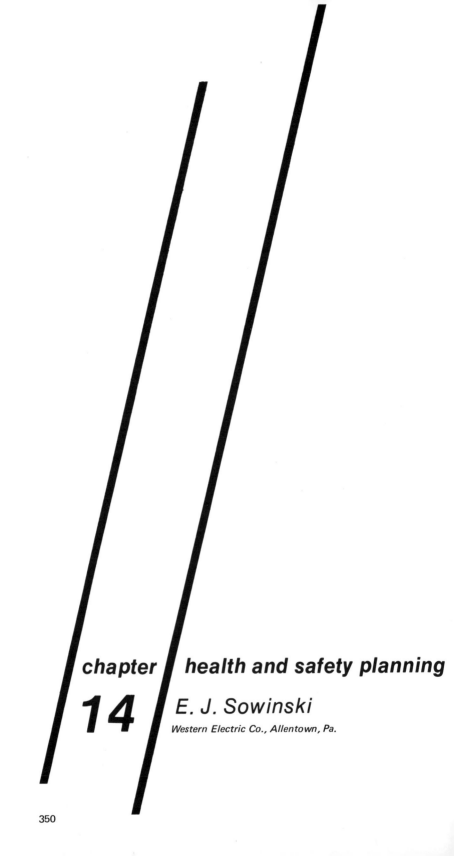

# chapter 14 / health and safety planning

## E. J. Sowinski

Western Electric Co., Allentown, Pa.

Environmental control is necessary for the health and safety of employees just as it is required for the successful manufacture of microelectronic devices. The competent planner in the field of health and safety follows certain broad guidelines in approaching problems. These are: (1) the recognition of environmental factors and stresses associated with work operations; (2) the evaluation of the magnitude of these factors and stresses in terms of ability to impair man's health and safety; and (3) the prescription of control methods to reduce or eliminate such factors and stresses.

The categories of hazards or stresses most frequently of concern in the microelectronics industry are: (1) chemical, in various forms of solids, liquids, and gases; (2) energy forms such as electromagnetic and ionizing radiation, and noise; and (3) ergonomic, such as body position in relation to safety.

It is important when evaluating and controlling hazards to recognize whether the hazard is acute and immediately dangerous to life and health, whether the hazard can cause chronic degeneration or produce an acceleration of the aging process, or whether it will cause only significant discomfort or inefficiency. To do this the health engineer utilizes qualitative and quantitative measurement techniques, experience gained from previous evaluations, and formal educational knowledge.

This chapter is intended as an overview to give perspective to planning necessary to maintain a high quality working environment. Sources of additional information are listed in the reference section. It should be kept in mind that the specific industry being considered will dictate the relative importance of each topic.

## chemical hazards

Previous chapters have described the demand for chemicals in the manufacturing of microelectronic devices. Plans for the supply, use, and disposal of chemicals must guard against the hazards presented by these materials. In this regard toxicity and flammability are the two hazards of major concern.

### toxicity

The toxicity of any chemical relates to its ability to bring about biochemical and/or physiologic derangements in a living system. Generally, for a chemical to be harmful to human health it must come in contact with or pass into the body. The three modes of body entry are ingestion, skin absorption, and inhalation. The maintenance of healthful working conditions requires the protection of employees from exposure to toxic chemicals by these three routes.

The effect of exposure to a toxic material may be one of severe poisoning, occurring acutely after very brief exposure (e.g., hydrogen cyanide poisoning from high concentrations), or from chronic exposures over days, weeks, months, or years (e.g., lead or benzene poisoning). On the other hand, exposure may produce chronic illness the symptoms of which are obscure and/or ambiguous.

**Ingestion and Skin Contact.** Adverse systemic health effects occurring from ingestion or skin contact of chemicals are relatively rare compared to exposure by the path of inhalation. In the case of ingestion the prohibition of eating or storing food in areas where toxic chemicals are used alleviates the possibility of intoxication by this route. Similarly, in the case of skin contact effective control is relatively easy to maintain via personal protective equipment (e.g., gloves, safety glasses, etc.), and systemic effects are limited by the skin barrier. The commonly used acids of microelectronic processes—sulfuric, hydrochloric, phosphoric, chromic, and hydrofluoric—can present severe splash hazards; their potential corrosive action on skin, clothing, and machinery should be well recognized. Alkaline cleaning and soaking tanks also present corrosive problems in microelectronics manufacturing. The chemicals commonly employed are caustic soda, soda ash, trisodium phosphate, rosin, sodium resinates, wetting agents, and emulsifiers. The splash hazard of alkaline baths is mainly dependent on employee work habits, the temperature of the bath, and its degree of alkalinity. Concentrated caustic sodas can produce deep and painful destruction of tissue. Even weak alkalies can soften outer skin layers, emulsify skin fats, and cause skin irritation. Personal protective equipment (rubber gloves, safety glasses, and rubber or plastic apron) should be used where splash hazards with corrosive chemicals exist.

Other examples where skin contact can be a major health concern include vapors of certain phenols, esters, aromatic nitro compounds, and amines. Slight amounts of hydrogen sulfide and dangerous amounts of hydrogen cyanide may be absorbed through the skin from contaminated air. Effective control for vapors which can be skin absorbed must involve skin protection and exhaust ventilation (discussed below).

**Inhalation.** A major percentage of vapor and particulate materials inhaled into the lungs is exhaled, but the amount retained is excreted by passage through the blood stream. Here it is in a position to reach vital organs and exert toxic action. Interestingly enough, a man inhales approximately 30 to 40 pounds of air per day, whereas he may eat no more than a couple of pounds of food. As a result, a major effort for the health engineer concerns the evaluation and control of gases and airborne suspensions.

Gases can exist in two basic forms: compressed and noncompressed. Many commonly used compressed gases are under a pressure as high as 2,600 psi. Noncompressed gases can originate as vapors from liquids and as reaction products of two or more chemicals.

Solvents and acids exhibit vapor pressures at ordinary temperatures and can give off toxic gaseous vapors. A prime example of toxic gases which can emanate as a reaction by-product are the oxides of nitrogen from bright dipping processes. Hydrogen cyanide, a highly toxic gas, also can emanate as a reaction by-product if mineral acids are mixed with cyanide salt solutions. Every precaution must be instituted for segregation of acids and cyanides in use, supply, and disposal. In all cases threshold limit values (defined later) should be consulted as a guide for acceptable exposure levels to airborne materials. Whenever possible any industrial process should be engineered to reduce inhalation exposures to an absolute minimum.

Airborne suspensions having the potential for inducing health effects via inhalation are referred to as aerosols. The term aerosol is used to describe dispersions of solids or liquids which possess sufficient stability to remain airborne for appreciable lengths of time and present inhalation hazards. Aerosol deposition within the respiratory system varies with the size, shape, and density of the aerosol particle. The percentage penetration of particles into the smallest lung air spaces—called alveoli—rises from essentially zero at 10 microns to a maximum at and below about one micron. Particles larger than 10 microns are essentially all removed in the nasal chamber and therefore have little probability of penetrating to the lungs where systemic absorption can occur. Upper respiratory collection efficiency drops off as size decreases and becomes essentially zero at about one micron. The efficiency of particle penetration to the alveoli is essentially 100% at a size of about 0.5 micron. The respiratory hazard for aerosol particles relates directly to their size. For gases and vapors, respiratory hazard relates to airborne concentration and exposure time.

**Threshold Limit Values.** Since inhalation is the major mode of personnel exposure to toxic chemicals the control of airborne contaminants is of prime concern to health and safety planning. The term "threshold limit value" refers to airborne concentrations of a substance at which it is believed that constant work exposure will not cause adverse effects on nearly all workers. They are time-weighted concentrations for a 7 or 8 hour workday and 40 hour workweek and are used as guides in the control of health hazards. They should not be used as fine lines between safe and dangerous concentrations. Time-weighted averages permit excursions above the limit provided they are compensated by equivalent excursions below the limit during the workday.

For mixtures of similar materials (i.e., solvents or acids) where additive health effects are expected, Eq. (14-1) is commonly utilized:

$$\frac{C_1}{TLV_1} + \frac{C_2}{TLV_2} + \frac{C_3}{TLV_3} + \cdots + \frac{C_n}{TLV_n} = 1 \qquad (14\text{-}1)$$

where $C$ = airborne concentration of contaminant and TLV = threshold limit value. If the number one is exceeded upon addition of the series $C/TLV$, the threshold limit value for the mixture is exceeded. In this regard Table 14-1 lists some TLV's for common solvents.

The relationship between threshold limit and permissible excursion is a rule of thumb and in certain cases may not apply. The amount by which

TABLE 14-1. Threshold Limit Values for Common Industrial Solvents

| Substance | TLV Vapor Concentration ppm* |
|---|---|
| Dichloromethane | 500 |
| Methanol | 200 |
| Trichloroethylene | 100 |
| Xylene | 100 |
| Benzene | 10 |

*Parts vapor per million parts air by volume.
Source: Occupational Safety and Health Standards, Federal Register (Oct. 18, 1972).

threshold limits may be exceeded for short periods without injury to health depends upon a number of factors, such as the health of personnel exposed, the nature of the contaminant, whether very high concentrations (even for short periods) produce acute poisoning, whether the effects are cumulative, the frequency with which high concentrations occur, and the duration of such periods. The evaluation of such factors involves experienced judgment in arriving at decisions as to whether a hazardous condition exists.

Because of wide variation in individual susceptibility, a small percentage of workers may experience discomfort from some substances at concentrations at or below the threshold limit. A smaller percentage may be affected more seriously by aggravation of a pre-existing condition. Tests are now available that may be used to detect hypersusceptible individuals to a variety of industrial chemicals.[1] These tests may be used to screen out the hyper-reactive worker during job placement routines.

## flammability and explosibility

The personnel hazards of flammable and explosive materials require no explanation; however, the chemical properties defining these conditions deserve a brief description. Many industrial chemicals encountered in microelectronics manufacture are flammable, but the critical temperature at which they ignite varies considerably. The lowest temperature at which a vapor-air mixture has reached the minimum concentration for ignition by application of a flame is called the flash point. Chemicals with a flash point below $150°F$ ($65.5°C$) are considered dangerously flammable. Solvents such as acetone, benzene, and toluene have flash points considerably below room temperature.

Certain mixtures of air and solvent vapor are capable of explosion, and the limits of concentration between which explosion occurs will vary with each solvent. The lower explosive limit is defined as the concentration of solvent vapor in the vapor-air mixture corresponding to the flash point; the upper explosive limit is that concentration which is no longer flammable.

The relative explosive hazard of chemicals can be assessed by correlating a number of factors—their boiling point, flash point, evaporation rate, upper and lower explosive limits, auto-ignition temperature, heat of combustion, and vapor pressures.

Many unique hazards are presented by liquid chemicals with regard to fire and explosion. Ethers such as ethyl ether, dioxane, and tetrahydrofuran tend to absorb and react with oxygen from the air to form unstable peroxides which can explode when they become concentrated by evaporation or distillation. The use of perchloric acid also presents an explosion hazard since this acid can decompose violently upon contact with organic materials.

## chemical control

Toxic materials seem to be a necessity for many manufacturing technologies in the microelectronics industry; consequently, methods of detection and preventative means must be available so that environmental concentrations are maintained at minimum levels. As used here, the term environment not only includes specific working environments inside a factory but also the out-of-doors. Planning for environmental control is most important for the satisfactory use of new chemicals; in turn, appropriate control must be designed for all hazardous materials.

### introduction of new chemicals

To provide continuing environmental protection it is becoming increasingly important to know the predicted effects of new chemicals on both personnel and the environment. This concept requires knowledge of chemical by-products as well as the toxic properties of parent compounds. The chemical nature of by-products formed from any chemical process depends on the properties of the parent compound, the manner in which it is used, and environmental stresses. Typical environmental stresses include chemical adsorption in ductwork, surface reactions, exposure to radiant energy, and biological degradation.

A practical industrial approach to identify potentially hazardous by-products should include information from the manufacturer and a review of the literature concerning the chemistry and analysis of the parent compound. Special attention should be given to all possible reactions, stabilities, and rates of decomposition. Methods for characterizing the parent compound and its decomposition products with documentation of spectra and physical characteristics should be obtained. Next, sampling and analytical methods for detecting the parent compound and its by-products in the environment should be established. Initial application of analytical methods should be included in pilot processes of new technologies. In this manner potentially hazardous by-products can be revealed, the substitution of less hazardous parent chemicals can be explored, and appropriate control measures can be established prior to routine manufacturing usage.[2]

**Substitution for Hazardous Chemicals.** Substitution of a less hazardous material for one which is more hazardous in terms of toxicity or flammability is the most effective restraining measure against potentially hazardous industrial situations. Specifics concerning this planning concept depend entirely on the new process in question, particularly on the environmental hazards of by-products.

One important aspect of control planning that is often overlooked is the effect of repeated exposures, usually by inhalation, to low concentrations of substances over long periods of time. For example, it may be known that excessive exposure to carbon tetrachloride is harmful, but since it is non-flammable and a good spectral solvent it might be used repeatedly with a person relying on his sense of smell as a warning to excessive exposure. The presently accepted threshold limit value for carbon tetrachloride is 10 ppm, a concentration well below the odor threshold for most persons. It is entirely possible for a person to inhale enough carbon tetrachloride vapors to cause adverse health effects even though a time is hardly recalled when the odor was objectionable. Because of this the use of carbon tetrachloride is avoided in microelectronic applications. Similar conditions can occur with many other chemicals and manufacturing by-products. Substitutes should be considered and employed where feasible.

### evaluation of chemical exposures

An extensive technology has developed with regard to the evaluation of chemical exposures. The traditional approach is the measurement of air concentrations of toxic materials utilizing sampling devices and direct reading instruments which have been developed for the purpose (see Chap. 16).

A different sampling approach which has some popularity involves the use of miniature samplers which can be attached directly to an individual whose exposure is being studied and which operates with lightweight pumps and battery pack.[3]    These devices can provide useful experimental data because they follow a worker as he moves about.

In many situations it is nearly impossible to get reliable time-weighted averages by air sampling, and a completely different approach must be utilized. Such an approach involves biological monitoring in which samples of blood or urine are analyzed as indicators of exposure.[4]    Less commonly utilized in this regard is breath, sweat, tissue, feces, or other biological materials. This approach is based on the premise that, when inhaled, certain materials or their metabolites are found in blood or other tissues at varying levels which bear some relationship to total individual exposure. Regardless of the method selected for monitoring, the purpose remains the same—the determination of time-weighted levels of specific materials with the intention of comparing analytical results to threshold limit values or pollution standards.

### design concepts

**Chemical Handling.**  Special provisions should be designed into the handling, storage, and disposal of toxic and flammable chemicals accepted for manufacturing use. Wherever it is feasible, flammable liquid storage, repackaging, and dispensing should be done in a separate building. Bulk storage should be in remote areas. Provision should be made for easy distribution of materials within the factory so that excessive storage in production areas can be prevented. Within production areas chemical storage should be in ventilated, fire resistive cabinets and safety containers. Provision should also be made for the collection and disposal of flammables and hazardous chemicals. The nature of these problems is such that individual analysis of expected problems is required before solutions can be found. In most cases a system of collection by either bulk containers or pipeline and remote destruction is a practical solution.

Provisions should be made for storing compressed gas cylinders of flammable or toxic materials in a separate, well ventilated building in which protection from the weather is provided. A well ventilated area should be established for the separate storage of peroxides and acids away from flammable solvents.

Compressed gas cylinders can be handled safely if several general rules pertaining to receiving, storage, transportation, usage, and disposal are followed. These include: (1) know cylinder contents and its properties; (2) fasten cylinders securely in use, transit, or storage; (3) transport cylinders only on a wheeled cart; (4) use cylinders only with equipment suitable for the contents; (5) do not use cylinders without a regulator; (6) close cylinder valves when not in use, and mark cylinder when empty. Regardless of the pressure rating of a cylinder the pressure of the gas in the cylinder depends to a great extent on its physical state. Gases which are liquefied in the cylinder such as carbon dioxide, propane, ammonia, and others will exert a vapor pressure as long as any liquid remains.

**Ventilation.** Operations involving the generation of airborne toxic and flammable materials must be controlled with efficient local exhaust ventilation. To meet the needs of safety and economy local exhaust systems must effectively remove airborne toxic and flammable materials and at the same time exhaust a minimum volume of air. Make-up air must be supplied to ventilated areas to replace the air removed by exhaust systems so that such systems work properly; exhausted air should not be recirculated.

Laboratory type hoods are the most commonly used means of removing gases, dusts, mists, vapors, and fumes from work sites in the microelectronics industry; however, they are often misused and frequently specified when not needed. For exhaust systems to be effective they must be located so toxic contaminants are removed away from the breathing zone of the worker as depicted in Fig. 14-1. An exhaust hood must confine contaminants within the hood, remove them from the work site through ductwork, and disperse them so they do not return into the building through the fresh air supply system. In cases where the exhausted materials involve air pollution appropriate treatment of exhaust is required. Filters, collectors, condensers, scrubbers, or other cleaning equipment can be installed to prevent the direct discharge of

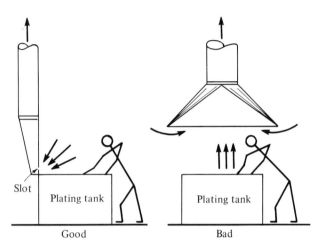

Good                          Bad

Direction of air flow

*Figure 14-1 Location of exhaust systems. (Courtesy* Industrial Ventilation, A Manual of Recommended Practice, *American Conference of Governmental Industrial Hygienists, Lansing, Michigan, 11 Ed., 1970.)*

chemical exhaust pollutants into the atmosphere (see Chap. 15). In cases where toxic chemicals must be used in clean rooms or clean benches, exhaust rates must be capable of maintaining vapor concentrations at safe levels during normal operations as well as in the event of accidental spills. In this regard it is important to recognize that clean rooms and many clean benches involve sizable percentages of recirculated air in which contaminants may concentrate.

An adequate velocity of air entering an exhaust hood at its face is the basic requirement for capture and control of contaminants generated within the hood air space. A minimum entering face velocity of 100 linear feet per minute for normal hood openings is recommended. Required exhaust rates depend on the toxicity of the material to be contained, its rate of evolution, and the nature of the enclosing hood (see Table 15-9).

Successful performance of exhaust systems depends on such factors as the velocity of air moving through the hood, entrance shapes, thermal loading, mechanical action of particles, exhaust slot design, and obstructions in the hood. Also, corrosion resistance, cleanability if contaminated, and in some cases the collection of contaminants such as radioisotopes, determine successful hood performance.

Monitoring the operational performance of exhaust systems is desirable especially where highly toxic or flammable materials are used. A pilot light can be provided to show if an exhaust fan motor has been turned on, or a pressure gage can be installed to indicate that the fan is drawing air.

**Fire and Explosion Control.** Effective control of fire and explosion is based on knowledge of the chief sources of ignition. These include: (1) flames—open lights, fires, burning material; (2) sparks—including those from static electricity and electrical equipment; (3) heated materials—glowing metals, overheated metal parts, hot cinders.

Control of these sources is mainly a matter of proper design and installation of electrical equipment and prohibition of naked lights or incandescent materials in areas where flammable solvents are used. Electrical equipment should be constructed to meet requirements of the National Electrical Code. If at all possible operating conditions should be altered so that electrical equipment is not subjected to hazardous atmospheres. This usually can be accomplished with planning so that switches, lights, motors, and electric contacts are not located in a hazardous area but instead are placed in remote or purged locations.

Static electricity tends to be generated by the flow and agitation of solvents in pipes and tanks, especially in dry atmospheres. Metal pipes, tanks, and machinery should therefore be efficiently grounded.

Further precautions against fire and explosion from flammable chemicals include:

1. Substitution of noncombustible for combustible materials. For example, trichlorotrifluoroethane (Freon) can be substituted for acetone in many microelectronic solvent cleaning operations.
2. Addition of noncombustible to combustible materials. For example, trichloroethylene is noncombustible, and if added in proper proportions to combustible solvents, it can render the resulting mixture nonflammable.
3. Regulation of temperature so that the vapor pressure will give concentrations well below the lower explosive limit.
4. Prevention of escape of vapor into the in-plant atmospheres by means of local exhaust ventilation and efficient general exhaust ventilation.
5. Segregation of operations, including storage for flammable and explosive materials.

The necessity for adequate ventilation where flammable gases and vapors are handled cannot be too strongly emphasized. This must include laminar air flow clean benches and clean rooms widely encountered in microelectronics manufacturing. Flammable vapor concentrations can occur due to the relatively large volumes of recirculated air normally encountered in these systems. Their design must include exhaust rates capable of maintaining nonflammable vapor concentrations during normal operations as well as in the event of accidental spills of flammables.

Automatic fire and explosion detection and/or protection equipment should be installed in areas in which special hazards are presented. For example, hydrogen gas detecting units can be installed to indicate hydrogen concentrations approaching explosive limits. Such devices are utilized in heat treating areas where extensive use is made of hydrogen furnaces. The instruments are generally adjusted to set off an alarm when hydrogen concentrations reach approximately 1%. Since the lower explosive limit for hydrogen is 4.1%, this allows for a reasonable safety factor. The most widely used protection method is the automatic water sprinkler. Sprinkler systems for fire protection are available with many different kinds of heads for automatic activation at different ranges.

## energy hazards

### noise and vibration

Prolonged exposure to intense noise can cause hearing impairment in exposed individuals. Noise is commonly defined as unwanted sound, and its major parameters are: 1) frequency in cycles per second or hertz (Hz); and 2) intensity or sound pressure in decibels (dB).

The frequency range of audible sounds for healthy young ears is usually considered to extend from 20 to 20,000 Hz. The frequencies comprising speech are found principally between 250 and 4,000 Hz. This frequency range, referred to as the critical speech frequency range, is considered most important to man since hearing losses in this range would handicap the individual in most daily activities.

The magnitude of pressure variations constituting a sound provide a measure of its strength or intensity. The intensity or pressure variations producing audible sound are quite small. Normal atmospheric pressure is $10^6$ dyne/cm$^2$, while the faintest sounds which can be detected by the ear are produced by pressure variations of approximately $2 \times 10^{-4}$ dyne/cm$^2$.

The ear is able to respond without difficulty over a wide pressure range. To eliminate the difficulties arising from handling large numbers it is customary to employ the decibel system for sound pressure measurement. The system is logarithmic in nature. The formula for computing sound pressure level in decibels (dB) is

$$dB = 20 \log_{10} \frac{P_1}{P_0}$$

where $P_1$ is the pressure of the sound being measured and $P_0$ is the reference pressure. In industrial hygiene work $2 \times 10^{-4}$ dyne/cm$^2$ is the accepted reference pressure.

Specifications of the maximum conditions of noise exposure which are tolerable to the ear, called "damage risk criteria," are valuable in serving as guides for: (1) hearing conservation programs in industry; (2) noise control procedures and techniques as applied to machinery and work environments; and (3) equitable rulings in court cases involving compensation for noise induced hearing loss. Many different damage risk criteria for noise exposure have been proposed. The most recent federal legislation provides for a con-

TABLE 14-2.   Permissible Noise Exposures

| Duration of Personnel Exposure per Day, Hours | Sound Level dBA* |
|:---:|:---:|
| 8 | 90 |
| 6 | 92 |
| 4 | 95 |
| 3 | 97 |
| 2 | 100 |
| 1 | 105 |
| $\frac{1}{2}$ | 110 |
| $\frac{1}{4}$ or less | 115 |

*dBA: refers to sound pressure level measured in dB on a sound level meter tuned to an A weighting network.
Source: Occupational Safety and Health Standards Federal Register (Oct. 18, 1972).

tinuing, effective hearing conservation program in all cases where noise levels exceed those outlined in Table 14-2.[5]

The control of noise can be accomplished by four basic methods. These include source control, control by enclosure, control by sound absorbing materials, and control by ear protection. The control of excessive noise at the source is the most effective method of control, but it usually is the most difficult to plan because of the lack of general rules. For example, identical machines may require different methods of noise control because of different production installations or different economic considerations. As alternatives the use of enclosures or sound absorbing paneling may suffice. The control techniques must be designed for the specific equipment in question.[6] In some cases the use of ear protectors provides the only practical solution. The use of cotton or other material stuffed into the ears should be discouraged. Most effective personnel control is effected by either good fitting ear plugs or ear muffs.

Vibration is linked with noise since the two often have common origins. The specific effects of vibration are not well defined; it is known that vibration can adversely influence binocular acuity. People exposed to vibration for long periods of time frequently complain of headaches, exhaustion, and fatigue. Excessive vibration should be avoided in manufacturing areas. Particularly useful in this regard are vibration-free benches for use with microscopes to eliminate the transmission of background vibrations.[7]

### lasers

The use of laser energy is a relatively new phenomenon in the field of physical science. Because of known and unknown biological damage which laser energies can induce, stringent safety precautions are necessary to prevent accidents which might cause serious permanent damage.

The name "laser" comes from the technical definition: *l*ight *a*mplification by *s*timulated *e*mission of *r*adiation. The beam of light is coherent since it emerges as rays that are parallel with the same wavelength. It is these unique coherent qualities that make the laser so narrowly beamed and powerful. Laser beams can be created from some 100 different gases and liquids and can be produced for continuous waves or in pulses, depending on the lasing material. Colors range from orange to blue through the invisible infrared spectrum.

Industrial applications of laser systems are being developed constantly; examples include microdrilling and welding, space communications, meteorology, survey and alignment instruments, holography (3-D photography), surgery, and as educational tools.

The primary hazard of lasers involves potential eye damage. This hazard is complicated because of the short duration of some laser pulses (micro or milliseconds). There may be no pain or discomfort when the beam strikes the retina of the eye. Retinal burns may take several days to become detectable under examination. To further complicate the hazard potential from laser beams, if the beam is reflected from a shiny surface the energy of the beam is reflected and has nearly the same amount of energy as the incident beam.

In addition to the eye injury, other potential hazards associated with the use of high energy lasers are electrical shock, skin burns, ozone and nitrous oxide production, oxygen depletion, ignition of combustible solvents, explosion of evacuated optical systems and localized food contamination due to deposition of vaporized metals or other toxic materials.

Several agencies have published safety standards for laser operation.

TABLE 14-3.  Thresholds for Biologic Damage by Laser Radiation (6,943Å Ruby Laser) at the Retina of the Eye

| Regulation or Code | Threshold Values | | |
|---|---|---|---|
| | *Q-Switched (Pulse Duration)* | *Non-Q-Switched (Pulse Duration)* | *Continuous Wave* |
| U.S. Air Force Regulation No. 161-24, Jan. 1967 | 0.125 J/cm² (10–100 ns) | | |
| British Ministry of Aviation Code of Practices for Laser Systems | 0.07 J/cm² (30 ns) | 0.85 J/cm² (200 s) | |
| University of California, LRL Hazards Control Quarterly Report 16 and 17, 1964 | 0.08 J/cm² (30 ns) | 0.7–0.8 J/cm² (200 s) | |
| Uniform Industrial Hygiene Codes or Regulations for Laser Installations ACGIH–1968 | 0.07 J/cm² (30 ns) | 0.85 J/cm² (200 s) | 6.0 W/cm² (white light) |

Source: Moore, W., Biological Aspects of Laser Radiation: A Reivew of Hazards, Public Health Service, Bureau of Radiological Health, Jan. 1969.

Thresholds for biologic damage at the retina of the eye are given in Table 14-3.[8] The threshold values are presented for three separate modes of operation: Q-switched, non-Q-switched, and continuous wave. The distinguishing feature of the Q-switched laser is the rate of energy delivery. For example, power densities ranging from 1 to 100 MW/cm$^2$ can be delivered in pulses ranging from about 5 to 100 nanoseconds (ns). All lasers, regardless of power rating, should be considered potentially hazardous and they should not be used indiscriminately.

A protective program for lasers should center around facility design and medical surveillance. The design of laser systems should follow as closely as possible a closed system technique, and high reflectant surfaces on and adjacent to target areas should be avoided. When used in manufacturing areas lasers should be enclosed and interlocked to eliminate the accidental exposure of operating personnel to direct or reflected beams. Periodic eye examinations and records for all operating personnel also are a necessity. The use of goggles or glasses is generally discouraged except in special research situations where other alternatives are impractical.

### ionizing radiation

Ionizing radiation refers to electromagnetic radiation capable of producing ions directly or indirectly in its passage through matter. Ionizing radiation includes alpha particles, beta particles, and X and gamma rays. Of the four varieties of ionizing radiation, X and gamma rays are of major health significance. This is due to their greater potential for deep penetration into soft tissue. Note that many radioactive isotopes that emit beta particles also emit gamma rays of deeper tissue penetrating ability.

The basic term for radiation exposure dose is the *roentgen*. It relates to the dose of X-ray or gamma radiation which is capable of producing one electrostatic unit of electricity in 1 cm$^3$ of air. Other terms of importance in radiation health are as follows:

*REM* (Roentgen Equivalent Man)—The quantity of radiation of any type which produces the same biological damage in man as that resulting from the absorption of one roentgen of X or gamma radiation.

*Exposure Dose*—A measure of amount of radiation at a given point.

*Absorbed Dose*—This describes what is taken from a beam of radiation at a point.

For safe operation, personnel should never remain unnecessarily in the vicinity of an X-ray tube in operation even if measurements have shown that shielding meets requirements. A good general rule is that exposure of individuals to radiation should always be kept to the lowest practicable level. Maximum permissible doses for an individual are usually considered to include all doses from all types and energies of radiation during the time period of interest delivered to the body region of interest. In this regard Table 14-4 outlines maximum permissible doses established by the State of Pennsylvania and currently used at the present time by many state health departments.

TABLE 14-4.  Ionizing Radiation Standards

| Part of Body | Maximum Dose per Calendar Year | Maximum Dose per Quarter Year |
|---|---|---|
| Whole body, head and trunk, gonads or lens of eye | 5 rems | 3 rems |
| Skin of whole body | 30 rems | 10 rems |
| Hands and forearms, feet and ankles | 75 rems | 25 rems |

Source: Pennsylvania Department of Health Regulations, Chapter 4, Article 433, Subchapter D (Jan. 30, 1970).

Major sources of ionizing-radiation in microelectronics manufacturing are the X-ray tubes used in X-ray diffraction or X-ray fluorescence equipment. In all cases such tubes should be completely surrounded, except for the emergent beam opening, with protective material of adequate thickness to reduce the radiation to tolerance levels or below at all points which can be occupied by personnel.

## human engineering and industrial safety

In addition to the established methods of combating industrial accidents, such as safety-education programs and special safety equipment, human factors engineering, or ergonomics, is employed. The intent is to improve worker efficiency and to reduce accidents by relating the mechanical design of equipment to the biological and psychological characteristics of the operator. This implies that machines and working areas must be built around the operator rather than placing a worker in a setting without regard for human requirements and capacities. Unless working environments are so constructed, it is not fair to attribute accidents to human failure.

### designing for human capabilities

Equipment and working methods need not be developed for individuals, but rather optimum conditions should be used for groups of workers. The best arrangements can thus be set forth with reference to general human capabilities. Data on the physical characteristics of working populations are useful in determining clearances and allowances in the working area. Examples are work bench dimensions based on sitting or standing height, or minimum head clearance based on sitting height or stature. In this regard it should be remembered that many occupational groups may be highly selected and differ from other groups and from the general public as well. Consequently, the population that will use a given machine or apparatus should be measured directly for a given task if it is at all possible to do so.

In addition to static measurements additional information on the dimensions of the human body under dynamic conditions is important. Controls, tools, or materials that must be handled often or accurately should be located within normal reach zones. Secondary items or controls used only occasionally may be located beyond normal reach zones but within maximum reach. In some cases modifications of normal dynamic conditions must be made. For example, consider the special atmosphere systems (see Fig. 11-3) commonly used in microelectronic manufacture. In enclosures of this type an operator is no longer permitted the freedom of movement that is possible on an open-bench work position. Such restriction of movement creates a relatively static work position for an operator and highlights the importance of a three-fold interrelationship of man, the task, and the enclosure in human factors engineering. Design studies have revealed the following criteria which minimize operator fatigue:[9]

(1) In drybox operations, no effective work can be performed within the first 5 inches of the enclosed bench edge because of restrictions imposed by the arm ports.

(2) An operator cannot be expected to work effectively beyond a point 15 inches inside a drybox enclosure.

(3) The ideal work area is generally a point somewhere between eight and and ten inches inside the enclosure, midway between the arm ports.

Figure 14-2a exhibits the dimensions of a basic drybox enclosure commonly used in the microelectronics industry for both sitting and standing operator positions. Figure 14-2b displays a comparison of the open bench with a drybox enclosure.

### accident prevention

In addition to designing for human capabilities there are at least five approaches that can be taken with respect to the prevention of accidents. Each can meet with varying degrees of success depending upon the appropriateness of the approach in a specific instance and upon the skill of those applying it.

**Selection and Training.** Strict adherence to this approach requires that initially the job be described in detail and individuals selected and trained to do the job. Those who cannot reach certain criteria of effectiveness within some specified period of time should be shifted to activities more suitable to their talents.

**Education.** This term refers to programs in safety education and is not concerned with the teaching of specific skills necessary to do the job. The purpose generally is to make individuals more "safety conscious."

**Admonishment or Punishment.** These terms refer to the withdrawing or suspension of rewards or privileges, hopefully so that the same mistake will not be made again.

**BASIC ENCLOSURE**

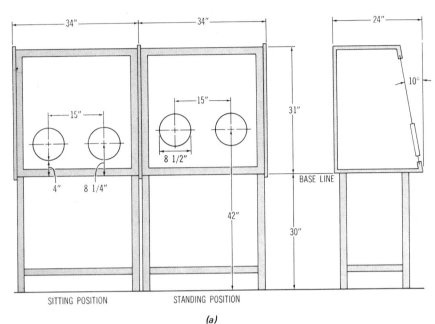

SITTING POSITION          STANDING POSITION

*(a)*

**VERTICAL WORK AREA**

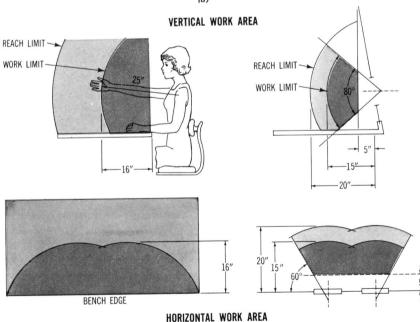

**HORIZONTAL WORK AREA**

*(b)*

*Figure 14-2 Ergonomic criteria for special atmosphere systems. (a) Basic enclosure. (b) Vertical and horizontal work areas.*

**Rewards for Accident-Free Behavior.** In a sense this is the antithesis of the previous technique. The "zero defects program" of the U.S. Defense Department is a variant of this technique which includes all errors, whether accident related or not. A significant reduction in accidents awaits an alert and progressive management that makes this item as central to their thinking as such topics as profits, efficiency, time schedules, etc. An example of this line of thinking is the safety motto of the Bell System which states: "No job is so important and no service is so urgent that we cannot take time to perform our work safely."

**Identification of Accident-Prone Individuals.** Much has been written about the accident-prone individual. Undoubtedly there are such individuals; however, one is on perilous ground when he attempts positively to identify such a person.

**The Man-Machine System.** More than any single concept, the man-machine system design concept is the one around which effective accident prevention must rally. All human factors engineering for accident prevention can be thought of as dealing with the appropriate allocation of tasks between man and machine in which the two components—man and machine—are in such intimate cooperative relationship that the engineer is obliged to view them as a single system.

## summary

This chapter has presented a general review of the factors involved in recognizing, evaluating, and controlling health and safety hazards in the microelectronics industry. The categories of hazards include: (1) chemical, with reference to toxicity and flammability; (2) energy, including lasers, ionizing radiation, and noise; and (3) ergonomic, such as body position in relation to safety.

Distinctions have been drawn as to whether or not a hazard is acute and immediately dangerous to health or whether it will only cause discomfort or inefficiency. Control aspects have been discussed with reference to degree of hazard.

## references

1. Stokinger, H. E., Mountain, J. T., and Dixon, J. R., *Arch. Envir. Health* **13**, 296 (1966).
2. Sowinski, E. J., and Suffet, I. H., Chapter 3 *Pollution: Engineering and Scientific Solutions* (Proceedings of the First International Meeting of the Society of Engineering Science held in Tel Aviv, Israel, June 12–17, 1972) ed. E. S. Barrekette, Plenum Publishing Corp., New York, 1972.
3. Mine Safety Appliance Company, 201 North Braddock Avenue, Pittsburgh, Pennsylvania 15208.

4. Goldwater, L. J., and Hoover, A. W., *Arch. Envir. Health* **15**, 60 (1967).
5. Occupational Safety and Health Standards, Federal Register, Oct. 18, 1972.
6. Cox, J. R., Chapter 18, Industrial Hygiene and Toxicology, Vol. I, Interscience, N.Y., 1958.
7. Barry Vibration Controls, 700 Pleasant St., Watertown, Mass. 02172.
8. Moore, W., Biological Aspects of Laser Radiation: A Review of Hazards, Public Health Service, Bureau of Radiological Health, January 1969.
9. Doxie, F. T., and Ullom, K. J., The Western Electric Engineer XI, 24, 1967.

# bibliography

## general overall field

1. Patty, Frank, editor, Industrial Hygiene and Toxicology, Vol. I, General Principles, 1958; Vol. II, Toxicology, 1963.
2. The Industrial Environment, its Evaluation and Control. U.S. Public Health Service, Div. of Occupational Health, Public Health Service Publ. No. 614, 1965.

## chemistry (analytical sampling)

1. Jacobs, M. B., The Chemical Analysis of Air Pollutants, Interscience Publishers, Inc., New York, 1960.
2. Jacobs, M. B. and Scheflan, L., Chemical Analysis of Industrial Solvents, Interscience Publishers, Inc., New York, 1953.
3. Air Sampling Instruments. For Evaluation of Atmospheric Contaminants, American Conference of Governmental Hygienists, 1014 Broadway, Cincinnati, Ohio, 2nd Ed., 1962.
4. Willard, H., Merritt, L. L., Jr. and Dean, J. A., Instrumental Methods of Analysis, D. Van Nostrand Co., Inc., 3rd Ed., 1958.

## engineering and specialties

1. Hemeon, W. C. L., Plant and Process Ventilation, The Industrial Press, New York, 2nd Ed., 1963.
2. Industrial Ventilation. A Manual of Recommended Practice, American Conference of Governmental Industrial Hygienists, P.O. Box 453, Lansing, Michigan, 11th Ed., 1970.
3. Brandt, A. D., Industrial Health Engineering, John Wiley and Sons, New York, 1947.
4. Drinker, P. and Hatch, T., Industrial Dust, Hygienic Significance, Measurement and Control, McGraw-Hill Book Co., 2nd Ed., New York, 1954.
5. Respiratory Protective Devices Manual, American Industrial Hygiene Association and American Conference of Governmental Industrial Hygienists, 1963.
6. Industrial Noise Manual, American Industrial Hygiene Association, 14125 Prevost, Detroit 27, Mich.
7. Handbook of Noise Measurement, General Radio Co., West Concord, Mass., 1963.
8. McCord, C. P. and Witheridge, W. N., Odors, Physiology and Control, McGraw-Hill Book Co., New York, 1949.
9. Blatz, H., Introduction to Radiological Health, McGraw-Hill Book Co., New York, 1964.

## toxicology

1. Sax, N. Irving, Dangerous Properties of Industrial Materials, Van Nostrand Reinhold Co., New York, 2nd Ed., 1963.

2. Browning, E., Toxicity of Industrial Organic Solvents, Chemical Publishing Co. (revised), 1953.
3. Browning, Ethel, Toxicity of Industrial Metals; Butterworth's, London, 1961.

## journals

1. American Industrial Hygiene Association Journal, 14125 Prevost, Detroit, Mich.
2. The Annals of Occupational Hygiene, The British Occupational Hygiene Society, Pergamon Press Ltd., 44–01 21st St., Long Island City, New York.
3. Archives of Environmental Health, American Medical Assoc., 535 N. Dearborn St., Chicago, Ill. 60610.
4. British Journal of Industrial Medicine, British Medical Association, Tavistock Square W. C. 1, London.
5. Health Physics, The Health Physics Society, Pergamon Press, 44–01 21st St., Long Island City, N.Y. 11101.
6. Journal of the Air Pollution Control Association, 4400 Fifth Ave., Pittsburgh, Pa.
7. Journal of Occupational Medicine, 3110 Elm Ave., Baltimore, Hoeber Medical Div.
8. Toxicology and Applied Pharmacology, Society of Toxicology, Academic Press, New York.

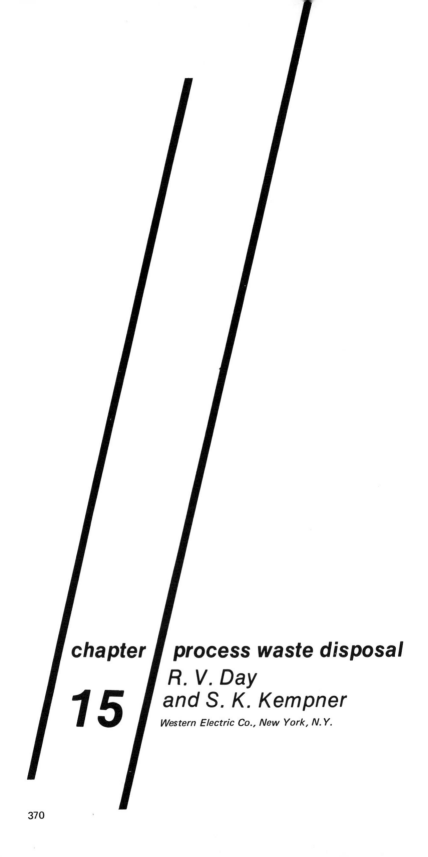

chapter **process waste disposal**

*R. V. Day*
*and S. K. Kempner*

**15**

*Western Electric Co., New York, N.Y.*

The manufacture of microelectronic equipment generates liquid and gaseous pollutants, and the treatment and removal of these waste products are essential to provide safe operating conditions in the factory and in the external environment. This chapter primarily deals with the identification, collection, and treatment of liquid and airborne waste products so that purification occurs to fulfill the commitment to the community and to comply with governmental standards.

The opening portion of this chapter deals with the sources of liquid waste, including the wastes from water softening, deionizing, plating, and etching processes. The liquid treatment process is then described beginning with the collection and segregation of the liquid wastes into separate sewer systems. The nature of the waste and the effluent standards are explained as a basis for planning the treatment process. After this has been established, the design concept and methods for the treatment of acid-alkali systems, chrome systems, cyanide systems, combined systems, and sludged disposal are reviewed.

The latter portion of this chapter deals with the steps that can be taken to remove fume and vapor contaminants from the exhaust streams before they are vented to the atmosphere. Starting with the significance of fume and vapor control the discussion then proceeds into the emission abatement equipment that can be installed on the plating and etching, paint spraying and baking, and degreaser exhausts. Equipment such as fume scrubbers, fume incinerators, carbon bed absorbers, and other solvent recovery units are described.

## liquid waste control

### preliminary planning

**Sources.** Many of the chemicals used for microelectronic processes eventually end up as a waterborne waste problem. As explained in Chap. 8, there is a large demand for soft water and deionized water in microelectronics. At the present time zeolite softening techniques are employed using NaCl to regenerate the zeolite material. The waste from the regeneration process contains high concentrations of soluble salts requiring proper disposal. The deionization process utilizes both strong alkali and strong acid in the regeneration of the ion exchange resins. This regeneration process is relatively inefficient, and high concentrations of soluble salts result in the effluent.

Etching processes produce waste that presents unique waste treatment problems. Etching consists of the removal of silicon, gold, copper, and other metals from the various substrates, and the etchant employed must be capable of dissolving the metals and maintaining the metals in solution. The removal of the metal from the waste stream becomes very difficult because the primary purpose of the etchant is to maintain the metal in solution. Conversely, the treatment process must be capable of reversing or neutralizing the basic action of the etchant. There are two waste streams generated from these processes—concentrated spent etchant and dilute rinse water containing a low concentration of the etchant. In some cases the concentrated waste can be considered for recovery of both etchants and metals to avoid a waste disposal problem. The dilute waste is acidic and contains a variable quantity of metal and acid depending on the individual operations.

Preceding many plating operations are metal cleaning processes which produce toxic wastes. The cleaning or metal preparation waste consists of both acid and alkali materials and cyanide compounds. The spent concentrated cleaning bath must be disposed of periodically as well as the continuous flow of dilute rinse water following each cleaning operation.

The plating or metal finishing techniques employed by a specific shop can be numerous and widely different in the toxic wastes produced. Cyanide is used in many plating operations (copper, zinc, cadmium, gold, etc.). Nickel and chrome plating are also widely used in microelectronics shops. The waterborne waste from these operations is limited to the diluted rinse waters following the plating operations, and the volume is variable because of the different rinsing techniques employed by each shop.

**Collection of Waste.** The various process wastes must be segregated to allow for treatment by the different disposal techniques. For instance, deadly poisonous cyanide gas would be liberated if cyanide waste and acid wastes were mixed in the same sewer system. Thus, the treatment techniques and the effluents dictate the complexity of the sewer collection system as well as the type of construction materials required for corrosion resistance.

For a typical microelectronics factory separate sewer systems may be installed for: (1) concentrated acids, (2) concentrated alkali, (3) diluted acid-alkali, (4) concentrated cyanide, (5) diluted cyanide, and (6) chrome-wastes. The construction material may vary from steel pipe for certain types of cyanide effluents to various reinforced plastics for more corrosive wastes. Unplasticized polyvinyl chloride (PVC) is probably the most common construction material used for corrosive wastes at temperatures less than $150°F$.

**Effluent Requirements.** The first step in the design of the waste treatment plant is the determination of the requirements established for the receiving stream or municipal sewer. These requirements are established by the regulatory agency having jurisdiction at the point of discharge of the waste and are set by a municipal sewer ordinance or the stream standards of a state agency.

The Water Quality Control Act of 1965 has required each state to establish stream standards for all interstate waters. These standards must be established to protect the present use of the stream as well as anticipated future use. In some streams a zoning type approach has been applied where the upstream water quality requirements are extremely high and where a certain degree of pollution is tolerated in the lower or tidal reaches of the stream.

TABLE 15-1.    Typical Standards for a Class A Stream

*Chemical Characteristics:*

| | |
|---|---|
| pH | 5–9 |
| Oil | 0 |
| Dissolved Oxygen | 3.0–5.0 ppm |
| Arsenic | 0.05 ppm |
| Barium | 1.0 ppm |
| Cadmium | 0.01 ppm |
| Chromium | 0.05 ppm |
| Copper | 0.02 ppm |
| Iron | 1.0 ppm |
| Lead | 0.05 ppm |
| Manganese | 0.5 ppm |
| Silver | 0.025 ppm |
| Zinc | 5.0 ppm |
| Chloride | 250.0 ppm |
| Cyanide | 0.00 ppm |
| Cyanate | 0.00 ppm |
| Fluoride | 1.0 ppm |
| Nitrate | 45 ppm |
| Sulfate | 250 ppm |
| Phenol | 0.001 ppm |

*Physical Characteristics:*
1. Not noticeably increase the turbidity of the stream
2. Not noticeably increase the color of the stream
3. Not have objectionable taste
4. Shall be free of offensive odors
5. Not have temperature in excess of 150°F

Table 15-1 presents a typical stream standard that would be applicable to protect the water for use as a drinking water supply and other high quality uses. Note that very strict standards are established for the metals normally encountered in plating operations. This necessitates that a high degree of treatments be employed. Stream standards also establish the effluent requirements for the municipal sewage treatment systems; thus, municipalities must enact appropriate sewer use codes to comply with these standards. Table 15-2 presents typical limits for waste discharge to a municipal sewer system.

A comparison of many sewer ordinances indicates an inconsistency in the regard to the limits established for metals and other compounds found in plating operations. Some codes state that no toxic metals will be discharged while other codes permit relatively high concentrations. Most ordinances require relatively low concentrations due to the high toxicity of metals to the biological treatment processes. Biological treatment processes are not capable of complete removal of the metals, and therefore, a municipality must limit the concentration in order to comply with stream standards.

The regulatory agency having jurisdiction may have some specific requirement in regard to installation of industrial waste treatment systems. In some areas the agency may require a batch treatment in lieu of a flow-through

TABLE 15-2.  Typical Limits or Standards for Waste Discharged
in a Municipal Sewer Ordinance

| Constituent | Limit | |
|---|---|---|
| Free Acid | None | |
| pH | 5 to 10 | |
| Fat and Oil | 100 | ppm |
| Cadmium | 2 | ppm |
| Chromium (total) | 3 | ppm |
| Copper | 1 | ppm |
| Lead | 0.3 | ppm |
| Nickel | 1 | ppm |
| Zinc | 2 | ppm |
| Cyanide | 1 | ppm |
| Suspended Solids | 500 | ppm |
| Dissolved Solids | 2000 | ppm |

type system, back-up chemical feed equipment, or duplicate instrumentation. In addition, there may be regulations regarding detention time or the furnishing of metering and sampling facilities that must be determined prior to preparation of plans and specifications.

**Waste Characteristics.**  The next step is the determination of the actual quantity and quality of the waste by a complete inventory of all operations. This inventory includes an estimate of work flow through the shop, concentration of cleaning and plating baths, estimated rinse water flow, estimated drag-out, dump schedules of concentrated plating and cleaning tanks, and other data. This information will determine the sewer system required to adequately handle the waste.

At this point consideration should be given to possible recovery of the chemicals, metals, and water to reduce the amount of wastes to be treated. Many recovery procedures are well established and have recovered the associated capital investment in a relatively short time period. Ion exchange techniques for the recovery of specific metals such as chrome or gold and the use of evaporation facilities for recovery of chrome or cyanide solutions have been employed for many years. Similarly, recovery techniques for waste etchants from printed circuit board operations can be instituted to recover the large quantity of metal and etchant that often are wasted.  A separate recovery system would be required for each specific process to avoid contamination by different process effluents being served by the same recovery unit.  Nonetheless, a waste treatment system is usually required because recovery systems will service only a part of the factory effluents.

## *treatment systems*[1-7]

**Design Concept.**  Compliance with effluent codes mandates the need for automatic controls with each treatment system.  These controls pace the addition of chemicals to ensure complete treatment in the most economical manner. Due to the variability of the waste a manual control system would cause fluctuation between overtreatment and incomplete treatment, with overtreatment usually predominating. Automatic control systems are of proven quality to produce the desired effluent characteristics.

The selection of either a batch type or a flow-through system is dependent on the waste flow unless the regulatory agency so specifies a required system. If the waste flow is greater than 50 gpm a batch treatment system is usually not economical since the tankage becomes excessive. A batch treatment system requires a minimum of three duplicate systems to provide maximum flexibility of operation; one tank being filled, one tank being treated, and the third tank being emptied. The flow-through or continuous system is used for the large flows but must always be preceded by an equalization and surge tank.

**pH Adjustment.** This process is basically concerned with acidic and alkaline solutions and elimination of most of their acidic or alkaline characteristics. Acids and alkalies are the converse of each other; the hydrogen ions (low pH) of an acid will react with the hydroxyl ions (high pH) of an alkaline solution to form water. The acid anion (mostly the chloride or sulfate ions) and the alkali cation (mostly the sodium ion) form a salt which normally remains in solution. Table 15-3 shows the basic equations involved in the pH adjustment process.

**TABLE 15-3.  pH Adjustment**

| | |
|---|---|
| Self Neutralization | $H^+ + OH^- \longrightarrow H_2O$ |
| Neutralization with Caustic Soda | $HCl + NaOH \longrightarrow NaCl + H_2O$ |
| Neutralization with Lime | $2HCl + Ca(OH)_2 \longrightarrow CaCl_2 + 2H_2O$ |

It is desirable to collect the strong acid waste and the strong alkali waste separate from the dilute waste in order to make the pH adjustment process controllable. These strong wastes are then pumped at a constant rate to the treatment system. The best features of the neutralization process with optimum pH adjustment control are achieved with a continuous flow-through system as shown in Fig. 15-1. Three tanks are used with each tank furnished with agitation capability. The initial pH adjustment should be made in the first tank while the final adjustment would be made in the third tank. Figure 15-1 shows that the strong acid and strong alkali wastes are mixed with the dilute waste just ahead of the first neutralization tank. (For a batch type treatment system the neutralization tanks should be designed to mix the waste and neutralizing agent effectively to make the process controllable.)

The system should be as fully automatic as is practical and be provided with chemical feed equipment and the instrumentation needed to continually control additions of the neutralizing agent. The treatment of an acid waste is

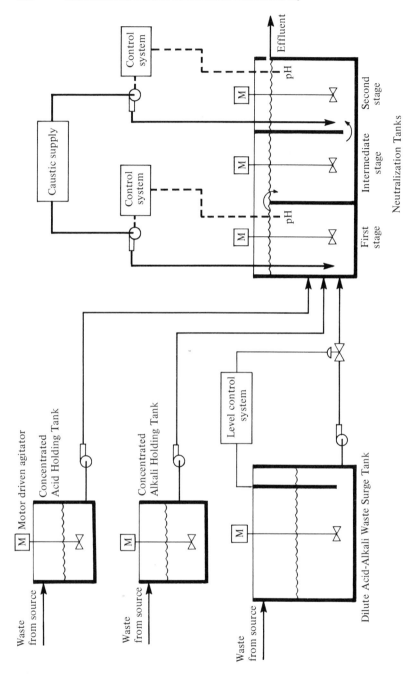

Figure 15-1 Typical waste treatment systems—Acid-alkali wastes: flow-thru neutralization system.

recommended to normally maintain a pH of zero mineral acidity in the first stage of neutralization and a pH in the second stage of neutralization as required in the subsequent combined waste treatment system.

The choice of the alkali to be used for neutralization is dependent on local conditions and the actual requirements of the alkali. Being low in cost, lime is preferred for waste which requires extensive treatment. Caustic soda is employed for waste having a relatively low treatment requirement because of its relative ease in handling.

Lime is the currently accepted alkali used for removal of fluoride bearing wastes. The partially soluble $CaF_2$ is formed and removed as a sludge, its solubility in water being 7 mg/l. as F. Present stream standards of 1-1.5 mg/l. means that sufficient water from the stream or sewer must be allocated to achieve proper dilution of the fluoride bearing effluent. A better method of fluoride removal is an important need of the microelectronics industry, since HF is a commonly used etchant.

**Chrome Treatment.** This process is basically concerned with hexavalent chromium and its complete reduction to trivalent chromium. In subsequent processes the trivalent chromium will be precipitated as the insoluble chromic hydroxide.

To achieve complete reduction to trivalent chromium requires an acid environment with a reducing agent. The rate of the reduction process increases as the pH decreases, and the reduction does not to go to completion at a pH of 5.0 or greater. The reducting reaction shown in Table 15-4 is quite rapid in the pH range of 1.5-3.0 and is completed in a few minutes. An

**TABLE 15-4.   Chrome Treatment**

| | |
|---|---|
| Chromic Acid | $CrO_3 + H_2O \longrightarrow H_2CrO_4$ |
| Sulfur Dioxide Solution | $SO_2 + H_2O \longrightarrow H_2SO_3$ |
| Reducing Reaction | $2H_2CrO_4 + 3H_2SO_3 \longrightarrow Cr_2(SO_4)_3 + 5H_2O$ |

additional retention period in the presence of a slight excess of sulfite ensures the complete reduction of the hexavalent chromium.

Figure 15-2 presents a schematic flow diagram of a continuous flow-through type system. The system consists of a reaction tank furnished with agitation and facilities to control the addition of acid and the reducing agent followed by a retention chamber to ensure the complete reduction of chrome. Preferably the trivalent chromium should not be precipitated in this system since the effluent will be discharged directly to the combined waste treatment system for further treatment.

The reduction of chrome can be accomplished using either sulfur dioxide gas, one of the sodium salts of sulfur dioxide, or ferrous sulfate as the reducing agent. The pH of the waste can be maintained at the proper level with acid, usually sulfuric acid. The reduction with ferrous sulfate can be ac-

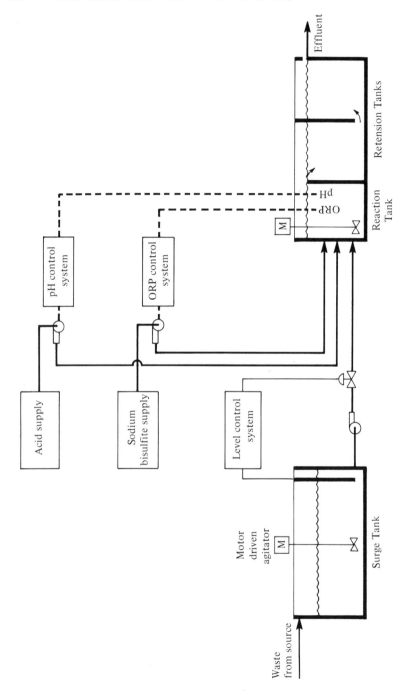

*Figure 15-2 Typical waste treatment systems—Chrome wastes: flow-thru reduction system.*

complished over a wide pH range, but this process produces an excessive quantity of an iron bearing sludge. The use of sulfur dioxide or one of its sodium salts reacts at a pH range of approximately 2.0 to 3.0 and produces a waste containing a minimum quantity of sludge.

The system should be provided with chemical feed equipment and instrumentation to continually control the process. The addition of the reducing agent is controlled to maintain a sulfite-trivalent chromium oxidation-reduction potential level, i.e., a small sulfite residual that ensures the completion of the reduction reaction. The instrumentation also controls the rate of acid additions to maintain a pH suitable for the reaction.

**Cyanide Treatment.** There are several oxidation methods that have been employed for cyanide destruction: (1) biological, (2) chemical, (3) electrolytic, (4) incineration, and (5) radiation. The chemical method employing the alkaline chlorination procedure is the most widely employed because: (1) the method lends itself to automatic control; (2) the process can be controlled to stop the reaction at the cyanate level; and (3) the process ensures 100 percent destruction of the cyanide.

This process is basically concerned with the oxidation of cyanide to carbon dioxide and nitrogen in an alkaline environment using chlorine as the oxidizing agent in two stages of treatment. In the first stage, the chlorine oxidizes the cyanide to cyanogen chloride, and this is then converted to cyanate. In the second stage the chlorine oxidizes the cyanate to nitrogen and carbon dioxide. Table 15-5 presents the basic equations of the oxidation process.

**TABLE 15-5.  Cyanide Treatment**

| | |
|---|---|
| Chlorine Solution | $Cl_2 + H_2O \longrightarrow HOCl + HCl$ |
| Hypochlorite | $HOCl + HCl + 2NaOH \longrightarrow NaOCl + NaCl + 2H_2O$ |
| 1st Stage Oxidation | $NaOCl + NaCN + H_2O \longrightarrow CNCl + 2NaOH$ |
| Hydrolysis Reaction | $CNCl + 2NaOH \longrightarrow NaCNO + NaCl + H_2O$ |
| 2nd Stage Oxidation | $3NaOCl + 2NaCNO + H_2O \longrightarrow$ $2NaHCO_3 + N_2 + 3NaCl$ |

In the first stage of treatment chlorine reacts with the cyanide radical (CN) at any pH to form cyanogen chloride (CNCl). This compound is so volatile and toxic that it is essential that it be converted by being hydrolyzed to the nonvolatile and less toxic cyanate as quickly as possible. It is important to complete such conversion before the second stage reaction occurs because the release of the end-product gases, carbon dioxide and nitrogen, in the second stage reaction tends to increase the liberation of cyanogen chloride. The rate of conversion of the cyanogen chloride to cyanate is dependent upon the pH of the waste, its conversion or hydrolysis being practically nil at pH 7.5 or less, fairly rapid at pH 8.0-8.5, quite rapid at pH 9.0-9.5, and exceedingly rapid (a matter of minutes) at a pH of 10 or above.

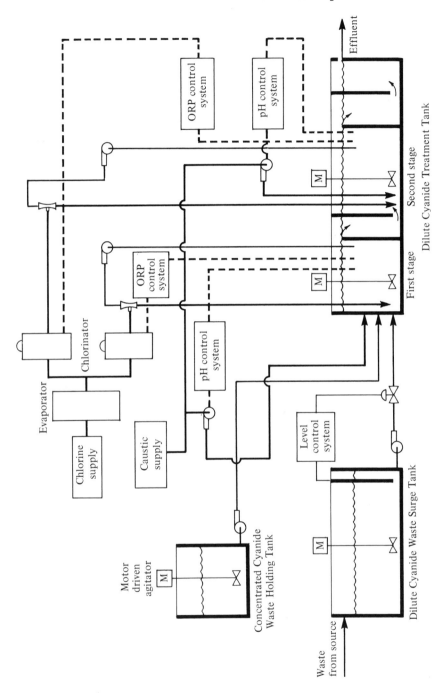

Figure 15-3 Typical waste treatment systems—Cyanide wastes: flow-thru oxidation system.

In the second stage of treatment chlorine reacts with the cyanate radical (CNO) to produce nitrogen gas and carbon dioxide. Part of the carbon dioxide reacts with carbonate or hydroxyl alkalinity to produce bicarbonates. The oxidation reaction is infinitely slow at a pH above 10, becomes increasingly rapid as the pH drops below 10, and reaches an optimum at pH 7.5.

Figures 15-3 and 15-4 illustrate flow-through and batch type treatment systems for cyanide destruction. The concentrated and the dilute wastes should be segregated so that the former can be pumped at a controlled rate together with the latter to the treatment system. The oxidation of the cyanide can be accomplished using either liquid chlorine or a hypochlorite solution as the source of chlorine. The pH of the waste must be maintained at the proper level in each stage of the process by the use of either acid or alkali.

Figure 15-3 illustrates a recirculated flow-through system designed to perform the oxidation in two separate stages and consists of two sets of reaction tanks placed in series. The first tank will be employed for the first stage of oxidation, cyanide to cyanate, while the second series of tanks provides for the second stage of oxidation, cyanate to nitrogen and carbon dioxide. Positive displacement or retention tanks should be installed following each stage of treatment to ensure the completion of the reactions prior to the discharge of the waste to the next stage of treatment.

The system should be provided with chemical feed equipment and instrumentation to control the system automatically. The rate of chlorine application is controlled by an oxidation reduction potential (ORP) instrument to maintain a chloramine-cyanate level, i.e., a small chloramine residual in the first stage of treatment. Another ORP instrument controls the rate of chlorine application to the second stage to maintain a free chlorine (ORP) level, i.e., a small free chlorine residual in the second stage of treatment. Instruments should also be provided to automatically control the pH in each stage of treatment.

Figure 15-4 illustrates a batch type treatment system designed to perform the complete oxidation of cyanide. The system is a stepwise titration of the cyanide with chlorine to produce the desired results. The titration is controlled automatically by ORP and pH instrumentation to attain the desired results. The first and second stages should be separated by the proper control of the pH. The flow diagram also shows the instrumentation required to automatically fill and empty the treatment tank.

**Combined Waste Treatment.** The removal of toxic heavy metals is common to all three waste treatment processes. The metals are present as insoluble hydrous compounds, oxides, and hydroxides in a finely divided particulate suspension. Table 15-6 shows the general equations using lime to convert any remaining metal ions into hydroxides preparatory to coagulation. It is preferable to remove metals in a common treatment system consisting of co-

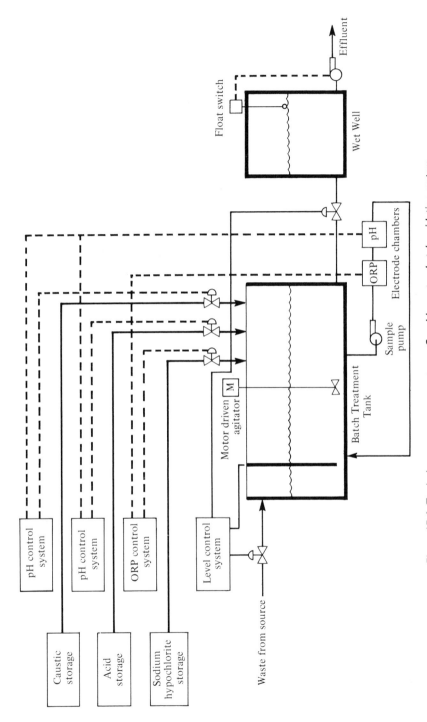

*Figure 15-4 Typical waste treatment systems—Cyanide wastes: batch oxidation system.*

**TABLE 15-6.  Combined Waste Treatment—Metal Removal (Lime as Source of Alkali)**

| | |
|---|---|
| Iron | $2FeCl_3 + 3Ca(OH)_2 \longrightarrow 2Fe(OH)_3 + 3CaCl_2$ |
| Zinc | $ZnCl_2 + Ca(OH)_2 \longrightarrow Zn(OH)_2 + CaCl_2$ |
| Copper | $CuCl_2 + Ca(OH)_2 \longrightarrow Cu(OH)_2 + CaCl_2$ |
| Chromium | $Cr_2(SO_4)_3 + 3Ca(OH)_2 \longrightarrow 2Cr(OH)_3 + 3CaSO_4$ |
| Nickel | $NiCl_2 + Ca(OH)_2 \longrightarrow Ni(OH)_2 + CaCl_2$ |
| Tin | $SnCl_4 + 2Ca(OH)_2 \longrightarrow SnO_2 + 2CaCl_2 + 2H_2O$ |

agulation and flocculation equipment followed by a high rate solids removal unit.

The three treated waste streams should be combined in a rapid mix tank for addition of coagulants and final pH adjustment as shown in Fig. 15-5. The metals are then completely precipitated from the solution to the point of minimum solubility or maximum insolubility. The pH range at this point in the processing is limited by the narrow range over which the hydroxides of trivalent chromium and zinc are more or less totally insoluble. This pH is on the order of 8.2–9.2 with an optimum of about 8.7. In deference to the most favorable pH for the use of ferric iron salts as a coagulant the pH should never be less than 9.0.

Referring to Fig. 15-5, the properly flocculated waste will pass downward through the sludge blanket in the bottom of the reaction well (the center section of the solids removal unit) and then flow upward through the sludge blanket in the solids removal compartment (the outer periphery). The flow through the sludge blanket affords the final clarification. As the waste flows upward to the overflow flume it will be continually entering an ever-widening area, causing the flow velocity to be progressively reduced. The decreasing velocity will cause the remaining floc particles to separate from the waste stream and fall back into the sludge blanket. The clarified effluent will flow over V-notch weirs into the collecting flume and then by gravity to the effluent meter pit for flow measurement and final disposal.

**Sludge Disposal.** The sludge that is withdrawn from the solids removal unit via a mechanical scraper contains about 1 percent solids and has the consistency and appearance of a very dilute mud slurry. The sludge can be disposed of by the use of a sludge lagoon or burial, but these methods usually are too expensive or not available at the site. The sludge drying beds, centrifuges, and rotary drum vacuum filters have been utilized to concentrate the sludge. Each of those can produce a relatively dry sludge having the consistency of wet clay suitable for landfill.

### effectiveness of treatment

The installation of the properly designed treatment processes and the proper operation will provide an effluent having the following characteristics:

1. One that is free of hexavalent and trivalent chromium.

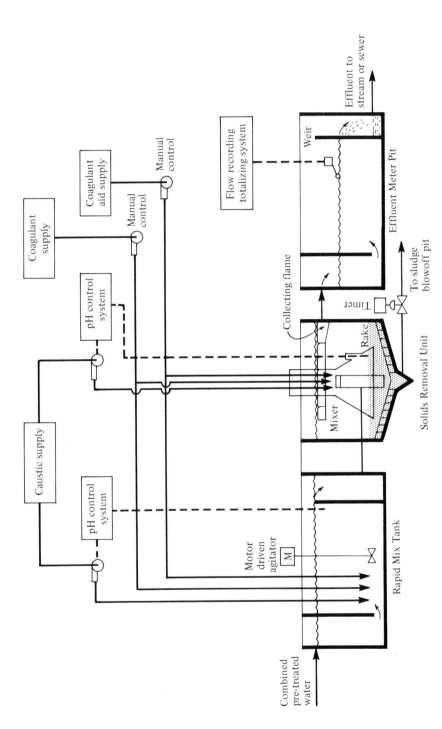

*Figure 15-5  Typical waste treatment systems—Combined wastes: flow-thru solids removal system.*

2. One that contains no cyanides except the iron cyanide complexes (the alkaline chlorination process does not destroy the less toxic iron cyanide complexes; it converts the ferrocyanide to ferricyanide).
3. One that has a pH of about 9.0 required for the efficient removal of the metal hydroxides (a final pH adjustment tank can be installed to maintain any required pH in the effluent).
4. One that will contain on the average not more than 5 mg/l. of suspended solids, mostly insoluble metallic compounds whose total metallic content is about 2.5 mg/l. and consisting of a mixture of all the various metals initially present in the raw wastes.

## vapor and fume control

### significance

Until recently, ventilating systems which remove the noxious fumes from work areas and exhaust them to the outside air were considered to be the total solution to the air pollution problems of electronics manufacture. They protected the employee by keeping vapor exposures well below the threshold limit values discussed in Chap. 14. They also provided protection for microelectronic products which might be sensitive to low level concentrations of certain chemical vapors. However, experience has shown that these corrosive fumes can be harmful in the outside environment. Building equipment in nearby roof penthouses has been damaged, and the paint finishes of automobiles parked in adjacent lots have been marred. In one case it was proven that corrosive fumes re-entering the building via the plant make-up air had caused product damage.[8] These same corrosive vapors will enter most clean rooms unless vapor control facilities are provided (see Chaps. 10 and 11). Finally, any manufacturer must consider the effect of chemical exhaust fumes on the community environment. Table 15-7 shows the Pennsylvania Ambient Air Quality Standards, as of October 20, 1969. These standards illustrate typical criteria being established for the protection of the community environment. While the values listed are not direct limits on stack emissions, they do prove that fume and vapor control beyond simple dilution is a necessity.

What kind of additional fume and vapor control is necessary for microelectronics? The answer to this question can be limited to acid-alkali and hydrocarbon emissions since they represent the major fume and vapor wastes of microelectronic processes.

### acid-alkali fumes

The chemical fumes from the various etching, cleaning, and plating processes can be in aerosol or gaseous form as shown in Table 15-8. For example, cyanides often enter the exhaust system as solid particles. This occurs when the surface of the foam created by air agitation of the plating bath dries out, leaving cyanide salts to be picked up by the fast moving exhaust air.

TABLE 15-7.  Pennsylvania Ambient Air Quality Standards Single Point Measurements,[a] Oct., 1969

| Substance | Annual | or | 30-day | or | 24-hr | or | 1-hr |
|---|---|---|---|---|---|---|---|
| 1. Suspended particles | $(65 \ \mu g/m^3)^b$ | | – | | $195 \ \mu g/m^3$ | | – |
| 2. Settled particles | $0.8 \ mg/cm^2$ | | $1.5 \ mg/cm^2$ | | – | | – |
| 3. Lead | – | | $5 \ \mu g/m^3$ | | – | | – |
| 4. Beryllium | – | | $0.01 \ \mu g/m^3$ | | $-^c$ | | – |
| 5. Sulfates (as $H_2SO_4$) | – | | $10 \ \mu g/m^3$ | | $30 \ \mu g/m^3$ | | – |
| 6. Sulfuric acid mist | – | | – | | $-^c$ | | $-^c$ |
| 7. Fluorides | | | | | | | |
| (soluble as HF) | – | | – | | $5 \ \mu g/m^3$ | | – |
| 8. Sulfur dioxide | $0.02 \ ppm^b$ | | – | | $0.10 \ ppm$ | | $0.25 \ ppm$ |
| | $(53 \ \mu g/m^3)$ | | | | $(267 \ \mu g/m^3)$ | | $(668 \ \mu g/m^3)$ |
| 9. Nitrogen dioxide | – | | $-^c$ | | – | | – |
| 10. Oxidants | – | | – | | – | | $0.05 \ ppm$ |
| 11. Hydrogen sulfide | – | | – | | $0.005 \ ppm$ | | $0.1 \ ppm$ |
| 12. Carbon monoxide | – | | – | | $25 \ ppm$ | | – |

[a]Maximum values, not to be exceeded.
[b]Geometric mean.
[c]The Council of Technical Advisors on Ambient Air Quality Standards is directed to consider such standards.
Source: Bureau of Air Pollution, Commonwealth of Pennsylvania.

TABLE 15-8.  Typical Fumes and Vapors

Aerosols—Airborne mists or fine dust transported by the exhausted air.
  1. Caustic cleaners—NaOH, KOH, etc.
  2. Metallic cyanides—$Zn(CN)_2$, $Cu(CN)_2$, etc.
  3. Acids—$HCl$, $H_2CrO_4$, $H_2SO_4$, $HNO_3$, etc.

Gases—Vapors emitted because of low boiling points, heated solutions, or air agitation.
  1. Halide gases     HCl, HF
  2. Oxides of nitrogen  $NO$, $NO_2$
  3. Acetic acid      $CH_3COOH$
  4. Chlorine        $Cl_2$
  5. Ammonia      $NH_3$

**Collection.** The hoods, ductwork, and fans required for chemical exhaust systems are standard designs well explained in *Industrial Ventilation, A Manual of Recommended Practice.*[9] The basic approach is to establish an exhaust air velocity at the process work position which is sufficient to capture the contaminated air being given off by the process solution. This velocity will vary over a wide range, depending upon the conditions existing at the work area. Significant conditions affecting capture velocities at a process include spray velocities, hood configurations, process machinery arrangements, and temperatures in plating or cleaning tanks. Table 15-9 shows the typical range of capture velocities, depending upon conditions that exist.

**TABLE 15.9.   Range of Capture Velocities**

| Condition of Dispersion of Contaminant | Examples | Capture Velocity fpm |
|---|---|---|
| Released with practically no velocity into quiet air. | Evaporation from tanks; degreasing, etc. | 50–100 |
| Released at low velocity into moderately still air. | Spray booths; intermittent container filling; low speed conveyor transfers; welding; plating; pickling | 100–200 |
| Active generation into zone of rapid air motion. | Spray painting in shallow booths; barrel filling; conveyor loading; crushers | 200–500 |
| Released at high initial velocity into zone of very rapid air motion. | Grinding; abrasive blasting, tumbling | 500–2000 |

From "Industrial Ventilation–A Manual of Recommended Practice, 11th Ed., 1970. Committee on Industrial Ventilation, American Conference of Governmental Industrial Hygienists, Lansing, Mich.

In addition to capture velocity, consideration should be given to the type of chemical fumes and the materials of construction for the exhaust system. If possible, each class of chemical vapor should be exhausted separately. The combination of different chemicals in a single duct must be carefully reviewed to prevent reactions which could result in more objectionable pollutants. For instance, cases have been noted where an easily scrubbed gas such as HCL becomes difficult to scrub when combined in a single exhaust system with $NO_2$ vapors. The duct and fans should be resistant to corrosion. This usually means a plastic or plastic-coated material. Polyvinyl chloride (PVC) is used commonly for ductwork since it is both resistant to acid-alkali attack and inexpensive. However, PVC is limited to applications where the temperature is less than $150°F$. Exhaust fans for acid-alkali systems are generally constructed of fiberglass or plastic coated steel. PVC fans are not recommended because they are structurally weak and subject to failure. When fiberglass is specified, a special coating is necessary for hydrofluoric acid atmospheres.

**Scrubbers.**   Acid-alkali fumes generally can be scrubbed (absorbed) with water. The highly soluble halide group has unique solubility characteristics when dissolved in water. Water will absorb over 35 percent hydrogen chloride by weight before the absorption process is affected by the vapor pressure of hydrogen chloride gas. In the concentrations that exist at the plating areas, contact absorption is unaffected by the concentration of solute in the scrubbing liquid (water).

Equilibrium limiting gases such as $SO_2$ and $NH_3$ need a reactive chemical added to the scrubbing liquid if they are to be absorbed in a similar fashion. For instance, tests on one particular scrubber indicated an efficiency of 6 percent with $NH_3$ using plain water as the scrubbing liquid. When a small amount of sulfuric acid was added to the water an efficiency of 90 percent was attainable.

Chemically reactive gases similar to $NO_2$ require additional research. It appears that something other than water alone is needed to affect the chemical reaction so that an absorptive reaction similar to HCL can be affected. The scrubbers commercially available at present are not capable of attaining efficiencies much greater than 50 percent with $NO_2$ on a single pass design.

The theory of the absorption of gases or vapors in a liquid is basic to the chemical engineer and has been thoroughly treated in numerous textbooks. It must be borne in mind that the air pollution control engineer is concerned not with a technology created for product purification or high concentrations

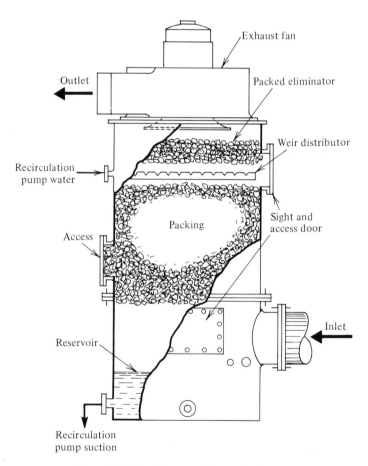

Figure 15-6  Vertical-counterflow packed tower.

but with a technology dealing with low concentrations in an air stream. An over-simplification would state that the efficiency of a specific scrubber depends to a large extent upon the amount of water-air surface area that it can create. This large amount of surface contact area is necessary so that there is sufficient intimate contact between the exhaust air stream and the scrubbing water to effect a mass transfer of the pollutant from the air stream to the water. Different scrubbers have different mechanisms for generating this water-air surface area. A spray tower does it with a tremendous number of tiny droplets of water; a plate tower causes a frothing of millions of air bubbles; the packed tower provides an enormous amount of wetted surface area over which the exhaust gases must pass.

Normally the concentration of the pollutant has an effect on the mass transfer rate, i.e., the greater the difference in concentration the greater the driving force and transfer efficiency. In the counter-current scrubber such as the vertical packed tower (Fig. 15-6) where the direction of the air stream is counter to that of the water flow, the cleanest air is scrubbed by the clean water and the dirtiest air by the dirty water. Here the driving force is at a maximum throughout the scrubber. In the co-current or cross flow scrubber (Fig. 15-7), where the air and the water flow in the same direction or at right angles to each other, this is not the case.

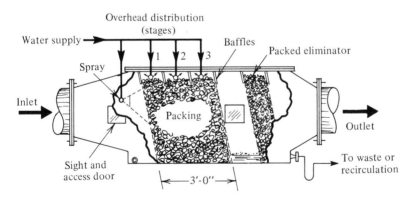

*Figure 15-7  Deep-bed horizontal gas scrubber.*

Another important parameter affecting the efficiency of a scrubber is the internal face velocity of the air stream. A plot of efficiency vs. water rate for a vertical packed tower with HCl gas (Fig. 15-8) shows an increase in efficiency with an increase in air flow. This points out the inadvisability of over-sizing a unit to accommodate a possible future need for increased capacity. For best efficiency the vertical scrubber should operate fairly close to its flooding velocity—the velocity of the air stream which will carry the scrubbing liquid or water out the exhaust duct rather than permitting it to flow down over the packing in a normal fashion. A plot of efficiencies vs. water rate for

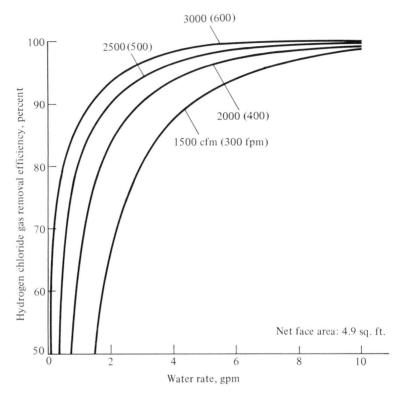

*Figure 15-8 Performance of vertical scrubber at 20 mg/m$^3$, HCl.*

a horizontal or cross-flow scrubber (Fig. 15-9) shows the opposite relation-ship. The efficiency of this type of scrubber can be improved by decreasing the face velocity of the exhaust air stream.

In selecting a scrubber the manufacturer's data must be carefully scrutin-ized so that the proper type is selected for a specific pollutant. A comparison of the efficiency vs. water rate for six different types of scrubbers in cleaning hydrogen chloride gas from an exhaust air stream is shown in Fig. 15-10.[10]

### hydrocarbon emissions

The organic chemicals used in the microelectronics industry are generally less corrosive to materials, but the flammability hazard is correspondingly more significant. Thus the materials used in the hydrocarbon fume exhaust systems be of sheet metal and steel construction, but the hoods and horizontal ducts require built-in fire protection. The toxicity potential of organic solvents varies; some common solvents like trichloroethylene and xylene have TLV's of 100 ppm while others like acetone and Freon have TLV's of 1000 ppm. The air pollution hazard from organic solvents appears to be less of a problem. For a microelectronics plant many solvent cleaning work positions consist of

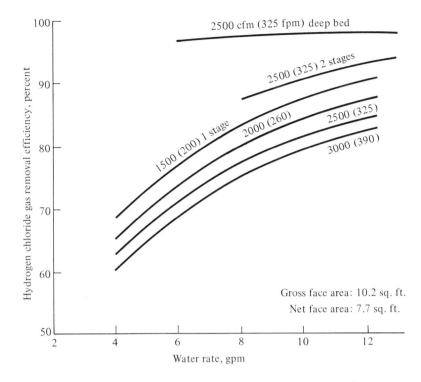

*Figure 15-9   Performance curves of horizontal scrubber at 20 mg/m³, HCl.*

small tanks containing one gallon or less of a solvent such as acetone. In cases of vapor degreasing where larger volumes of solvent are involved reclamation by absorption or refrigeration becomes economically practical. Where flammable solvent concentrations in the exhaust reach levels of 1500–2000 ppm flame incineration can be employed. In these cases the resultant reduction of solvent in the exhausted air reduces the air pollution potential.

**Solvent Emission Control.** Where vapor degreasing with some type of chlorinated hydrocarbon is employed, the cost of the solvent is generally high enough to justify air pollution control equipment in the form of a solvent recovery unit on cost savings basis. One such unit is a two bed activated charcoal absorber similar to equipment described in Chap. 10. During operation one bed absorbs the solvent from the exhaust air while the other bed is steam stripped. Automatic timing controls periodically reverse the operation to the alternate carbon bed. Stripping or regeneration is accomplished by passing steam through the bed. The heat of the steam drives off the solvent into a condensing still where the steam and solvent are separated and the solvent returned to the degreasing unit.

A second type of recovery unit for a vapor degreasing tank is a simple refrigeration coil. In a vapor degreaser the solvent in the tank is heated to

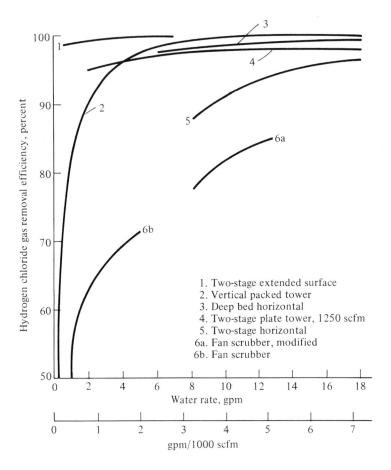

Figure 15-10 A comparison of scrubber effectiveness at 2500 SCFM, 20 mg/m³, HCl.

about 190°F to form the degreasing vapor. To prevent excessive loss of this vapor, cooling coils with water at normal plant temperature are placed inside the tank around its perimeter to condense the vapor and return it to the tank. This cooling is effective, but a certain amount of the vapor escapes. By placing refrigeration coils at a temperature of about −20°F in the tank above the normal cooling coils additional condensation is accomplished, and a substantial reduction in solvent loss and/or air pollution is effected.

Where the concentrations of nonchlorinated solvents are high enough, flame incineration is a positive means of solvent emission control. (Chlorinated hydrocarbons should not be incinerated because HCl gas is formed in the process.) The contaminated exhaust air is passed through a fume incinerator where it mixes with the combustion gases and comes in direct contact with the flame—usually at temperatures from 1100 to 1400°F. Where large volumes of air are exhausted from a solvent work station the incinerator fuel requirements become a meaningful consideration. For instance, 5000 SCFM of exhaust air heated to 1200°F require roughly 8,000 SCFH of natural gas. Therefore, the recovery of this waste heat should be considered before installing incineration equipment.

Incineration temperatures in the area of 600 to 900°F are feasible by using a catalyst—often a platinum alloy—to aid the combustion process. When using a catalyst, the potential for poisoning of the catalyst by the offending vapors should be determined prior to the installation.

**Paint Spraying and Baking.** Finishing areas where parts are sprayed with paint and then baked or dryed in an oven are sources of solvent or hydro-carbon emissions. Paint or lacquer is composed of roughly 25–50 percent solid material dissolved in roughly 50–75 percent liquid solvent. In the course of the spraying and drying operation about 90 percent of this solvent eva-porates and is carried out the exhaust stack. (A water wash spray booth re-moves only the solids from the exhaust air, the vaporized solvents passing right through.)

Because labor regulations require high face velocities at paint spray booth areas, the exhaust air from a spray booth generally has a very low contami-nant concentration in a high volume air stream. Thus the size requirements for any type of pollution abatement equipment, whether it be a carbon bed absorber or a fume incinerator, become quite severe. It is safe to say at this writing that it is impractical to install solvent emission controls on spray booth exhausts. One of the main objections to hydrocarbons in the atmosphere is that certain ones (classified as reactive) are subject to photochemical reactions with $NO_2$ in the presence of sunlight to form photochemical air pollution which causes eye irritation and vegetation damage. The answer to the spray booth emission problem appears to be the reformulation of paints, thinners, and lacquers to a product whose vapors are not subject to photochemical reaction.

## summary

The proper collection and disposal of process waste is the last stage of the chemical processes so important to the making of electronic products. The same precision used to make the product also must be applied to controlling liquid and gaseous pollutants because of the potential harm to employees, the products, and the community environment. There are many ways to accom-plish this control, with the more common techniques having been described in this chapter. Looking to the future it is axiomatic that better waste disposal techniques will have to be developed to handle the increasing quantity and variety of waste by-products of a rapidly changing microelectronics manu-facturing technology.

# references

1. Anon., "No Pollution Here! Plant Burns its Cyanide Wastes," *Plant Engineering,* June, 1961.
2. J. J. Novotny, "Incinerator Burns Liquid Waste Safely," *Plant Engineering,* December, 1964.
3. R. F. Byron, J. Danaczko, Jr., A. L. Dixon, and L. M. Welker, "Radiation Decomposition of Waste Cyanide Solutions," U.S. Patent No. 3,147,213.
4. J. K. Easton, "Electrolytic Decomposition of Concentrated Cyanide Plating, Wastes," *Plating,* 53, 1340 (November, 1966).
5. L. B. Sperry, and M. R. Caldwell, "Destruction of Cyanide Copper by Hot Electrolysis," *Plating,* 36, 343 (April, 1949).
6. N. S. Chamberlin, and J. B. Snyder, Jr., "Technology of Treating Plating Wastes," 10th Industrial Waste Conference Purdue University (1955).
7. N. S. Chamberlin, and R. V. Day, "Technology of Chrome Reduction with Sulphur Dioxide," 11th Industrial Waste Conference Purdue University (1956).
8. L. N. Carbonaro, "Investigate Causes for Corrosion in the Buildings of the Hawthorne Works Areas and Recommended Methods for Control," Unpublished Report, Western Electric Co., Chicago, 1968.
9. *Industrial Ventilation, A Manual of Recommended Practice,* 11th edition, Am. Conference of Governmental Industrial Hygienists, Lansing, Mich., 1970.
10. S. K. Kempner, E. N. Seiler, and D. H. Bowman, "Performance of Commercially Available Equipment in Scrubbing Hydrogen Chloride Gas," Journal of the Air Pollution Control Association (March 1970).

# bibliography

1. A. C. Stern, *Air Pollution,* Volumes I, II, III, Academic Press, New York, 1968.
2. *Air Pollution Manual,* Parts I & II, American Industrial Hygiene Association, Detroit, 1968.
3. *Air Pollution Engineering Manual,* U.S. Dept. of Health, Education & Welfare, Cincinnati, Ohio, 1967.
4. G. M. Fair, J. C. Geger, and D. A. Okun, *Water and Wastewater Engineering,* Vol. 2, John Wiley & Sons, Inc., New York, 1968.
5. *Sewage Treatment Plant Design,* American Society of Civil Engineers, New York, 1959.

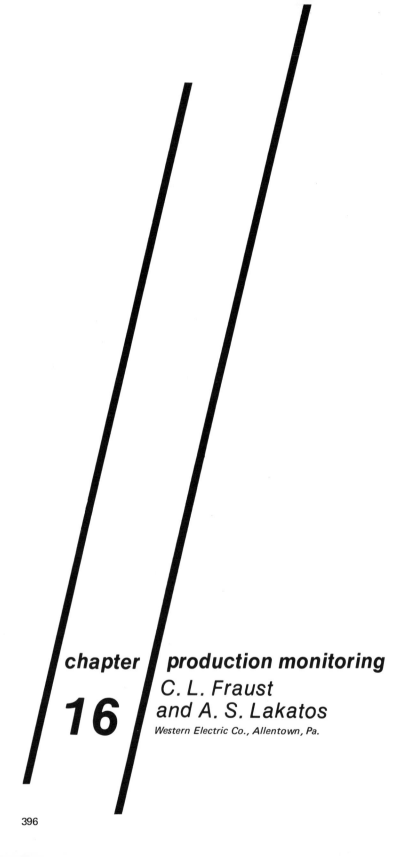

**chapter**

**16**

**production monitoring**

*C. L. Fraust*
*and A. S. Lakatos*

Western Electric Co., Allentown, Pa.

Now that the reader has become acquainted with the problems of microelectronic device manufacture as well as the complexities of environmental control procedures employed to produce uncontaminated devices, it is time to consider techniques to ensure that these control facilities are maintained. By this time it should be apparent that the environment greatly influences device processing since microelectronic devices are sensitive to contamination of every size, shape, and form. It should also be apparent that the expense of environmental control imposes severe restrictions on wholesale implementation of control facilities. Because of this, good engineering judgment is of prime importance. Where possible, decisions as to the necessity of specific forms of environmental control have resulted from monitoring methods that indicate detrimental levels of contamination deemed harmful to microelectronic devices. Since any exposure to ambient conditions will lead to contamination, an important phase of this work involves decisions as to the relative risk entailed by exposure of devices to the atmosphere for specified lengths of time.

Monitoring is the process that enables the engineer to obtain information related to the continuing performance of an environment. The environment may take the form of a single work position or may be the open floor factory ambient. Monitoring may also be provided for direct measurement of contaminating agents or indirect measurement of process parameters. Typical of the latter are various electrical tests performed to ensure the integrity of devices at various production stages. Diligent use of monitoring programs will detect product contamination problems and the type of control needed. Monitoring programs enable the engineer to evaluate existing control facilities and determine its effectiveness as well as its maintenance requirements.

The purpose of this chapter is to familiarize the reader with the various techniques employed in providing monitoring programs in microelectronic device manufacturing. Three phases of monitoring will be considered. First, a discussion of electrical monitoring schemes will be presented. Since the device itself represents the premier manifestation of contaminants, it is fitting that monitoring schemes start at this point. Next, representative forms of direct monitoring associated with manufacturing will be discussed. This section will also treat monitoring of environmental control facilities. Finally, monitoring schemes applicable to studying general plant environments will be explored. This information is presented to indicate how one analyzes for various types of common contaminants that may be present in the general ambient.

## *electrical monitoring of product*

In Chap. 2 it was seen that reduced yield in microelectronic circuits can usually be related to process contamination in one form or another. The types of defects and failures caused by contaminants can generally be classified into three types: surface effects, bulk effects, and contact and interconnection problems.   Much work has been done in the semiconductor field in analyzing device failures and relating them to the appropriate failure mechanisms.[1]   This effort has provided the device engineer with a long list of relationships between the degraded electrical behavior of a device and its probable cause.   Monitoring the electrical characteristics of a product thus seems to afford many useful methods for the measurement of contaminant related effects; the more common techniques will be presented and will draw heavily on the material covered in Chap. 2.

## types of measurements

Table 2-1 relates the major failure modes in microcircuits to their probable contaminant cause.   Based on this information, effective electrical controls can be established by in-process monitoring of such product electrical parameters as: 1) breakdown voltage; 2) leakage current; 3) low current gain ($h_{fe}$); 4) saturation voltage ($V_{CE(SAT)}$); and 5) capacitance-voltage relationships. Several of these parameters are normally used for in-process monitoring of the device characteristics as a control of the manufacturing operations themselves.

Measurements of the breakdown voltage and leakage current of a junction can be readily accomplished following the photoresist and oxide etch step which precedes the next diffusion operation in the process.  Thus, the integrity of the isolation junction can be verified following the base photoresist and etch operation since contact can be made with electrical probes to adjacent isolation regions through the open base areas.  The base-collector junction can be tested prior to emitter diffusion provided the collector contact or deep collector area has also been opened up in the oxide as is normally the case.  Finally, the completed transistor including the base-emitter and collector-base junction can be tested prior to metallization when all the contact areas are opened in the oxide.

Since the reverse characteristic of a junction is susceptible to almost every type of contaminant commonly found in the manufacturing process, it is an especially powerful tool to be used in control procedures.  The sensitivity of a reverse biased junction to contaminating effects is readily seen by considering the very high fields and low currents characterizing operation in this area and the fact that the contaminants and defects are strongly influenced by these high fields.  Typical field strengths in the depletion region of a reverse biased junction are in the neighborhood of $10^5$ volts/cm.[2]

**Breakdown Voltage, Leakage Current, and Gain.**  Breakdown voltage and leakage current monitoring will compare actual voltage-current relationships against ideal junction criteria.   The solid line in Fig. 16-1 shows the I-V characteristic for an ideal p-n diode.  Semiconductor devices are especially susceptible to ionic contaminants at the junction surface.  A positive charge in the oxide film will bend the Fermi level and enhance n type silicon toward $n^+$ at the silicon–silicon dioxide interface.[3]   This will result in a breakdown

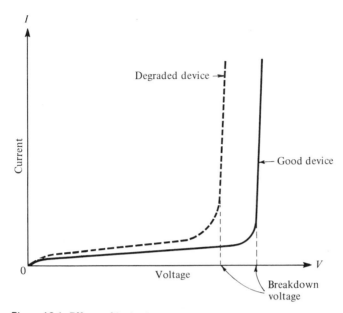

*Figure 16-1  Effects of ionic charge in oxide layer on breakdown voltage.*

voltage somewhat lower than that expected and is illustrated by the dashed line in Fig. 16-1. Since the isolation and collector-base diode characteristics are primarily dependent upon the $n$ doped side of the junction, the cleaning, diffusion, and oxide growth operations accompanying the formation of these junctions must be carefully controlled against ionic contaminants which can cause gross reduction in the theoretical breakdown voltage expected.

Channeling, as shown in Fig. 2-5, is characterized by a relatively high constant leakage current, on the order of microamps, extending out until the breakdown region is reached. Its cause is generally attributed to contamination by positive ions such as sodium and potassium in the oxide or oxide-silicon interface causing a shallow inversion layer at the surface which effectively increases the junction size. This may occur for both the collector-base and emitter-base characteristics. Prolonged exposure to the atmosphere and human handling are particularly abundant sources of this type of contamination.

Metal atoms, especially copper, have been shown to selectively precipitate in junction areas and dislocation regions causing a changed voltage-current relationship referred to as "softening." A similar "soft" characteristic can also be caused by ionic precipitates in the junction transition region.

A more general type of degradation is illustrated by Fig. 16-2. Basically, a resistive path in parallel with the junction is found to exist. This may be caused by surface contamination by any impurities which become conductive in the presence of moisture from the air. Improper cleaning and rinsing after a plating operation, acid etch procedure, etc., can leave such residues. An

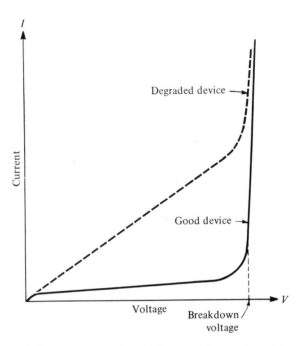

*Figure 16-2 Soft IV characteristic.  This is created by metal precipitates in junction areas and dislocation regions.*

oxide pinhole caused by particulate contamination at an earlier step may serve as a trap for contaminants causing resistive leakage in parallel with a junction.  A diffusion pipe caused by a pinhole in a prior photoresist step will have the same general characteristic.  Variations of this type of failure can occur.  For instance, the junction leakage may look normal to a particular voltage level, then become resistive out to the true breakdown voltage.  The cause is the spreading of the depletion region with increased voltage into an area with a defect similar to those listed.

Along with leakage and breakdown measurements, low current gain ($h_{fe}$) measurements are also useful for observing contaminant effects.  The same contaminants which increase junction leakage currents will also degrade $h_{fe}$ since the small signal currents will be masked by the leakages.

**Saturation Voltage.**  $V_{CE(SAT)}$ is a measure of the collector-to-emitter voltage of a transistor operating with both the collector-base and emitter-base junctions forward biased, i.e., the transistor is being turned on hard.  In this mode of operation the device is especially sensitive to contact resistance problems because any additional resistance in series with the junction will cause an IR drop which adds to the $V_{CE(SAT)}$ readings.  Device failures due to contact resistance are generally attributable to contamination during the formation of the contact areas.  Whether the contacting procedures involve deposition of metals by evaporation, sputtering, plating, or by actual bonding of metals, careful preparation of surfaces is required. Contamination may take the form of both organic and inorganic impurities and generally leads to resistance and adherence problems. Analysis of bonding failures has shown the presence of chloride ions and hydrocarbons at the bond interface, with the hydrocarbons attributable to poor removal of photoresist from a prior operation.[4]

**Capacitance-Voltage Relationships.** Unlike the previous techniques which use measurements on the actual device as controls, C-V measurements are generally taken on dummy wafers which are processed along with actual product. The technique involves the formation of a metal-oxide semiconductor (MOS) capacitor on the dummy control wafer. The structure of a typical MOS capacitor is shown in Fig. 16-3a.

The capacitance measurements should be performed at some frequency above 500 kHz so that minority carriers cannot follow the variation of the

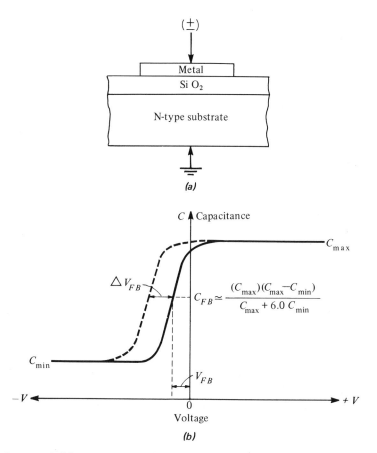

*Figure 16-3 (a) MOS capacitor. (b) Typical capacitance vs voltage characteristic.*

measurement signal and are held at the surface by the applied bias. Testing involves sweeping from some negative voltage through zero to some positive voltage, with an x-y recorder plotting capacitance vs. voltage. The maximum positive voltage is held for some period of time (approximately 30 sec), and the procedure is then reversed, sweeping from positive to negative voltage. Figure 16-3b shows such a C-V characteristic.

From this curve a number of parameters can be calculated including the effective dielectric thickness ($t_{ox}$), flat band capacitance ($C_{fb}$), flat band voltage ($V_{fb}$), surface state charge density ($N_{ss}$), and threshold voltage ($V_t$). Of these, the surface state charge density ($N_{ss}$) provides a particularly good indication of contaminant levels since it is a measure of space charges within the dielectric due to mobile ions, fast surface states, surface stage charges, and ionized traps. Ionic contamination thus adds to the surface state density. The effect of trapped ions can be seen by a horizontal translation to the left of the C-V trace as shown by the dotted line in Fig. 16-3b. As the bias voltage is increased toward the maximum positive value, mobile ion contaminants will tend to drift to the oxide-silicon interface. This will cause an additional accumulation at the silicon surface resulting in an increase in the threshold voltage and subsequent shift to the left of the C-V plot. Elevated temperatures will enhance this drift and produce more dramatic shifts. Measurements taken over a period of time have shown that shifts of less than 0.5 volt at 300°C are possible with good production controls while ionically contaminated wafers will shift well over 100 volts with this measurement technique. Therefore, C-V measurements can be used to monitor contamination of diffusion and oxide growth operations as well as the cleaning procedures accompanying them.

### limitations

Electrical monitoring of product for contamination is most effective when employed in conjunction with direct process controls and other contaminant measurement systems. The measurement results are indirect (the effect rather than the actual contaminant is shown) and only partially quantitative (the amount and specific type of contamination is not determined). Also, such tests are after the fact, i.e., the detected device failure means that the device must be junked. Nonetheless, electrical monitoring provides a useful source of data for contamination analyses.

## in-process monitoring

Monitoring of contamination may take several forms when performed during the production of microelectronic devices. For one thing the monitoring may be applied directly to a particular device. Other monitoring techniques are applicable in measuring a specific environment encountered by the device or its bulk material precursor. Monitoring may be continuous or periodic at various frequencies dependent upon the information desired. Monitoring may involve sophisticated instrumentation or may involve a series of simple direct observations. This, too, depends upon the particular information required.

## visual inspection

Continuously employed visual inspection at various stages of production is a very important form of monitoring. While examination is primarily to un- cover flaws, it is also possible to observe the presence of foreign matter such as particles, films, or residues on the surface of material. Pinholes can also be detected by visual techniques. An extension of this type of analysis is the automatic microscope discussed in Chap. 5. With this instrument it is possible to scan individual devices and quickly determine the number and size distri- bution of particles or pinholes on a device surface.

Another example of a visual technique is the transparent tape test to measure film adherence. In this test a piece of adhesive tape is placed on the surface and rapidly pulled off. Poorly adherent films will peel with the tape. A similar technique involves attaching leads to the surface of the substrate and noting if an applied pulling force produces peeling.[5]

## sealed device monitoring

Many microelectronic devices manufactured are hermetically sealed to isolate them from the environment. To ascertain whether the seal provided is ade- quate, the devices are monitored for the presence of minute leakage paths which may result in premature device degradation under operational condi- tions.[6]   The system involves a pressurized exposure of sealed devices to radioactive krypton, Kr-85, an inert radioisotope possessing a 0.52 MeV gamma ray and a 0.24 MeV average beta ray. If a leakage path is present, Kr-85 will flow through the opening into the device.

Detection of the Kr-85 inside the device is accomplished using a scintilla- tion counter. Knowing the specific activity of Kr-85, the counts can be related to the volume of the radioactive gas in the device. For a given ex- posure time and gas pressure the leak rate under standard operating condi- tions can be calculated for comparison against criteria established by product reliability data.

## organic film monitoring

As was pointed out in Chap. 6, exposure of a substrate or device to minute traces of vapors still leads to rapid formation of an organic surface film. To ensure that a surface is free of excessive amounts of this type of contaminant, several types of batch surface monitoring are performed.

A prime means of estimating contamination from organic films is measure- ment of the contact angle produced when a droplet of deionized water is placed on a surface that is normally wettable.[7,8]   Since hydrophobic organic films will render a surface less wettable, the contact angle may be related to the degree of surface contamination. (Chapter 6 describes the limits of appli- cation.)   Figure 16-4a illustrates the contact angle. To measure this quantity,

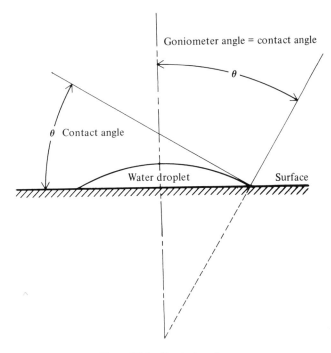

Figure 16-4a  Contact angle.

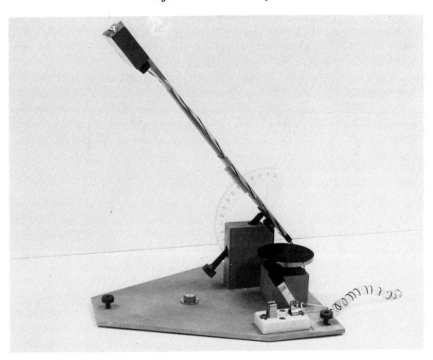

Figure 16-4b  Reflection goniometer.

a drop of deionized water is placed on each of several representative test surfaces. Then a beam of light is focused downward towards the center of the droplet. The reflection of this light source on the bubble surface is viewed through a small aperture or lens located adjacent to the light source. The light source and viewing device are mounted on an eccentric block connected to a moving arm goniometer so that movement of the light source produces parallel movement of the goniometer. By moving the light source one is able to track the motion of the reflection along the bubble surface and accurately determine the contact angle, defined by the intersection of the substrate surface and a tangent drawn to the point of intersection of the droplet and the substrate. Figure 16-4b illustrates this piece of equipment. From geometric considerations the angle turned by the goniometer arm is equal to the contact angle. The larger the contact angle, the more contaminated is the surface. Similarly, the larger the rate of increase, the more rapidly the surface is being covered with hydrophobic contamination.

Two other techniques—the atomizer test[9] and the water break test[10]— are available for monitoring surface cleanliness by surface-wetting ability. These techniques are more subjective in that they involve observing the pattern produced by wetting a surface either by spraying a fine mist onto the surface or by dipping the surface into deionized water. Comparisons with pictures of standards provide the basis for acceptance or rejection.

A final technique for detection of surface contamination is that of evaporative rate analysis.[11]   In this method a radioactive tracer, generally a C-14 tagged compound dissolved in a volatile solvent, is employed. A drop of the solvent is placed on the test surface, and a Geiger-Mueller tube is used to detect the beta emissions from the surface. Ideally, the rate of evaporation of tracer and solvent, enhanced by passage of nitrogen over the surface, is rapid for a clean surface. In the presence of organics, however, the evaporation rate is retarded, and the radiation counts will remain high relative to those for clean surfaces. Factors leading to nonideal behavior are surface porosity, surface area, particles, and contaminating films that are miscible with the tracer material.

## water system monitoring

**Resistivity.** In Chaps. 6 and 7 it was shown that many microelectronic devices are dependent upon extremely high purity water (up to 18 megohm-cm) for chemical bath formulations, rinsing, and wet oxidation processes. The quality of the water can be guaranteed by continuous resistivity (or conductivity) measurements using an immersed probe and a Wheatstone bridge circuit. Automatic operation with compensation for temperature is possible with some types of instrumentation.

These measurements are used to determine when it is time to regenerate the deionizing columns which produce the high purity water (see Chap. 8).

At a rinse position water resistivity measurements also serve to indicate when the rinsing operation has been completed. During a rinsing operation ionic impurities are washed off the process parts being cleaned, thus decreasing the resistivity of the water. For example, if as little as 1 ppm of NaCl is present in otherwise pure water, the resistivity will drop from 18 to 0.45 megohm-cm.[12] Resistivity monitoring of rinse water is a highly sensitive monitor of ionic contamination that can be done on a continuous basis.

There are other forms of contaminants in high purity water for which periodic in-process monitoring techniques have been developed. The silting index[13] and filter plugging tests[14] for particles have already been discussed in Chap. 5. The discussion here will be limited to silica analysis, specific ion analysis, and microbiological monitoring.

**Silicates.** A material that presents water problems in microelectronic device production is silicon—specifically in the form of insoluble silicates. Ordinarily, deionization columns will remove the great majority of |soluble silicates. Effective filtration of colloidal silica suspensions, however, is difficult since their particle size can be considerably less than 0.1 micron. Periodic monitoring of process water for the presence of colloidal and soluble silicates is desirable to relate its significance to process performance and product failure modes.

The analysis technique for silicates is outlined in ASTM Standard Method D859-68.[15] The method involves complexing silica with ammonium molybdate to form a yellowish-green complex and converting this to a blue complex by the addition of 1-amino-2-naphthol-4 sulfonic acid. This method is sensitive to concentrations of silica in the range 0.02–1 mg/1. While the method is specific for soluble silicates, it has been shown that for low concentrations of silica, the hydrous silica is converted to a form that is complexed with ammonium molybdate.[16] The blue complex obeys Beer's law and can be quantitatively analyzed using a spectrophotometer and standard silicate solutions.

**Ion-Selective Analysis.** The monitoring of purified water supplies for specific inorganic materials such as sodium and potassium is sometimes required for contaminant source identification. This is generally accomplished either by atomic absorption spectroscopy (see Chap. 3) or by ion-selective electrodes. Atomic absorption techniques are basically of a batch type. Samples of water are examined periodically for a variety of trace impurities using a variety of hollow cathode lamps, each specific for a particular element. Quantitative analyses are performed by comparing results with those obtained from standards made up to contain known amounts of the element in question.

A technique applicable to continuously monitoring for specific elements is that of ion selective electrodes. Among the materials detectable by this type of technique are sodium, potassium, calcium, lead, cadmium, bromide, chloride, fluoride, cyanide, iodide, nitrate, perchlorate, and sulfide.[17] The electrodes employ either solid or liquid membranes across which there is a potential difference between the sample and a reference. The response to varying concentration obeys the Nernst equation as follows:[18]

$$E = E_a + \frac{RT}{nF} \ln a = E_a + \frac{RT}{nF} \ln \gamma (X)^n \qquad (16\text{-}1)$$

where

$E$ = measured potential
$E_a$ = reference potential
$R$ = universal gas constant
$T$ = absolute temperature
$n$ = charge of ion
$F$ = Faraday constant
$a$ = ion activity
$\gamma$ = single ion activity coefficient
$X$ = ionic concentration

**Biological Contaminants.** Conventional water treatment at city treatment plants involves chlorination to a residual chlorine concentration for biological control.[19] Under certain conditions some organisms will survive this treatment, and it is possible for them to become entrenched in the deionized water system. Microorganisms such as bacteria, bacterial spores, and fungal spores readily become airborne from soil deposits and can enter a factory environment with ventilation air. Deionized water rinses may become contaminated from settlement or diffusion of organisms into the rinse tank. It is most likely that microbial, especially bacterial, contamination arises from accidental or careless contact of human hands with the tanks or from other human initiated causes. Decontamination is generally best accomplished with hypochlorite solution. Halogenated hydrocarbons, ethylene oxide, and formaldehyde are other examples of sterilizing agents.[20]

The problems associated with microbial contamination go beyond the presence of viable organisms. The prime reason for this is the ability of viable organisms to reproduce in an exponential manner if given a favorable environment for growth. A typical doubling time for this growth is only 20 minutes.[21] Both dead and live organisms may cause physical contamination, much the same as any form of particulate contamination. The chemical composition of the organisms may also be of importance. While these organisms are simple compared to higher forms of life, they are composed of water, carbonaceous material (fats, proteins, carbohydrates, enzymes, vitamin structures), and inorganics. The main elements found in these lower forms of life are carbon, hydrogen, oxygen, nitrogen, sulfur, and phosphorus. Other elements found in some organisms include potassium, iron, cobalt, magnesium, manganese, calcium, sodium, molybdenum, copper, and vanadium.[21,22]

A variety of techniques are available for sampling for microbial contamination. Basically each technique involves first trapping organisms, then culturing them in nutrient broth, and, finally, visual counting of colonies produced. The theory behind this type of sampling is that organisms will, in the main, be collected as individual cells. These cells are cultured and undergo cellular division so that a colony of organisms is formed which may now be readily counted. The number of colonies produced is used to represent the number of organisms originally present on a one to one basis. This assumption is not completely valid, since organisms may be present initially in agglomerates or filaments.

Two techniques are recommended for analysis of samples believed to contain bacteria.[23,24] The first of these is the standard plate count. In this procedure 1 ml of sample is placed in a sterile petri dish. Sterile nutrient agar is added. For highly contaminated water it may be necessary to dilute the sample to ensure that discrete colonies will be produced rather than a mat of organisms. The covered petri dish is then inverted, placed in an oven maintained at constant temperature for a specified incubation period (usually 48 hours at 25°C or 24 hours at 35°C), and then counted with the aid of a microscope.

The second technique involves filtering 10 ml of water through a 0.45 micron membrane filter. To the filter holder on the underside of the membrane, nutrient agar is introduced onto an absorbent pad either by pipette or by syringe. The filter holder is then capped, inverted, and placed in an incubator as described earlier. After incubation the filter may be stained to highlight colonies. Results are reported as the number of colonies found after incubation per ml of water examined.

### monitoring storage facilities

**Need.** Storage facilities are extremely important considerations in any formal scheme of environmental control. Since it is not always possible to process material completely in one continuous flow scheme, it becomes necessary to store this material at various stages of production until the next process may be implemented. While some difference in style and material may be noted among storage boxes, the basic method of operation is the same. The storage box is generally a cubicle with access doors for entering and removing stored material. A dry environment is maintained using a compressed air or nitrogen gas supply of proven quality. The sides and the doors of each facility should be designed to prevent the influx of particles, organic vapors, and moisture into the dry box during insertion and removal of product.

Tests have shown that particulate levels and vapor levels in a storage facility have been degraded to levels found in the ambient air surrounding the facility because of frequent use. Each time the dry box was opened particles and vapors were introduced inside. When the box was sealed again, the supply gas of proven quality in a properly operating dry box should have flushed both vapors and particles from within. The effect of opening and closing a dry box is illustrated in Fig. 16-5 for the example of particles.

In general, it is desirable to supply enough gas of proven quality to provide approximately five turnovers in one minute. This may be accomplished by continuously supplying gas at the required flow rate or by supplying a smaller amount of gas except when opening the box. Then a sufficient flow of gas is supplied to produce the required flushing.

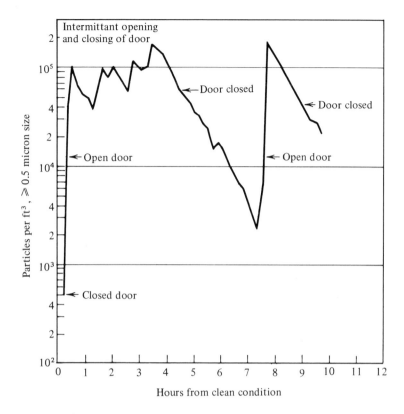

Figure 16-5  *Effect of opening and closing a dry box door.*

### clean room and clean bench monitoring

Clean room and clean bench facilities afford the user a means of controlling particles in the work environment to within specified limits. Detailed discussions of these facilities may be found in Chap. 11. Monitoring the clean facility periodically ensures that continuing particle control is achieved during production use. Particle measurements with a light scattering particle counter will alert operating personnel to maintenance problems with the clean air supply (damaged filters and poor air distribution). When a filter becomes plugged, the resistance to flow through it increases. This in turn leads to a lower air velocity through the filter. A plugged filter condition is readily determined by monitoring the velocity profile across a room or across the face of a filter with an anemometer.

Monitoring of clean rooms may also be used to indicate the effects of disturbances such as the movement of workers within the clean room, heat producing ovens and furnaces, and mechanical equipment. The implementation of this type of monitoring has been greatly improved by the development of equipment to interface light scattering particle counters with high speed digital computers.[25]   Data from particle counters can be directly collected onto paper tape or magnetic tape so that the large amounts of data collected may be analyzed with a computer. Using such a system the time savings in monitoring data handling and analysis is astronomical. Without the burden of manual data handling, it is now possible for the engineer to devote significantly more time to the more important aspects of evaluation of particle monitoring data. Typical results obtainable with computerized data handling are found in Fig. 16-6, a portion of computer run-off of one minute particle counts collected in a class 1,000 clean room. From this analysis it was possible to see the influence of traffic patterns of workers as was discussed in Chap. 13.

Several other monitoring procedures are applicable to clean room and clean bench operation. It is necessary to monitor both the temperature and the humidity for clean rooms as well as the level of organic vapors in the

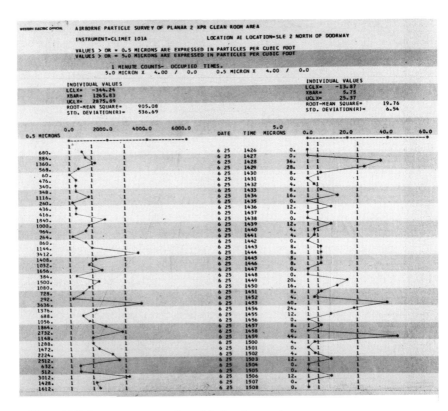

*Figure 16-6  Computer output of particle data from a clean room.*
*Notes: Lines on graphs are drawn in manually.*
    *XBAR is the mean number of particles for each size group.*
    *ROOT-MEAN SQUARE is the standard deviation of each particle distribution.*
    *STD. DEVIATION (R) is the standard deviation of the range.*
    *LCL and HCL are XBAR ∓ 3 STD DEVIATION (R), respectively.*
    *Time is listed as Military Time.*

room.   Temperature and humidity monitoring is generally continuous while organic vapors may be monitored either periodically or continuously depending on the type of information desired.   If vapor control is part of the clean room or clean bench facility, monitoring the organic vapor levels is necessary to determine when replacement of regeneration of the carbon filter is required.   This may be readily done with a total hydrocarbon analyzer, which is an instrument for continuously monitoring organic vapors using a hydrogen flame ionization detector.[26,27]

### other monitoring systems

Two other examples where monitoring is employed for controlling production facilities are worthy of mention.   The first deals with monitoring heat treating furnaces for contaminants.   In these furnaces ceramic headers and other piece parts are treated in an atmosphere of hydrogen and nitrogen.   Samples are collected periodically in evacuated cylinders for analysis with a gaseous mass spectrometer for water vapor and other gaseous impurities which might cause localized changes in the reducing atmosphere.

The second deals with the use of special atmosphere chambers.   Such systems are used in transistor manufacture where discrete devices are hermetically sealed in metal cans to isolate them from the environment in which they will operate.   Welding of this can onto the header is performed in an enclosed glove box (Fig. 11-3) being supplied with high purity nitrogen.   The oxygen content of the box should be continuously monitored to ensure that oxygen is not leaking into the system.   This can be done with a variety of instruments, and detection down to the ppm level is possible.   Such a system can also include an electrical interlock to shut down the welders when the oxygen levels exceed a predetermined level (typically 1%).

### monitoring the general plant environment

For the most part the microelectronic factory environment is largely an integrated complex of many independent operations.   While a good portion of the environment is controlled by facilities such as clean rooms, clean benches, local exhaust hoods, storage boxes, and cascade rinse baths, the predominant part of the plant environment is uncontrolled.   Because of this lack of control it becomes necessary to institute a different form of monitoring from that used to monitor the control facilities mentioned above.   The difference lies in the fact that monitoring for control facilities is primarily a check on the facility.   On the other hand, the information obtained from general monitoring schemes may provide a basis for deciding upon the necessity for environmental control for a particular process.   This information may also indicate the nature of the contaminating agents present and the possible relationships to specific manufacturing problems.

The techniques for monitoring constituents from three major classes of industrial contaminants now will be reviewed.   The three classes include particles, inorganic chemicals, and organic chemicals.

### monitoring particulate matter

Particle measurement systems have been treated fully in Chap. 5 of this text. The reader is referred to that chapter for detailed information concerning specific techniques. Particles are distinguished from other classes of contaminants in that their mode of action is primarily physical or due to their physical presence. This does not preclude the fact that the chemical nature of particles may be significant and require separate consideration.

The primary techniques employed in monitoring airborne particulate contamination are listed in Table 16-1. This type of contamination monitoring can be divided into two distinct categories: that which requires manual[28-36] analysis and that which is performed automatically.[37-42] Anyone who has had the misfortune of having to count particles visually with a light microscope will be quick to recognize the distinct advantages offered by light-scattering particle counters. Using this type of equipment, it is possible to monitor continuously both open floor and clean room areas to determine daily variations and long term variations in particle counts. Particle counting systems interfaced with a digital computer are most valuable for monitoring open floor areas. It is possible to observe trends such as differences between various factory locations, work shifts, and times of the year using the above techniques. An example of the latter is shown in Fig. 16-7, which shows $\geq 0.5$ micron and $\geq 5.0$ micron particle data over a period of almost a year.

While the trend in monitoring has been heavily in the direction of automatic systems, there is still substantial use for other techniques. This is especially true with regard to monitoring of larger particles—25 microns and larger in diameter—since the collection efficiency for most of the commercial light-scattering counters drops off for large particles. This is not to say that if a large particle enters the illuminated volume of the counter it will not be counted, merely that it is difficult to entrap the particles in the inlet stream. Therefore, it is sometimes expedient to use methods such as wet impingement,[28,29] vacuum filtration,[30-33] and cascade impaction[34-36] in order to collect these larger particles for analysis by microscopy. If information beyond that of determining number and size of particles is desired (shape, weight, mass mean diameter, etc.), it is necessary to be able to collect particles for analysis. Collection of particles also permits their chemical analysis.

### inorganic material

Inorganic material is probably the most deleterious form of contamination for microelectronic devices. Their effects in many instances are noticeable at levels below a part per million by weight. Great care must be taken in measuring atmospheric concentrations of materials of this type. The instrumentation requires extreme sensitivity. Sampling and analysis techniques must be specific in most instances for specific chemicals. A variety of procedures summarized in Table 16-2 are available. For reasons of brevity the following discussion will be limited to some of the more important techniques for col-

**TABLE 16-1.  Airborne Particle Detection**

| Sampling Technique | Collection Vehicle | Mode of Analysis | Size Range (μm) | Information Obtained |
|---|---|---|---|---|
| Settlement | Various filter media | Gravimetric | > 5 | Weight/unit area-time |
| Impaction | Cascade impactor | Gravimetric | > 5 | Weight/unit volume, size distribution |
| High volume vacuum filtration | Various filter media | Gravimetric | > 5 | Weight/unit volume |
| Settlement | Membrane filters | Microscopic | > 5 | Number/unit area-time, size distribution |
| Vacuum filtration | Membrane filters | Microscopic | > 5 | Number/unit volume, size distribution |
| Wet impingement | Greenburg-Smith and Midget impingers | Microscopic | > 5 | Number/unit volume, size distribution |
| Direct sampling | Light Scattering Particle Counter | Light Scattering | 0.5–25.0 | Number/unit volume, size distribution |
| Direct sampling | Condensation Nuclei Counter | Condensation of moisture on particles, light scattering | 0.01–0.5 | Number/unit volume |
| Direct sampling | Whitby Analyzer | Diffusion and collection of charged particles | 0.01–0.5 | Number/unit volume, size distribution |

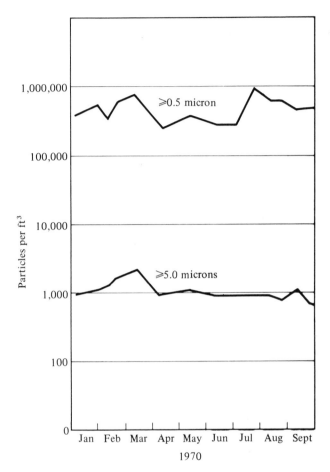

*Figure 16-7 Seasonal trend of particle concentration in a microelectronics factory.*

lection and analysis of inorganic chemicals of concern to microelectronic manufacturing.

**Metals.** The analytical instrumentation for detection of metals has been discussed in Chaps. 3 and 4. The prime instruments for this type of work are the electron microprobe, X-ray fluorescence spectrometer, emission spectrometer, spark source spectrometer, atomic absorption spectrometer, scanning electron microscope, and the ion microprobe. With the exception of the atomic absorption spectrometer these instruments are usually used for detection of metals present on device surfaces or within the material. However, these instruments are generally not sensitive enough for direct application to atmospheric analysis and in many instances are not sensitive enough for surface analyses.

Concentration techniques are generally required for atmospheric analysis

| Sampling Technique | Collection Medium | Material Collected | Method of Analysis | Sensitivity |
| --- | --- | --- | --- | --- |
| High volume vacuum sampler | Various filter media | Metals, non-metals | X-ray fluorescence spectroscopy | Microgram |
| High volume vacuum sampler | Various filter media | Metals | Neutron activation | Picogram |
| Filtration | Membrane filters | Cations | Atomic absorption spectroscopy | Nanogram |
| Filtration | Membrane filters | Cations, anions | Ion-selective electrodes | Nanogram |
| Filtration | Various filters | Metals, non-metals | Ring oven | Sub-microgram |
| Direct | Water, base | Cations, anions | Conductivity | Nanogram |
| Absorption | Ammonium molybdate | Acidic materials | Titration | Microgram |
| Direct | | Silicates | Colorimetry | Nanogram |
| Direct | | Oxygen | Gas chromatography | Nanogram |
| Direct | | Oxygen | Potentiometry | Microgram |
| Direct | | Oxygen | Amperometry | Microgram |
| Direct | | Water vapor | Dew point | Microgram |
| Direct | | Water vapor | Thermal conductivity | Microgram |
| Absorption | Sodium tetrachloromercurate | Sulfur dioxide | Colorimetric | Nanogram |
| Direct | Sulfuric acid-hydrogen peroxide | Sulfur dioxide | Conductivity | Nanogram |
| Direct | Various halides | Sulfur dioxide | Amperometric | Nanogram |
| Direct | | Sulfur dioxide | Flame photometry | Nanogram |
| Absorption | Sulfuric acid, acetic acid, N-(1-naphthol)-ethylendiamine dihydrochloride | Nitrogen dioxide | Colorimetric | Nanogram |
| Direct | | Oxides of nitrogen | Chemiluminescence | Sub-nanogram |
| Absorption | Buffered potassium iodide | Ozone (oxidants) | Colorimetric | Microgram |
| Direct | Buffered potassium iodide | Ozone (oxidants) | Microcoulometric | Nanogram |
| Absorption | Potassium iodide and sodium hydroxide | Ozone (oxidants) | Coulometric | Microgram |
| Direct | | Ozone | Chemiluminescence | Sub-nanogram |

to increase the amount of material available for analysis. The amount required varies from instrument to instrument, but generally one requires at least several nanograms or as much as a microgram for detection (see Table 3-1).

The primary methods for concentrating airborne atmospheric samples to measure levels of metallic contaminants are exactly the same as for collection of particulate matter. Since the bulk of this material will be found as metal oxide or salt, it is possible to concentrate these materials onto filter media or into solution by impingement.

An excellent technique for general airborne particulate analysis involves collection of samples with high volume samplers. These pieces of equipment enable the investigator to collect relatively large amounts of sample in reasonably short periods of time. Analysis has been performed using X-ray fluorescence with excellent sensitivity.[43,44] Quantitative analysis may be performed by using a coprecipitation technique to prepare standards on membrane filter media.[45] Neutron activation methods may also be applied for sample analysis with extreme sensitivity for heavy metal atoms.[46] However, the nature of the neutron activation systems is such that most analytical laboratories would not be in a position to perform this analysis from both an equipment and a radiation hazard standpoint. This type of analysis is generally performed by specialized laboratories.

Atomic absorption has been shown to be quite suitable for analysis of atmospheric samples.[47] While the method is intended for use with cellulose acetate filters it is equally applicable for samples collected by wet impingement. First, the membrane filter sample is placed in a platinum crucible, dissolved in acetone, and ashed in a furnace to remove volatile organics. Analogous to this, the water solution from the impinger samples may be evaporated in a platinum crucible and then ashed. Then the residue is dissolved in concentrated nitric acid on a hot plate with the residue being diluted to 25 ml with deionized water. Atomic absorption analysis may now be performed using suitable standards. Another procedure[48] involving high volume sampling on glass fiber filters and atomic absorption analysis also appears well-suited for atmospheric measurement of metallic contaminants.

Aside from the spectrometric techniques for analysis of metals, a large number of wet chemical methods are available for specific metals. Unfortunately, most wet methods require a larger sample than is generally feasible to supply. Approximately milligram quantities are required.[49]

**Reactive Gases.** Included in this group of inorganic materials are oxygen, water vapor, hydrochloric acid vapor, sulfur dioxide, oxides of nitrogen, and ozone.

Determination of oxygen is important in areas where inert or reducing environments are required. For laboratory analysis, air samples can be collected in glass bottles at the point of application and transferred to the instrument location. Gas chromatography can be used for oxygen analysis.[50,51,52]  Using low or subambient temperatures, high surface area columns, and thermal conductivity or ionization detectors (for ultrahigh sensitivity), it is possible to detect oxygen in gaseous mixtures. Several portable continuous monitoring instruments are also available for on-the-spot monitoring. These instruments can generally function at the 1 ppm (by volume) level or greater by means of potentiometric or electrochemical reaction. A potentiometric instrument responds to a potential difference produced across two platinum electrodes separated by a material permeable to oxygen.[53,54]  The voltage produced is inversely proportional to the logarithm of the ratio of the partial pressure of oxygen in the air examined to the partial pressure of atmospheric oxygen.  In an electrochemical instrument oxygen is depleted by cathodic reaction in an alkaline electrolyte.[55]  This depletion of oxygen produces hydroxyl ions from the anode, thus establishing a current that is proportional to the oxygen partial pressure.

Water vapor is commonly monitored by dew point[56] or thermal conductivity[57] measurements. The dew point of a gas is the temperature at which moisture will condense.  By cooling an air sample or a surface upon which condensation can occur, it is possible to determine this value for a given water vapor partial pressure.  Knowing the dew point, one can readily calculate the water vapor concentration from the ideal gas law or obtain it directly from dew point tables.[58]   Thermal conductivity detectors make use of thermistors.  Their resistivity characteristic varies with temperature.  Over one heated element is passed dry air for reference and over the other, the moist air.  By virtue of the difference in thermal conductivity of moist and dry air, the moist air element will be cooled at a different rate than the dry air element.  The difference in current produced across these elements is then measured by a Wheatstone bridge arrangement; the current required to balance the circuit is proportional to the moisture content of the moist air.

An effective technique for the specific analysis of hydrochloric acid vapors involves bubbling air through deionized water. Silver nitrate is then added to precipitate the chloride ion as silver chloride. For dilute suspensions of silver chloride, turbidometric analysis using a spectrophotometer may be employed.[59]  This reaction has been made use of in the development of a continuous monitoring microcoulometer.  This instrument continuously titrates the collection solution and records the amount of silver ion required to maintain the initial level of silver ion.[60]

The reactive gases sulfur dioxide, nitrogen dioxide, nitric oxide, and ozone are most commonly thought of in connection with air pollution.  In conjunction with reactive hydrocarbons and ultraviolet radiation, these materials are presumed to be building blocks for photochemical smog and oxident gases.[61]

The leading method for wet determination of sulfur dioxide is the West-Gaeke method.[62,63,64]   This is a colorimetric technique that involves sampling air in midget impingers containing sodium tetrachloromercurate.  Dichlorosulfitomercurate is produced, which turns red-purple upon addition of acid bleached pararosaniline and formaldehyde. The final complex is pararosaniline methylsulfonic acid. Determinations are made spectrophotometrically by comparison with standards prepared with a sodium metabisulfite solution. While this is basically a laboratory technique, several continuous monitoring instruments are available for continuous colorimetric determination of $SO_2$ in ambient locations.[65]

Another widely used technique for continuous monitoring of sulfur dioxide is a conductometric technique.[64,65] Air containing $SO_2$ is scrubbed through an electrolyte solution of sulfuric acid and hydrogen peroxide. The conductivity of the reagents is measured with two platinum electrodes. Sulfur dioxide reacts with hydrogen peroxide to produce sulfuric acid. By measuring the conductivity of the scrubbed air supply with a second set of platinum electrodes, it is possible to relate increased conductivity to sulfuric acid produced from sulfur dioxide. Coulometric and flame photometric techniques are also available for continuously monitoring sulfur dioxide.[65,66]

$NO_2$, or its dimer $N_2O_4$, is most commonly determined colorimetrically using the Griess-Saltzman reaction.[67] Air containing nitrogen dioxide is scrubbed in an impinger containing an aqueous solution of sulfonic acid, acetic acid, and N-(1-Naphthyl)-ethylenediamine dihydrochloride. The pink color produced can be compared spectrophotometrically with prepared standards for quantitative analysis. Nitric oxide may also be determined using this method by passing the air stream through potassium permanganate solution. This step converts the NO to $NO_2$, whence it can be determined either separately or together with $NO_2$ by the above method. Continuous monitoring instruments employing this method with countercurrent scrubbing of the gas stream are available. A modification of the Griess-Saltzman reagent was proposed by Lyshkow for increased $NO_2$ absorption with better color development.[68] In recent years a technique has been developed to monitor $NO_2$ and NO by measuring the chemiluminescence from the reaction of NO and ozone. In systems of this type $NO_2$ must be catalytically reduced to NO for analysis.[69]

The prime technique for determining ozone or other oxidants in air involves liberation of iodine from a buffered potassium iodide solution.[70,71] Ozone will react with potassium iodide to produce iodine and potassium hydroxide. Analysis can be performed spectrophotometrically by comparison with iodine standards if they are run within one hour of sample collection. A continuous oxidant monitor is available that utilizes a microcoulometric version of the above technique.[72,73] A technique using chemiluminescence has also been described to monitor ozone.[74]

### organic contamination

Organic chemicals comprise the largest group of chemicals in the plant environment. This is due to the widespread usage of organic materials both inside the plant as well as outside. Organics may be present as particles (solids), liquids, vapors, and gases. The distinction between vapors and gases is that vapors are gaseous phases of normally liquid materials. Product surfaces are subject to contamination by organic materials by a number of mechanisms. First of these mechanisms is deposition in particulate form from the atmosphere. Next is adsorption of vapors from the environment, the presence of these vapors being attributable primarily to solvent operations as well as pollution from outside. Then there is deposition of nonvolatile residue from chemical processes involving contaminated solvents. Also possible is incomplete solvent or acid stripping, as well as residual solvent on the surface from incomplete evaporation after stripping.

Surface monitoring has been discussed previously; what remains is to describe applicable techniques for determination of organic materials in the atmosphere. The material that follows reviews the most widely employed monitoring techniques for this type of analysis.

Collection of organic particles from the atmosphere is accomplished in the same manner as with other particle forms. For this type of work filter media such as membrane filters are preferred because changes in sample composition may occur in solution if impinger samples are taken. Once samples have been collected, analysis is generally performed by a combination of selective extraction, column chromatography, and gas chromatography.[75,76]

Samples are extracted with ether, water, acid, and base in such a way as to result in four primary fractions; water insolubles and solubles, acids, bases, and neutral components. Neutral components are then separated into aliphatics, aromatics, and oxygenated portions using column chromatography. Further distinction as to specific members of these groups can be made using gas chromatography, or a gas chromatography-mass spectrometer coupling.

The primary techniques for collection of organic vapors are listed in Table 16-3. The problem of analyzing organic vapors in air has been the subject of widespread investigation for many years. The analytical aspects of organic vapor determinations have been greatly enhanced with the development of quantitative gas chromatography. However, for unknown constituents such as those present in most environments the use of this instrument is subject to the limitations of trial and error techniques for qualitative identification. The gas chromatograph is primarily a highly sensitive instrument that has the capability of separating components from a mixture. By coupling the gas chromatograph with a high sensitivity, fast scanning mass spectrometer, it is now possible to scan and identify chromatographic peaks as they are eluted from the column packing.

The remaining requirement is to collect samples efficiently in such a manner that sufficient sample is available in a form suitable for analysis. Normally, at least nanogram quantities of trace components are needed for good analysis. For direct analysis of air, such as by direct sampling of air with plastic bags,[77,78] this requires at least ppm (by volume) concentrations. Unfortunately, it is desirable to be able to determine fractional ppm concentrations and even ppb concentrations. Hence, it is necessary to concentrate air samples several orders of magnitude to adequately analyze samples by

TABLE 16-3. Methods for Collecting Organic Vapors from Air

| Technique | Collection Media | Description |
|---|---|---|
| Plastic bags | "Saran," "Mylar," "Teflon" | Air forced into bag. Syringe used to remove aliquots for analysis. |
| Freeze-out | Cryogenic liquids | Air passed through cold trap causing condensation of vapors. Subsequent heating will release vapors (after removal of cold trap). |
| Absorption | Organic solvents | Air is scrubbed in a midget impinger. Aliquots of the resultant solution used for analysis. |
| Absorption | Gas chromatography Column Packings | Air is passed through media to absorb organic vapors: Desorption generally accomplished by heat. |
| Adsorption | Various solid media | Air is passed through media to adsorb organic vapors. Desorption generally accomplished by heat or solvent extraction. |

current analytical procedures. To this end, a number of sampling procedures have been developed, each finding a certain amount of acceptance and usage.

Probably the earliest technique employed was freeze-out.[79,80] This entails passing air through a cold trap maintained at a specified subambient temperature. Entrapped vapors are condensed in these cold traps and can be collected for later analysis. It is also possible to selectively trap out vapors by using a series of cold traps arranged in order of decreasing temperature. In this way components with high boiling points are trapped out before those having lower boiling points.

Another technique that has found usage for collection of volatile organic vapors is absorption into solution with an appropriate solvent.[81,82,83] In this technique a midget impinger equipped with a fritted glass bubbler is generally employed. Typical scrubbing solvents are water, isopropyl alcohol, hexane, and xylene. One problem associated with general usage of organic scrubbing mediums is the masking of gas chromatographic peaks by the solvent peak unless the impurities are known. If the identity of the impurities is known, it may be possible to adequately separate them from the bulk solvent.

Another absorptive technique used to sample atmospheres for vaporous contaminant employs gas chromatographic column packings.[84,85,86] Typical applications involve fixed bed collection as well as fluidized bed collection.

Trace vapors are selectively retained in the stationary phase of the column packing, the degree of retention being a function of a number of variables which include the type of stationary phase, the type of material being sampled, and the temperature of sampling. Desorption is generally accomplished by heating the collection column and driving the trapped material into a containment system, such as a gas sampling loop, or directly into the gas chromatographic column.

Another approach to sampling has been the adsorption of vapors onto solid surfaces.[87-97] Typical collection media in this class are activated carbon, silica gel, molecular sieves, and porous polymer beads. The advantage of solid adsorbents is high surface area, which allows for collection of large samples. Generally, activated carbon and silica gel are the preferred materials. Silica gel is better for collection of polar materials, but silica gel will adsorb significant amounts of polar water vapor. Activated carbon is a general purpose adsorbent having the highest surface area among the group listed. While activated carbon is essentially nonpolar in structure, the presence of trace surface impurities renders the material effective for adsorption of polar molecules and reactive gases. In contrast to silica gel activated carbon will preferentially displace water vapor in the presence of organic molecules. Silica gel is easier to desorb than activated carbon, probably because it does not have the same fine pore structure.

Desorption of samples collected on solid surfaces is generally accomplished either by heating the material off of the surface or by displacement of the collected vapors by a strongly retained solvent.

At present there are only two methods for continuously monitoring organic vapors in air. These are nondispersive infrared spectroscopy[98] and hydrogen flame ionization.[26,27] Nondispersive infrared analyzers can be used for total hydrocarbon analysis or made specific for particular components. Hydrogen flame ionization detectors are generally used for determining total hydrocarbons present. This type of instrument is considerably more sensitive than infrared instruments. Basically, the instrument measures the total number of ions produced by the combustion of organic vapors in a hydrogen flame.

Organic halides may be measured directly in the atmosphere with a halide meter.[99] This instrument functions through decomposition of volatile organic halides by sparking the air sample. Originally it was thought that the instrument responded to the characteristic emission of copper halide formed from the reaction of the copper electrode and the halide. It has since been shown that the actual mechanism is an intensification of the nitrogen spectra from the air caused by the presence of halides.[100]

A microcoulometer is also available which may be used to monitor organic chlorides.[60] Air containing organic chlorides is passed into a high temperature furnace. There the organic chlorides are oxidized to hydrochloric acid and detected as previously described.

## *significance of monitoring programs*

Monitoring is a very powerful tool when used properly. A program of monitoring can be instrumental in providing information pertinent to environmental fluctuations, information that may be related to manufacturing problems. However, monitoring for the mere sake of monitoring is of limited value, expensive, and time consuming. As was illustrated earlier, one is concerned with a wide variety of contaminants and there is no universal instrument for sampling and specific detection of contaminants. Hence, monitoring programs must be geared to specific measurement procedures.

There are no straightforward guidelines for instituting programs of monitoring; a good deal of engineering judgment must be used. Product engineers, being knowledgeable of the manufacturing processes, are in an excellent position to predict possible contaminating agents. Also, they are familiar with the possible weak links in the manufacturing chain. This knowledge will greatly enhance the proper decisions on the materials to be monitored and their appropriate measurement techniques.

The next item to be established is where monitoring should be instituted. To a large extent this is where engineering judgment is of prime importance. Microelectronic device manufacture consists of a series of operations starting with bulk materials and ending with a minute finished product. In many instances, well over a hundred discrete operations are involved. To monitor each step of a multi-operational process would be an impossible task. At most of these operations monitoring is not necessary. For example, it would not be particularly important to monitor a device or the environment in which it exists if the next processing step is an acid etch which would remove the upper layers of material. However, it may be highly significant to perform this operation subsequent to an impurity deposition, an oxide growth, a film deposition, or a plating operation when surface cleanliness is of prime importance. Numerous examples are available to substantiate the point that certain critical operations warrant specific monitoring procedures, these being stages when a certain degree of cleanliness is required.

Another consideration is storage time. There is little question that product contamination is time dependent. The longer a production material is exposed to a contaminant source, the more likely is the chance for contamination. A production scheme that does not allow for delays or storage times is most desirable. At present, with few exceptions, this is not a general industrial practice. Monitoring must be used to evaluate storage facilities as well as aiding in the establishment of environmental control facilities.

Once the material to be monitored and the stage at which monitoring is to be instituted is established, it is necessary to determine the frequency of sampling. In some instances continuous sampling is desirable. Examples of this type of monitoring are the measurement of general plant and clean room humidity, oxygen in welding atmospheres, and water vapor in dry air supplies. In other instances only periodic monitoring is required, or even individual samples may suffice. Clean bench monitoring is a good example of periodic monitoring to check for leaks in the HEPA filters as well as to ensure that the velocity across the clean bench is still as prescribed.

The net result of monitoring programs is a systematic means of obtaining feedback information that may be used to guarantee the effectiveness of environmental control facilities, determine where environmental control should be instituted, and finally, to determine the nature and quantity of contaminants present in the plant environs.

## references

1. J. Vaccaro, *et al.*, *Reliability Physics Notebook*, U.S. Department of Commerce, AD-624 769, 1965.
2. J. S. Smith and J. Vaccaro, "Physical Basis for Evaluating the Reliability of *p-n* Junction Devices," IEEE Transactions on Reliability, R-17, 20–27, 1968.
3. B. Kurz, "Degradation Phenomena of Planar Si Devices Due to Surface and Bulk Effects," Sixth Annual Reliability Physics Symposium Proceedings, pp. 47–65, 1968.
4. J. E. Cline and S. Schwartz, "Electron Microprobe Techniques for Failure Analysis of Silicon Planar Devices," Sixth Annual Reliability Physics Symposium Proceedings, pp. 193–200, 1968.
5. R. W. Berry, P. M. Hall, and M. T. Harris, *Thin Film Technology*, Van Nostrand Reinhold, New York, 1968, p. 134.
6. Consolidated Electrodynamics Corporation, *Operation and Maintenance Manual*, Radiflo Leak Detector, Pasadena, Calif.
7. J. W. Van Valkenburg, Jr., "Factors that Influence the Magnitude of the Contact Angle," Ph.D. Dissertation, University Microfilms, 11, 365, 1955.
8. T. Fort, Jr., and H. T. Patterson, "A Simple Method for Measuring Solid-Liquid Contact Angles," *J. Colloid Sci.*, 18, 217 (1963).
9. "Standard Method of Test for Hydrophobic Surface Films by the Atomizer Test," Book of ASTM Standards, Part 8, F21-65, 1969.
10. "Standard Method of Test for Hydrophobic Surface Films by the Water-Break Test," Book of ASTM Standards, Part 8, F22-65, 1969.
11. J. L. Anderson, D. E. Root, Jr., and G. Greene, "Measurement and Evaluation of Surfaces and Surface Phenomena by Evaporative Rate Analysis," *J. of Paint Technology*, 40, 320 (1968).
12. A. W. Michalson, "Monitoring Water Quality with the Solu Bridge," *The Analyzer*, 8, 18 (1967).
13. "Standard Method of Test for Silting Index of Fluids for Processing Electron and Microelectronic Devices," Book of ASTM Standards, Part 8, F52-69, 1969.
14. C. W. Baldwin, "TI Gets Super Clean Water for Semiconductor Works. Part II– Demineralization Problems," *Plant Engineering*, 17, 144 (1963).
15. "Standard Methods of Test for Silica in Industrial Water and Industrial Waste Water," Book of ASTM Standards, Part 23, D 859-68, 1969.
16. D. T. Chow, and R. J. Robinson, "Forms of Silicate Available for Colorimetric Determination," *Anal. Chem.*, 25, No. 4 (1953).
17. "Guide to Specific Ion Electrodes and Instrumentation," CAT 1961, Orion Research Inc., 1969.
18. S. E. Manahan, "Electrochemical Methods," American Industrial Hygiene Association Conference, Detroit, 1970.
19. C. N. Sawyer, *Chemistry for Sanitary Engineers*, McGraw-Hill, New York, 1960, pp. 246–256.

20. *Contamination Control Handbook*, NASA CR 61264, Clearing House for Federal Scientific and Technical Information, Springfield, Va., pp. VI-16–27, 1969.
21. K. V. Thimann, *The Life of Bacteria*, The McMillan Co., New York, 1955, pp. 131–163.
22. O. Wyss, O. B. Williams, and E. W. Gardner, Jr., *Elementary Microbiology*, John Wiley and Sons, New York, 1963, pp. 131–145.
23. *Standard Methods for the Examination of Water and Wastewater*, American Public Health Association, Inc., New York, 1960, pp. 492–493.
24. "Standard Methods for Detection and Enumeration of Microbiological Contaminants in Water Used for Processing Electron and Microelectronic Devices," Book of ASTM Standards, Part 8, F60-68, 1969.
25. R. M. Seip, "Computerization of Airborne Particle Counts," Paper presented at the 9th Annual Technical Meeting of the American Association for Contamination Control, April 19–22, 1970, Anaheim, Calif.
26. A. J. Andreatch, "Flame Ionization Detector. Application to Industrial Hygiene and Air Pollution Studies," *Archives Env. Health*, 4, 317 (1962).
27. A. P. Altshuller, "Analysis of Organic Substances in the Atmosphere by Gas Chromatographs with Ionization Detectors," *Air Water Pollution*, 7, 87 (1963).
28. L. Greenburg and G. W. Smith, U.S. Bur. Mines Dept. Invest. No. 2392, 1922.
29. J. B. Littlefield and H. H. Schrenk, "Bureau of Mines Midget Impinger for Dust Sampling," U.S. Bureau of Mines Dept. Invest. No. 3360, 1937.
30. "Standard Method for Sizing and Counting Airborne Particulate Contamination in Clean Rooms and Other Dust-Controlled Areas Designed for Electronic and Similar Applications," Book of ASTM Standards, Part 8, F25-68, 1969.
31. A. Goetz, "Application of Molecular Filter Membranes to the Analysis of Aerosols," *Am. J. Public Health*, 43, 150 (1953).
32. M. W. First and L. Silverman, "Air Sampling with Membrane Filters," *AMA Arch. Ind. Health*, 7, 1 (1953).
33. D. A. Fraser, *AMA Arch. Ind. Health*, 8, 412 (1953).
34. K. R. May, "The Cascade Impactor: An Instrument for Sampling Coarse Aerosols," *J. Sci. Inst.*, 22, 187 (1945).
35. C. N. Davies and M. Aylward, "The Trajectories of Heavy Solid Particles in a Two-Dimensional Jet of Ideal Fluid Impinging Normally Upon a Plate," *Proc. Phys. Soc.*, B64:889, (1951).
36. W. E. Ranz and J. B. Wong, "Jet Impactors for Determining the Particle Size Distribution of Aerosols," *AMA Arch. Ind. Hyg. and Occup. Med.*, 5, 464 (1952).
37. F. T. Gucker, C. T. O'Konski, H. B. Pickard, and J. N. Pitts, Jr., "Photoelectronic Counter for Colloidal Particles," *J. Amer. Chem. Soc.*, 69, 2422 (1947).
38. P. L. Magill, "An Automated Way to Count Fine Particles," *Air Engineering*, 4, 31 (1962).
39. "Particle Counting Technology," *Contamination Control*, VIII:11, 1969.
40. F. J. Van Luik, Jr., and R. E. Rippere, "Condensation Nuclei, A New Technique for Gas Analysis," *Anal. Chem.*, 34, 1617 (1962).
41. G. F. Skala, "New Instrument for the Continuous Measurement of Condensation Nuclei," *Anal, Chem.*, 35, 702 (1963).
42. K. T. Whitby and W. E. Clark, "Electric Aerosol Particle Counting and Size Distribution Measuring System for the 0.015–1.0 μ Size Range," *Tellus*, XVIII, 573 (1966).
43. L. D. Blitzer, *et al.*, Unpublished Data, Bell Telephone Laboratories, Murray Hill, April 8, 1970.
44. R. W. DeMott, Unpublished Data, Western Electric Co., November, 1970.
45. C. L. Luke, "Determination of Trace Elements in Inorganic and Organic Materials by X-Ray Fluorescence Spectroscopy," *Anal. Chim. Acta.*, 41, 237 (1968).
46. B. R. Payne, Editor, *Radioactivation Analysis Symposium*, Butterworths, London, 1960.

47. J. Y. Hwang and F. J. Feldman, "Determination of Atmospheric Trace Elements by Atomic Absorption Spectroscopy," *Applied Spectroscopy*, 24, 371 (1970).
48. G. D. Carlson and W. E. Black, "Determination of Trace Quantities of Metals from Filtered Air Samples by Atomic Absorption Spectroscopy," Paper presented at the 63rd Annual Meeting of the Air Pollution Control Association, St. Louis, Mo., 1970.
49. A. C. Stern, Editor, *Air Pollution*, Vol. II, Academic Press, New York, 1968, pp. 157–170.
50. W. M. Graven, "Ionization by Alpha-Particles for Detection of the Gaseous Components in the Effluent from a Flow Reactor," *Anal. Chem.*, 31, 1197–99 (1959).
51. E. W. Lard and R. C. Horn, "Separation and Determination of Argon, Oxygen, and Nitrogen by Gas Chromatography," *Anal. Chem.*, 32, 878 (1960).
52. T. L. Chang, "Gas Chromatographic Methods for Mixtures of Inorganic Gases and Methane and Ethane," *J. Chromatography*, 37(1), A-26 (1968).
53. H. S. Spacil, "The Solid-Electrolyte Oxygen Sensor . . . Its Features and Uses," *Metal Progress*, 96, 106 (1969).
54. Thermo Lab Instruments, Inc., Glenshaw, Pa.
55. Beckman Instruments, Inc., Bulletin 0-4016B, Fullerton, Calif.
56. Cambridge Systems, Inc., Bulletin 880-B1, Newton, Mass.
57. Mine Safety Appliances Company, Bulletin 0709-1, Pittsburgh, Pa.
58. F. Rosebury, *Handbook of Electron Tube and Vacuum Techniques*, Addison-Wesley Publishing Co., Reading, 1965, p. 226.
59. A. C. Stern, Editor, *Air Pollution*, Vol. II, Academic Press, New York, 1968, p. 102.
60. Dohrmann Instruments Co., Bulletin AI-11, Mountainview, Calif.
61. P. L. Magill, F. R. Holden, and C. Ackley, Editors, *Air Pollution Handbook*, McGraw-Hill, New York, 1956, pp. 249–250.
62. P. W. West and G. C. Gaeke, "Fixation of Sulfur Dioxide as Disulfitomercurate (11) and Subsequent Colorimetric Estimation," *Anal. Chem.*, 28, 1816 (1956).
63. R. V. Nauman, P. W. West, F. Tron, and G. C. Gaeke, "A Spectrometric Study of the Schiff Reaction as Applied to the Quantitative Determination of Sulfur Dioxide," *Anal. Chem.*, 32, 1307 (1960).
64. A. C. Stern, Editor, Air Pollution, Vol. II pp. 55–78, Academic Press, New York, 1968.
65. C. E. Rodes, H. R. Palmer, L. A. Elfers, and C. H. Norris, "Performance Characteristics of Industrial Methods for Monitoring Sulfur Dioxide: I. Laboratory Evaluation," *J. Air Pollution Control Assoc.*, 19, 575 (1969).
66. Melpar, Inc. Technical Information, Falls Church, Va.
67. B. E. Saltzman, "Colorimetric Microdetermination of Nitrogen Dioxide in the Atmosphere," *Anal. Chem.*, 26, 1949 (1954).
68. N. A. Lyshkow, "Rapid and Sensitive Colorimetric Reagent for $NO_2$ in Air," *J. Air Pollution Control Assoc.*, 15, 481 (1965).
69. A. Fontijn, A. J. Sabadell, and R. J. Ronco, "Homogeneous Chemiluminescent Measurement of Nitric Oxide with Ozone," *Anal. Chem.*, 42, 575–9 (May, 1970).
70. C. E. Thorp, *Ind. Eng. Chem., Anal. Ed.*, 12, 209 (1940).
71. B. E. Saltzman and N. Gilbert, "Iodometric Microdetermination of Organic Oxidents and Ozone. Resolution of Mixtures by Kinetic Colorimetry," *Anal. Chem.*, 31, 1914 (1959).
72. G. M. Mast and H. E. Saunders, Research and Development of the Instrumentation of Ozone Sensing, *Instr. Soc. of America Trans.*, 1, 325–328 (1962).
73. L. Potter and S. Duckworth, "Field Experience with the Mast Ozone Recorder," *J. Air Pollution Control Assoc.*, 15, No. 5 (1965).
74. J. A. Hodgeson, K. J. Krost, A. E. O'Keefe, and R. J. Stevens, "Chemiluminescent Measurement of Atmospheric Ozone," *Anal. Chem.*, 42, 1795–1802 (Dec., 1970).
75. R. L. Shriner, R. C. Fuson, and D. Y. Curtin, *The Systematic Identification of Organic Compounds*, John Wiley & Sons, New York, 1964, p. 101.

76. A. C. Stern, Editor, *Air Pollution*, Vol. II, Academic Press, New York, 1968, p. 190–191.

77. A. P. Altshuller, A. R. Wartburg, I. R. Cohen, and S. F. Sleva, "Storage of Vapors and Gases in Plastic Bags," *Int. J. Air Water Pollution*, 6, 75 (1962).

78. A. L. Van der Kolk, "Sampling and Analysis of Organic Solvent Emissions," *Amer. Ind. Hyg. Assoc. J.*, 28, 588 1967.

79. A. C. Stern, Editor, *Air Pollution*, Vol. II, Academic Press, New York, 1968, p. 37.

80. E. R. Stephens and F. R. Burleson, "Analysis of the Atmosphere for Light Hydrocarbons," *J. Air Pollution Control Assoc.*, 17, 147 (1967).

81. R. H. Mansur, R. F. Pero, and L. A. Krause, "Vapor Phase Chromatography in Quantitative Determination of Air Samples Collected in the Field," *Amer. Ind. Hyg. Assoc. Quart.*, 20, 175 (1959).

82. B. Levadie and J. Harwood, "An Application of Gas Chromatography to Analysis of Solvent Vapors in Industrial Air," *Amer. Ind. Hyg. Assoc. Quart.*, 20, 21 (1960).

83. J. E. Dennison and R. P. Menichelli, "Determination of Trichloroethylene and Chloroform in Environmental Atmospheres," Unpublished Data, Western Electric Co., Engineering Research Center, 1969.

84. F. R. Cropper and S. Kaminsky, "Determination of Toxic Organic Compounds in Admixture in the Atmosphere by Gas Chromatography," *Anal. Chem.*, 35, 735 (1963).

85. A. Dravnieks and B. K. Krotoszynski, "Collecting and Processing of Airborne Chemical Information," *J. Gas Chromatography*, 4, 367 (1966).

86. A. Dravnieks and B. K. Krotoszynski, "Collecting and Processing of Airborne Chemical Information," *J. Gas Chromatography*, 6, 144 (1968).

87. P. W. West, S. Buddhadev, and D. A. Gibson, "Gas-Liquid Chromatographic Analysis Applied to Air Pollution Sampling," *Anal. Chem.*, 30, 1390 (1958).

88. E. J. Otterson and C. J. Guy, "A Method of Atmospheric Solvent Vapor Sampling on Activated Charcoal in Connection with Gas Chromatography," Paper presented at the Philadelphia Meeting of the American Conference of Governmental Industrial Hygienists, 1964.

89. C. L. Fraust and E. R. Hermann, "Charcoal Sampling Tubes for Organic Vapor Analysis by Gas Chromatography," *Amer. Ind. Hyg. Assoc. J.*, 27, 68 (1966).

90. F. H. Reid and W. R. Halpin, "Determination of Halogenated and Aromatic Hydrocarbons in Air by Charcoal Tube and Gas Chromatography," *Amer. Ind. Hyg. Assoc. J.*, 29, 390 (1968).

91. C. L. Fraust and E. R. Hermann, "The Adsorption of Aliphatic Acetate Vapors onto Activated Carbon," *Amer. Ind. Hyg. Assoc. J.*, 30, 494 (1969).

92. L. D. White, D. G. Taylor, P. A. Mauer, and R. E. Kupel, "A. Convenient Optimized Method for the Analysis of Selected Solvent Vapors in the Industrial Atmosphere," *Amer. Ind. Hyg. Assoc. J.*, 31, 225 (1970).

93. A. P. Altshuller, T. A. Bellar, and C. A. Clemons, "Concentration of Hydrocarbons on Silica Gel Prior to Gas Chromatographic Analysis," *Amer. Ind. Hyg. Assoc. Quart.*, 23, 164 (1962).

94. F. R. Cropper and S. Kaminsky, "Determination of Toxic Organic Compounds in the Atmosphere by Gas Chromatography," *Anal. Chem.*, 35, 735 (1963).

95. N. E. Whitman and A. E. Johnston, "Sampling and Analysis of Aromatic Hydrocarbon Vapors in Air: A Gas-Liquid Chromatographic Method," *Amer. Ind. Hyg. Assoc. J.*, 25, 464 (1964).

96. M. Feldstein, S. Balestrieri, and D. A. Levaggi, "The Use of Silica Gel in Source Testing," *Amer. Ind. Hyg. Assoc. J.*, 28, 381 (1967).

97. F. W. Williams and M. E. Umstead, "Determination of Trace Contaminants in Air by Concentrating on Porous Polymer Beads," *Anal. Chem.*, 40, 2232 (1968).

98. Mine Safety Applicances, Bulletin No. 0700-4, Pittsburgh.
99. C. D. Yaffe, D. H. Byers, and A. D. Hosey, *Encyclopedia of Instrumentation for Industrial Hygiene*, University of Michigan Press, Ann Arbor, 1956, p. 41.
100. G. O. Nelson, "The Halide Meter—The Myth and the Machine," *Amer. Ind. Hyg. Assoc. J.*, **29**, 586 (1968).

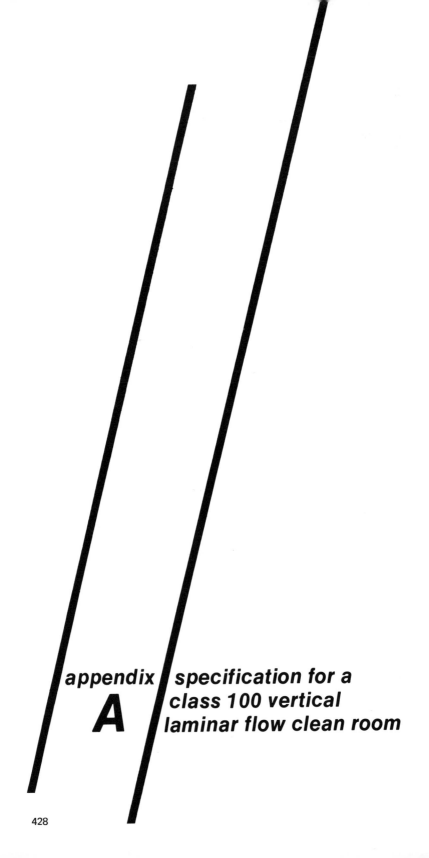

**appendix A** specification for a class 100 vertical laminar flow clean room

INDEX

Section 7   General Work Details
Section 8   Air Handling and Conditioning System
Section 9   Controls and Instrumentation
Section 10  Construction Standards
Section 11  Standard Equipment Specifications

(NOTE: Sections 7 through 11 are typical standard construction specifications, and therefore are not included in this Appendix.)

## 1.0  JOB DESCRIPTION–GENERAL

1.1 This specification covers the complete design, fabrication, erection and testing of a *Class 100 Vertical Laminar Flow Clean Room* encompassing approximately _____ square feet of clean space on the _____ floor of Building _____ at the _____ Company.

1.2 The general clean room design types considered acceptable for this project are:
(NOTE TO ENGINEER: Retain only those room design types suitable for the specific project.)

1.2.1 A clean room with: perforated raised floor; under-floor return air plenum; adjacent equipment room housing blowers, air conditioning units (less refrigeration), and prefilters; and, with one of the following air distributing means in the ceiling:

(a) Final filters mounted in a suspension system to form the clean room ceiling, (with lighting fixtures as specified in Section 5.9), and with the air supply plenum above sectionalized and/or baffled as required to meet the air flow characteristics specified. The distance from blowers to the far end of any plenum section shall not exceed 30 feet.

(b) Final filters mounted in the bottom face of modular multiple plenums, not over 4 ft. wide each, joined together to form the clean room ceiling, (with lighting fixtures as specified in Section 5.9). Each plenum to be served by an individual blower, with the distance from blower to far end of plenum not to exceed 30 feet. Volume control and air distribution within plenum must satisfy air flow characteristics specified.

(c) Final filters mounted in individual plenum units set into a "T-bar" type ceiling suspension structure, each with volume control and served by connection to a distributing ductwork-blower system. Each ductwork-blower system shall serve not over a 5 foot width of room nor a distance from blower to furthest filter unit exceeding 30 feet.

1.2.2 A clean room with: perforated raised floor; under-floor return plenum; and with supply blower-final filter module units supported by a ceiling "T-Bar" type structural system, or suspended and joined together to form the ceiling. Lighting fixtures to be as specified in Section 5.9. The space above the blower-filter modules to be a blower supply plenum conducting air to the blower inlets.

Air return from the under floor plenum to the blower supply plenum may be by:

(a) Return air risers spaced within the clean room and containing prefilters and air conditioning units (less refrigeration).

(b) Return air chase adjacent to the clean room containing prefilters and air conditioning units (less refrigeration).

1.2.3 A clean room with: perforated raised floor; under-floor return plenum; adjacent equipment room housing blowers, prefilters, air conditioning units, (less refrigeration), final filters in a central bank; and with a "T-bar" type ceiling containing air diffusing panels, (of a design specifically approved by the owner), that are served from an air supply plenum above that is sectionalized and/or baffled so as to meet the air flow characteristics specified. Lighting fixtures to be as specified in Section 5.9.

## 2.0 GENERAL PERFORMANCE REQUIREMENTS

### 2.1 *Cleanliness Level*

The performance criteria based on airborne particle counts are as follows:

2.1.1 *Unoccupied*—<10 particles per cubic foot of 0.5 micron size or larger with no particles detected of 1.0 micron size or larger. Measurement is at any point throughout the room area and at any elevation from 3 feet below the ceiling to 30 inches above floor level.

2.1.2 *At Rest*—<10 particles per cubic foot of 0.5 micron size and larger, with no particles detected of 1.0 micron size and larger. Measurements will be at a point 12 inches upstream (above) of any installed production equipment or work position and throughout the room at an elevation of 30 inches above floor level and no closer than 24 inches, horizontally, to any production equipment.

NOTE 1: For the design guidance of the Contractor the intended Normal Operation Conditions will be: a particle count level not exceeding the limits for Class 100 conditions (as defined by Federal Standard 209a) at a point 12 in. upstream of all operating positions and throughout the room at an elevation of 30 inches above floor level and no closer than 24 inches, horizontally, to any contamination generator.

NOTE 2: The terms "Unoccupied," "At Rest" and "Normal Operating Conditions" are used throughout this specification to indicate occupancy conditions as defined below:

*Unoccupied*—This will be the bare room before production facilities are installed and with room air system operating, following initial clean down.

*At Rest*—All production equipment will be in operation but no operating personnel will be present.

*Normal Operating Conditions*—All production equipment will be in operation and all operating personnel present.

### 2.2 *Particle Dispersion and Room Recovery*

In the Unoccupied, At Rest, and Normal Operating Conditions, lateral dispersion of particles generated at any point from 3 feet below the ceiling to 30 inches above the floor shall not extend horizontally beyond 24 inches from the generating point when measured at any elevation between the generation source and 30 inches above the floor, (provided the generation source and the test point are not under the influence of a heated convection column or a forced cross flow velocity).

The room air system shall be capable of providing recovery from contamination at any unobstructed air flow point within one minute of removal of contamination source.

### 2.3 *Air Flow Characteristics*

2.3.1 Air supply to the room shall be sufficient to maintain an average air velocity, based on projected floor area, that is not less than 80 fpm. Verification shall be per Para. 4.2.3 (where maximum limits for velocity readings are indicated) and Para. 4.2.4.

2.3.2 The air flow system shall be adaptable for adjustment to reduce the average velocity to as low as 50 fpm without violating air pattern parallelism per Section 2.3.3.

2.3.3 Unoccupied air flow patterns shall be parallel downward throughout the room with horizontal velocity components from an elevation 3 feet below the ceiling to a point 18 inches above floor level limited within the range necessary to satisfy Section 2.2.

### 2.4 *Room Pressurization and Leakage*

2.4.1 The over-all leakage rate of the entire room and air handling system in the unoccupied condition shall not exceed 1% of the total air supply with all doors closed.

2.4.2 The air handling system(s) and room construction quality shall be capable of sustaining and controlling, in the At Rest condition a positive pressure above plant atmosphere of 0.06 in. of water in the main clean spaces and 0.03 in. of water in the entryway with all doors closed and with all exhaust positions operating at design capacity. When entryways are opened, an outward air flow shall be maintained without affecting any exhaust position performance. Room pressurization conformance shall be verified in accordance with Section 4.

2.5 *Temperature and Relative Humidity*

2.5.1 A uniform temperature and a uniform relative humidity shall be maintained under normal operating conditions, at an elevation of 36 in. above floor level throughout the room and no closer than 24 inches horizontally to any production or operating personnel, (assuming radiation shielding between any heat source and measurement point).

2.5.2 Temperature and relative humidity control levels are specified in Section 3.0 together with permissible maximum control variation limits.

2.6 *Sound Level*

2.6.1 *Unoccupied Condition*—The sound level conditions within the clean room space due to the combination of transmission through the construction from the ambient and the operation of air handling and conditioning systems shall not exceed an NC (Noise Criterion) 55 curve by octave band analysis based on USA Std. SI. 6-1960.

For design purposes the sound level existing in the intended location meets an NC _____ curve with octave band levels as follows:

| Band Center Frequency (Hz) | Sound Pressure Level (db) |
|:---:|:---:|
| 63 | |
| 125 | |
| 250 | |
| 500 | |
| 1000 | |
| 2000 | |
| 4000 | |
| 8000 | |

2.7 *Vibration Levels*

The vibration isolation and damping characteristics of the room construction and air handling and conditioning equipment shall be such that the following limits are not exceeded on the clean room floor surface in any direction in the unoccupied condition:

2.7.1 *Vibration Velocity Measured at Floor Surface*—.0025 inches/sec. at frequencies between 500 and 15000 cycles per minute in any direction. For design purposes, the following vibration velocity exists on the _____ floor at the intended room location: _____ inch/sec., (equivalent to a peak to peak displacement of _____ mils at a frequency of _____ cycles per minute).

2.8 *Light Intensity*

The lighting installation shall be designed for "sustained light intensities" at a height of 30 inches above the floor and based on a "good" maintenance factor condition as follows: _____ .

## 3.0 AIR CONDITIONING LOADS AND OPERATING CONDITIONS

3.1 Control levels of temperature and relative humidity to be maintained within the clean room are as follows:

| Room | Temperature | Relative Humidity |
|------|-------------|-------------------|
| | (_____°F± _____) | (_____% RH ± _____% RH) |

3.2 *Heat Loads and Make-Up Air Requirements*

The heat load induced by the operating equipment incident to the process within the clean room, the number of operating personnel to be accommodated, and the volume of process exhaust to be accommodated are shown in the table below.

| Room | Personnel | Process KW | Process Exhaust* (CFM) |
|------|-----------|------------|------------------------|

*Pressurization air quantities necessary per Section 2.4 are in addition to process exhaust, and are to be 1% of main air supply quantity.

3.3 The Contractor shall provide air conditioning equipment to accommodate the heat gains from the above internal operating loads in addition to all heat gains incident to equipment installed as part of this contract to functionally operate the room, and the heat gains for make-up air to accommodate the process exhaust, air change for personnel, the leakage determined by design and workmanship incident to the room construction, and the pressurization make-up air quantity specified.

    (a) The air conditioning equipment shall also have sufficient reserve capacity to bring the room from shutdown at ambient conditions to the control temperature and relative humidity levels within 4 hours, with all loads present except personnel and lighting.

3.4 For purposes of computing air handling and conditioning equipment capacities and sizes,

    (a) The source for make-up air shall be *outside* the plant at:

          _____°F Dry bulb temperature

          _____°F Wet bulb temperature

          _____% Relative humidity

    (b) The ambient conditions surrounding the clean room shall be taken as:

| Summer | Winter | |
|--------|--------|--|
| _____ | _____ | °F Dry bulb temperature |
| _____ | _____ | °F Wet bulb temperature |
| _____ | _____ | % Relative humidity |

## 4.0 CERTIFICATION

4.1 *General*

The Contractor shall make all acceptance certification tests of his work as required by the Owner's Engineer during the progress of the work and at completion to demonstrate to the Owner the strength, durability, and fitness of the installation to satisfy the specification, and demonstrate the satisfactory functional and operating efficiency as well as adjustment of all equipment.

The Contractor shall supervise all tests and shall provide a competent engineer to conduct all tests in the presence of the Owner's Engineer or authorized representative.

The Contractor shall furnish all instruments, ladders, lubricants, test equipment, personnel, and any material required for the tests, except as agreed to by the Owner's engineer in writing.

The tests must be carried out, and the results approved in order that the system be "final" accepted by the Owner.

The tests shall be performed after the Contractor has made his initial operating and balancing adjustments, and satisfied himself that the installation is ready for acceptance certification testing.

NOTE: Instrumentation to be used shall be in accordance with the descriptions given in each case, and shall have a demonstrated accuracy and sensitivity suitable for the test procedure. All instruments shall be properly calibrated according to the manufacturer's recommendations, and so certified to the Owner's satisfaction at the time of test.

4.2 *Certification Tests—"Unoccupied Conditions"*

Upon completion of Contractor's initial balancing, adjustment and testing, and after the area is cleaned, the Contractor shall conduct the acceptance tests in the sequence presented in Sections 4.2.1 through 4.2.11, in accordance with the referenced sections of the American Association for Contamination Control Standard CS-6T "Testing and Certification of Particulate Clean" Rooms.

4.2.1 *Final Filter Installation Leak Test*

Final filter installations shall be tested for pinhole leaks to insure that the HEPA filters and their installations are properly sealed.

The filters and their installation shall be tested by the _____ _____ type HEPA filter Installation Leak Test using the _____ _____ Method in accordance with Para. _____ _____ of AACC Standard CS-6T. Acceptance shall be in accordance with Para. _____ of that standard.

4.2.2 *Induction Leak Test*

(a) *The purpose of this test* is to insure against particle intrusion into the clean space by induction through construction joints, (joints between individual supply modules assembled together, wall joints, duct joints, lighting fixtures in ceilings, etc.).

(b) This test shall be conducted in accordance with the provisions of Para. 8 of the AACC Standard CS-6T. Acceptance shall be in accordance with Para. 8.4 of that standard.

(c) Where necessary, repairs may be made by a procedure acceptable to the Owner.

4.2.3 *Test for Velocity and Air Flow Uniformity in the Clean Space*

(a) *The purpose of this test* is to establish the average air flow velocity and the uniformity of air flow within the work zone(s) of the clean room(s).

The work zone is that volume within the clean room which is designated for clean work and for which certification of laminar flow is required. The work zone is the zone throughout the entire horizontal cross-section of the room between entrance and exit planes located at 36 inches below the ceiling and 30 inches above the floor, respectively; (See Section 2.1.1 and 2.2).

(b) This test shall be conducted in accordance with the provisions of Para. 9.1 through 9.4 of the AACC Standard CS-6T.

(c) *Acceptance*

The design value of the average velocity in the work zone is given in Section 2.3.1.

The measured average velocity shall be equal to the design average velocity within (−)0.0 fpm, (+)30 fpm.

A minimum of 80% of the total number of velocity measurements shall be within plus or minus 20 feet per minute of the measured average velocity. The remaining 20% of the velocity measurements shall be within plus or minus 30 feet per minute of the measured average velocity.

Correction of conditions not meeting these requirements shall be a matter of agreement between Contractor and Owner.

4.2.4 *Test for Main Air Supply Quantity, Make-up Air Supply Quantity and Air Supply Reserve Capacities for Loaded Filters*

(a) The purpose of these tests is to establish within plus or minus 10% the quantity of main air supply to the clean room, the quantity of make-up air delivered by the make-up air handling system, and the reserve capacities of the air handling systems to accommodate a loaded HEPA filter condition.

(b) *Procedure—Main Air Supply Quantity*
This test shall be conducted in accordance with Para. 10.2 of AACC Standard CS-6T. Acceptance shall be as per Para. 10.2.3 of that standard.

(c) *Procedure—Make-up Air Quantity*
This test shall be conducted in accordance with Para. 10.3 of the AACC Standard CS-6T. Acceptance shall be per Para. 10.3.3 of that standard.

(d) *Procedure—Reserve Capacity of Air Handling Systems*
Determine the reserve capacity of the main air handling system(s) and the make-up air handling system(s) by comparing field data with manufacturer's fan curves for each fan as detailed in Para. 10.4.1a of the AACC Standard CS-6T.

(e) *Acceptance*
The reserve capacity of the main air supply and make-up air supply shall be a fan static pressure of 1 in. w.g. minimum above the test values at the design throughputs.

4.2.5 *Test for Air Flow Parallelism, Dispersion and Recovery Characteristics.*

(a) The purpose of these tests is to verify the parallelism of air flow throughout the work zone and the capability of the clean room to limit the dispersion of, or recover from the effects of, internally generated contamination.

(b) *Procedure—Parallelism and Dispersion*
Perform this test, (after completion of the air uniformity tests per Section 4.2.3), in accordance with Para. 11.3.1 through 11.3.7 of AACC Standard CS-6T. Instrumentation shall be per Para. 11.2 of that standard.

(c) *Procedure—Recovery Characteristics*
This test shall be performed in accordance with Para. 11.4.1 through 11.4.3 of AACC Standard CS-6T. Instrumentation shall be per Para. 11.2 of that standard.

(d) *Acceptance*
Dispersion at all test points shall not extend radially from the point of smoke source further than the limits specified in Section 2.2. Beyond this limit particle counts shall not exceed 100 per cu. ft. $\geqslant 0.5$ micron size.

Recovery from smoke generation shall at all points be within the recovery time limit specified in Section 2.2.

Corrections—If parallelism, dispersion control or recovery characteristics do not meet Owner's specifications, adjustments to air flow control devices (dampers, floor perforations, exit grilles, etc.) will be required. Following adjustments, air flow uniformity and air delivery tests must be repeated to demonstrate compliance with the specifications.

4.2.6 *Tests for Pressurization Control*

(a) The purpose of these tests is to verify the capability of the clean room system to maintain a positive pressure in the clean room above the surrounding ambient. This test should be performed after the airflow uniformity, air delivery and parallelism, etc., tests. (NOTE: This test will be repeated in the "At Rest" condition per Section 4.3).

(b) This test shall be performed in accordance with Para. 12.1 through 12.3 of AACC Standard CS-6T.

(c) *Acceptance*

Pressure differentials maintained with all doors closed shall meet the requirements of Section 2.4.2.

With doors open; inward air flow is unacceptable as evidenced by air flow direction and by particle counts inside the doorway rising above 100 particles per cu. ft. of 0.5 micron size and larger.

4.2.7 *Particle Count Test*

(a) The purpose of this test is to establish conformance to the requirements of Section 2.1.1 and serve as a secondary check on the HEPA filter installation leak tests.

(b) *Apparatus*

Particle Counter, per Para. 3.10.1 of AACC Standard CS-6T, except with capability to measure particle concentrations $\geqslant 0.5$ micron size, $\geqslant 1.0$ micron size, and $\geqslant 5.0$ micron size.

(c) *Procedure*

Perform this test after completion of the HEPA filter leak test, air uniformity test and pressurization tests.

Measure the count of particles $\geqslant 0.5$ micron size and $\geqslant 1.0$ micron size at the center of each 2 ft. X 2 ft. grid at a level of 30 in. above the floor.

In moving from grid to grid, move the particle counter while operating in such a manner as to detect possible "streaming" leaks through the filters that will cause localized areas of high particle count in the work zone.

Record all measurements together with their grid locations.

(d) *Acceptance*

No particle count greater than 10 per cu. ft. $\geqslant 0.5$ micron size shall be detected.

No particles at all $\geqslant 1.0$ micron size shall be detected.

4.2.8 *Test for Temperature & Humidity Control Capability*

(a) The purpose of these tests is to demonstrate the capability of the system to control temperature and humidity at the levels and within the limits specified in Sections 2.5 and 3.0.

(b) This test shall be conducted in accordance with Para. 13 of the AACC Standard CS-6T, with acceptance per Para. 13.4 of that standard.

4.2.9 *Tests for Lighting Level and Uniformity*

(a) The purpose of this test is to verify that the specified lighting levels and uniformity of lighting within the room have been met.

(b) This test shall be conducted in accordance with Para. 14.1 through 14.4 of the AACC Standard CS-6T with the working level established at 30 in. above floor level.

(c) *Acceptance*

Lighting levels throughout the room shall meet the requirements of Section 2.8. Corrections if required shall be a matter of agreement between Contractor and Owner.

4.2.10 *Test for Noise Level*

(a) The purpose of this test is to establish the air-borne sound pressure levels produced by the basic clean room mechanical and electrical systems as experienced within the room and at adjacent external occupied areas, and to verify that performance meets the buyer's specifications.

(b) This test shall be performed in accordance with Para. 15.1 through 15.4 of the AACC Standard CS-6T.

(c) The sound pressure levels reported shall not exceed those corresponding to a Noise Criterion (NC) 55 curve.

Conditions not meeting the specifications shall be corrected by the Contractor using approved sound attenuation materials, or by equipment changes as necessary, and as agreed to by the Owner.

4.2.11 *Tests for Vibration Characteristics*

(a) The purpose of this test is to establish the vibration levels produced by the clean room mechanical support systems, as experienced within the room and at adjacent work locations external to the room, and to verify that performance meets Owner's specifications.

(b) This test shall be conducted in accordance with Para. 16.1 through 16.4 of the AACC Standard CS-6T.

(c) *Acceptance*

Measured values shall not exceed the requirements stipulated in Section 2.7.

Corrections, if required, shall be a matter of agreement between Contractor and Owner.

4.3 *Performance Tests—"At Rest" Conditions*

Following the installation of operating equipment, the Contractor shall return within two weeks of written notification and rebalance and retest the complete installation in the "At Rest" condition. Readings shall be taken within each 10 foot square grid to verify room capabilities in regard to:

(a) The particle count requirement of Section 2.1.2, per procedures of Section 4.2, except at each 2 foot square grids.

(b) Particle dispersion and room recovery requirements of Section 2.2. These shall be demonstrated by smoke pattern studies performed in accordance with Section 4.2.

(c) The temperature and humidity control levels specified in Sections 2.5 and 3.0, per procedures of Section 4.2.

(d) Pressurization control capability to meet the requirements of Section 2.4 shall be re-verified per Section 4.2.

(e) Make-up air quantity to meet the requirements of Section 3.2 shall be re-verified per Section 4.2 using blower Hp and speed measurements together with fan manufacturer's curves.

4.4 *Operation and Maintenance Instructions*

4.4.1 Upon completion of the installation, the Contractor shall furnish six copies of operating and maintenance instructions. Operating instructions shall include: manufacturer's name, size and type, serial number and foreign print number for each item of equipment; normal starting procedures for the various elements of the system; detailed drawings; wiring diagrams, repair parts list and maintenance instructions for the various elements of the system. The Contractor shall provide all services necessary to properly instruct the owner's Engineer on the operation and maintenance of all systems.

4.4.2 Included in the six bound sets of operating and maintenance instructions shall be the following *certified* material:

1. Operating and maintenance procedure
2. List of recommended spare parts
3. Approved shop drawings
4. Certified material lists
5. Certified performance curves for all equipment supplied

4.4.3 Printed instructions and system control diagrams covering the operation and maintenance of each major item of equipment, shall be posted at locations designated by the Owner's Engineer, in a wooden frame under glass protection.

4.4.4  Upon completion of the work and after all tests are completed and the systems balanced and set, and at the time designated by the Owner's Engineer, a competent engineer shall be provided by the Contractor to instruct a representative of the Owner's maintenance department in the proper operation and maintenance of the airconditioning and air handling systems.

## 5.0  ENCLOSURE CONSTRUCTION DETAILS

5.1  *General*

Raised floor (when required), walls, ceiling, plenums, and ductwork shall be solidly constructed to minimize expansion and contraction at joints. The construction shall also minimize noise and vibration that may be transmitted throughout the structure. Ceiling height shall be a minimum of 9 feet in all rooms except equipment rooms.

5.2  *Floor*

5.2.1  *Base Floor*

The existing base floor is _____

_____

_____ .

The Contractor shall treat this floor before installation of the clean room floor as follows: _____

5.2.2  *Vertical Laminar Flow Clean Room Floor*—shall be wall to wall raised type so as to provide room for piping, electrical, and exhaust duct services as well as a return air plenum. The minimum height shall be sufficient to maintain plenum air velocities not over 1200 ft/min. maximum, with the capability of accepting duct work, piping, etc. to a maximum fill of _____ % of the flow area at any cross-section plane.

The floor shall be made up of 24 in. × 24 in. × 1¼ in. steel modular sections supported on adjustable height pedestals, (or other approved floor system with panels not over 48 in. wide × 10 ft. long with an approved number of 24 in. × 24 in. nominal size removable and relocatable access panels), and secured rigidly in place to prevent movement or vibration, yet easily removable for access to the lower plenum.

All floor panels shall be perforated. The perforation hole size and the number of panels equipped with air flow adjustment dampers (at least one half of panels) shall be designed to accomplish the balancing of the room for the required air flow conditions (Para. 2.3, 4.2.4), (or so designed as to provide automatic air flow control without dampers, regardless of process equipment placement). The pedestals, where used, shall be anchored to the floor either mechanically or with an acid, solvent and waterproof adhesive bond through a base of not less than 16 sq. inch.

The raised floor shall be capable of carrying a uniform live load of 250 pounds per square foot with a deflection not to exceed .040 inch. Each individual 24 in. × 24 in. (nominal) floor panel area shall be capable of supporting a concentrated load of 1,000 pounds per square inch with a maximum deflection of 0.100 inch. The floor panels shall be a steel frame construction with any nonsteel interior components fire-resistant treated and with a 1/8 in. fire-resistant, high-pressure plastic laminate, double wear tile surface of a color to be selected by Owner, or other approved finish. Edges shall be trimmed with a side leveled molding that will not delaminate when panels are removed and replaced, (or other suitable treatment). The finished floor shall be level with plus or minus 0.100 inch throughout the room, and plus or minus 0.062 inch within 10 ft.-0 in. A vinyl cover base (color to be selected by the owner) shall be installed along all the edges of all interior walls or partitions.

### 5.3 Walls

5.3.1 *Exterior enclosure walls* shall provide a continuous vertical finish between base floor and _____, as shown on Owner's drawings.

(a) Exterior enclosure wall shall be constructed as follows, (or approved equal):

(NOTE TO ENGINEER: Retain one or more of the following descriptions, or replace with other desired description.)

Hollow or solid core construction; modular design type, with interior and exterior surfaces having a high pressure laminate finish of Formica, Micarta, melamine, metal, or metal skin over gypsum panels.
Or:
Metal skin gypsum panels applied to both sides of metal studs with joints caulked with silicone caulking or sealed with appropriate gasketing-type metal trim strips specially designed for the purpose.
Or:
Gypsum panels applied to both sides of metal studs or a fire-retardant treated wood framing system. Joints shall be taped, cemented and smoothed.

(b) Color of wall finish or paint will be determined by owner at a later date.

(c) Windows in exterior enclosure walls shall be _____ above the clean room floor, located generally as shown on Owner's drawings. All windows shall be sealed with materials such as viton or neoprene gaskets to prevent air leakage into or out of the room(s), and mounted in anodized aluminum or painted steel frames. Windows shall be ¼ in. polished safety plate glass so placed in the frames that they are flush with the interior wall surface.

(d) All metal interior trim shall be anodized aluminum or painted steel.

5.3.2 *Interior of divider walls* shall extend from the clean room floor to the clean room ceiling and shall be constructed of same materials as exterior walls.

(a) Windows in interior walls shall be located generally as shown on Owner's drawings and shall be of ¼ in. polished safety glass in anodized aluminum or painted steel frames.

### 5.4 Amber Partitions

When required, as shown on Owner's drawings, amber partitions consisting of both solid and transparent sections shall be glazed with ¼ in. Rohm and Haas Plexiglas #2422 or approved equal.

### 5.5 Entryways and Pass-Throughs

5.5.1 Entryways shall be provided and generally located as shown on the Owner's drawings. The enclosed portion shall be constructed at the same level as, and conform to the general construction of the main clean area. The entryway shall provide an air seal sufficient to sustain the minimum pressurization above plant atmosphere specified in Section 2.4.2.

(a) When required for raised floor clean rooms, the entryway shall include raised floor sections, steps and railings as shown on Owner's drawings. Steps shall be rigidly connected to the main room raised floor structure and to the base floor. The stair structure shall consist of angular supporting framing, tread plates and riser closure plates with 3/16 in. thick rubberized non-skid tread cemented in place on tread plates, all designed for a loading on steps of 250 lbs. per ft$^2$. All trim on stairs and railings shall be anodized aluminum with #4 finish.

5.5.2 Pass-throughs shall be provided generally as located on Owner's drawings for process material transfer between rooms and to outside areas. Pass-throughs shall be of the inside dimensions, constructed of materials specified, and include interlocks and alarms as specified on drawings. Construction shall provide air seals to minimize air leakage through pass-throughs. Trim shall conform to that used in the room construction.

(a) When clean bench type pass-throughs are specified, these shall be provided by a certified clean bench supplier acceptable to the Owner and shall conform to Owner's clean bench specifications. The room contractor shall be responsible for their procurement and proper incorporation into the construction.

5.6 *Doors*

All doors shall be hollow metal generally located and sized as shown on Owner's drawings, with ¼ in. polished plate glass windows mounted 54 in. above floor, or solid core for equipment room. All doors shall be Kawneer, narrow stile "190", or approved equal. Double doors shall consist of a pair of doors, one fixed, one active.

Doors will have approximately 2 in. vertical stiles, 2½ in. top rail, and 4 in. bottom rails with center divider and push plate. Doors are to be complete with all hardware, ¼ in. polished plate glass, concealed overhead door closers with built-in hold-open device, latch lock with level handle inside, and completely gasketed for dust control, with rubber gaskets and automatic door seals against anodized aluminum thresholds.

Entrance doors shall be anodized aluminum doors and frames with fixed glass transom. Doors other than entrance doors shall be finished with similar color to the rooms which they serve.

Any special emergency exit door will have a latch lock with level handle inside and will open with key from outside.

Special attention is to be paid to all door construction and fitting to preserve air pressure.

Lock systems shall be compatible with Owner's master lock system.

5.7 *Vertical Laminar Flow Clean Room Ceiling*

5.7.1 The clean room ceiling shall consist of a structural support and suspension system, air distributing means, lighting system and sprinkler system.

5.7.2 The air distributing means in the ceiling may consist of one of the types delineated in Section 1.2.

5.7.3 Where final filters, or final filter-blower modules are installed in the ceiling, the support and suspension systems shall be structurally constructed so as to remain dimensionally stable and retain the necessary sealing requirements throughout the useful life of the final filters.

5.7.4 Where final filters or air distributing ceiling panels are used with a supply plenum arrangement above, the room ceiling and the plenum ceiling structural systems shall be capable of withstanding the maximum blower static pressure without deflecting more than _____ in. from the original level position.

5.7.5 The ceiling, as installed, shall be level throughout the room within _____ _____ in.

5.7.6 To insure that the ceiling area is occupied by a maximum of air distributing surface, support grids or "T" bars and wall sealing angles should occupy a very minimum portion of the total area. Framing members or T-bars should be held to a maximum width of approximately two inches.

(a) See Section 5.9 relative to lighting fixture design and installation in this regard.

5.8 *Ceiling Suspension*

5.8.1 From existing building structure

(a) Where the Owner's drawings so indicate; the ceiling system may be suspended from the existing building structure provided the weight to be suspended does not exceed _____ lb. per sq. ft.

(b) Suspension hangers must be at least _____ in. diameter rod or wire.

(c) Existing building structural members shall not be drilled for connections of suspension hangers.

5.8.2 *Separate Supporting Steel Structure*

(a) Where the Owner's drawings do not indicate suspension from the existing building structure, or where the load to be suspended exceeds the value in Section 5.8.1, the Contractor shall design, provide and install a separate ceiling suspension structure.

(b) The supporting structure shall in turn be supported by the clean room enclosure wall system if this be of the load-bearing design or by suitably located columns extending down to the base floor.

(c) The Contractor shall be responsible for the design of any required base plates to suitably transmit the loads to the base floor.

(d) Base plate loadings shall not exceed _____ lbs. per sq. ft.

(e) The structural system shall be designed and located so as not to load any existing base floor framing beam more than would result from a concentrated load of _____ lb. at the center of the beam span.

5.9 *Lighting*

5.9.1 *Fixtures in Clean Room*

(a) Lighting shall be provided by standard four-foot fluorescent lighting fixtures to meet the lighting levels specified in Section 2.8.

(b) Light fixtures and their ballasts shall be capable of being maintained from within the room. Ballasts shall be high power factor type carrying CBM certification and operate at 120 volts, 60 cps. Fixtures to have Buss GLR fuses and HLR fuse holders.

The lights are to be supported by the ceiling framing, with the wiring detail and mountings sealed into support frames against air leaks into the room. Materials shall be noncorrosive and have nonshedding surfaces. Selected fixtures as indicated by Owner's Engineer shall be used as emergency and/or night lights, connected in parallel, with one stub-up through ceiling.

(c) When required for photolithographic areas, gold fluorescent lamps shall be used in place of daylight fluorescent lighting. See Section 2.8 for lighting levels.

(d) Entryway fixtures shall be on a separate circuit.

(e) All fixtures shall be low voltage contactor-controlled with push button stations wall-mounted in the entryway. The Contractor shall provide and install all lighting wiring and controls from a _____/120 volt, 3 phase, _____ wire source with ground. Circuit wiring in lighting fixtures used as wireways must be #12THHN wire. A green equipment ground wire shall be run between fixtures and solidly connected to each fixture frame and to building steel. Contractor shall lamp all fixtures.

(f) Lighting fixtures shall be laid out in single rows and mounted so not to disturb the laminar flow.

(g) Flush mounted fixtures shall not be more than 12 in. wide, and shall have air tight housings with anodized aluminum eggcrate diffusers or hinged acrylic plastic diffusing lens covers. Fixtures shall be sealed into the ceiling framing to prevent induction leaks into the room, (except where the air above the entire ceiling is already final filtered).

(h) Pendant fixtures suspended below the ceiling shall not be more than 6 in. wide and shall have air foil shaped diffusing lenses.

(i) Lighting fixtures shall not be installed adjacent to walls.

5.9.2 *Fixture Layout in Equipment Rooms*

The Contractor will furnish one vapor-tight incandescent fixture between each pair of fans in the equipment rooms. These fixtures will maintain 30 foot-candles in areas that must be accessible for maintenance activity. Connect fixtures in parallel with one stub-up through ceiling.

## 6.0 FILTERS

### 6.1 *Final Filters in Main Air Handling System*

6.1.1 Final filters shall be:
HEPA, type C, Grade 1, 6 in. deep for ceiling mounting or 12 in. deep for remote location, per A.A.C.C. Tentative Standard CS-1T; i.e., fire resistant construction throughout, 99.97% minimum removal efficiency for 0.3 micron particles at rated flow, and probed to certify freedom from pinhole leaks. Final filters shall be supplied as part of the room construction and installed in holding frames such that the installation shall be free of pinhole leaks as evidenced by in place leak testing per Section 4.2.1.
Or:
HEPA, type B, Grade 1, 6 in. deep for ceiling mounting or 12 in. deep for remote locations per A.A.C.C. Tentative Standard CS-1T; i.e., fire resistant construction throughout, 99.97% minimum removal efficiency for 0.3 micron particles at rated flow, and tested for overall penetration at 20% of rated flow for freedom from significant pinhole leaks. Final filters shall be supplied as part of the room construction and installed in holding frames such that the installation shall be free of gross leaks as evidenced by in place leak testing per Section 4.2.1.

(NOTES TO ENGINEER: Retain one of the above descriptions. For rooms with final filters mounted in the ceiling, it is recommended that type "C" HEPA filters be specified. For rooms with final filters mounted remotely in ducts, plenums, equipment rooms, type "B" filters may be accepted at some cost saving. For filters mounted in the clean room ceiling, 6 in. deep units are recommended. For remotely located filters, 12 in. deep units are recommended to take advantage of their high capacitor at a slight increase in individual cost to obtain an overall cost saving.)

6.1.2 Installation
(a) When installed in a holding frame or filter blower module system for use with a plenum type supply, (either ceiling located or remotely located), or in a holding frame in a duct system, the filter frame shall be constructed leak-tight, shall be designed to accept the filter and hold it captive upon initial insertion, and shall provide means for tightening and maintaining the effective seal of the filter in the frame. Filters shall be individually removable with a minimum of loose parts and without disturbing adjacent filters or filter seals.
(b) When installed in the clean room ceiling in filter-blower modules or in units connected to supply ductwork systems, effective means shall be provided for sealing the modules or filter units into the ceiling structure to prevent induction of particles into the room.
(c) When installed in the room ceiling, filters shall be removable and insertable from inside the room, and shall be provided with guards over the exposed faces in the room. Guards shall provide mechanical protection for the filters and shall present a pleasing appearance to the room. Guards may be anodized aluminum flattened expanded metal, perforated stainless steel, (or approved substitute), having 60% minimum open area and sufficient mechanical strength to prevent sag.
(d) When installed in a duct system, filters shall be removable from the side of the duct, or from inside the duct through access doors, without dismantling the duct-to-holding-frame seal.

(e) When installed remote from the clean room ceiling, the number of filters shall be determined on the basis of the manufacturer's rated capacity values at 1 in. water gage pressure drop.

6.2 *Prefilters in Main Air Handling System*

(NOTE TO ENGINEER: For filters of the type specified in Section 6.2.1, it is recommended that 12 in. deep, 85% efficient units be specified. If the alternate prefilters of Section 6.2.2 are unacceptable, delete Section 6.2.2. It must be borne in mind, however, that permitting only prefilters of the type specified in Section 6.2.1 may eliminate some room design types.)

6.2.1 Prefilters in the main air handling system shall preferably be:

(a) Pleated, extended media, rigid frame cartridge type, _____ in. deep, and shall have a minimum efficiency of _____ % when rated by the ASHRAE Std. 52-68 Atmospheric Air Stain Test Procedure. Filters shall be of fire resistant construction meeting Underwriters' Laboratories UL-900 class 1 rating.

(b) The number of filters shall be determined on the basis of the manufacturer's nominal rated capacity values.

(c) The prefilters shall be supplied as part of the room construction and shall be installed in holding frames of leak-tight construction, and provided with means for obtaining and maintaining an effective seal of the filter into the frame.

(d) When installed in fan plenums (equipment rooms), the filter frame shall be designed to accept the filter and hold it captive upon initial insertion. Filters shall be individually removable with a minimum of loose parts, and without interrupting room operation.

(e) When installed in a duct system, filters shall be removable from the side of the duct, or from inside the duct through access doors, without dismantling the duct-to-holding frame seal.

(f) All filters shall be in place before the supply air, make-up air, (and exhaust) fans are operated and tested.

(g) All filters shall be protected during initial room start-up and clean-down by 1 in. thick medium efficiency fiber glass prefilter blanket material, Class 2 fire rating, held in place on the upstream face. This material shall be removed prior to acceptance testing.

6.2.2 Alternate Main Air Handling System Prefilters—Where the Contractor's design precludes the use of prefilters per 6.2.1, prefilters in the main air handling system may be:

(a) Preformed replaceable fiber glass cartridge type, 2 in. thick minimum, having a minimum efficiency, at rated capacity, of 45% using atmospheric dust, (or 94% using, by weight, 96% Cottrell precipitate and 4% lint), when tested by the ASHRAE Std. 52-68 Atmospheric Air Stain Test Procedure. Filters shall be of a fire retardant construction meeting Underwriters' Laboratories UL-900 Class 2 rating.

a) The number of filters shall be determined on the basis of the manufacturer's nominal rated capacity values.

b) Filters shall be provided as part of the room construction and shall be installed in holding frames of leak tight construction, and provided with means for attaining and maintaining an effective seal of the filter into the frame.

c) Where separate prefilter banks are provided, holding frames shall be permanent "J" cross section galvanized steel with medium support, medium retainer and stainless steel latch hardware. Filter shall be individually removable with a minimum of loose parts, and without interrupting room operation.

d) Where filters are installed in floor return locations or in combination with final filters in filter blower modules, they shall be individually removable with a minimum of loose parts and without disturbing the final filter seal or interrupting room operation.

e) All filters shall be in place before the supply air, make-up air, (and exhaust) fans are operated and tested. Filters shall then be removed and replaced after testing and acceptance by the Owner to provide the full filter dust loading capacity on completion of the construction and testing phases.

### 6.3 *Make-Up Air Handling System Particulate Filters*

6.3.1 Particulate filters in the make-up air handling system shall be of the same type, size and efficiency as specified in Section 6.2.1 and shall be subject to all the installation provisions of Section 6.2.1c through h.

(a) The number of filters shall be determined on the basis of the manufacturer's nominal rated capacity values.

### 6.4 *Activated Carbon Filters In Make-Up Air Handling System*

(NOTE TO ENGINEER: If the need exists to limit the level of fume and vapor concentrations within the room, and if the make-up air has fume and vapor concentrations above this level, then activated carbon filters are recommended in the *make-up air handling system.* In this case, complete the necessary information. If fume and vapor concentration in the make-up air is sufficiently low and make-up air quantities are high enough, then activated carbon would not be required. In this case, eliminate Section 6.4.)

6.4.1 Activated carbon filters shall be of the removable, perforated, activated carbon filled, multiple element gas and odor adsorption, 24 in. X 24 in. cell type for removal of the following materials to levels of _____ ppm in the effluent stream when presented with influent concentrations as indicated and when operated at the manufacturer's rated air flow capacity:

| *Material* | *Influent Concentration ppm in Air* |
|---|---|

6.4.2 Installation

(a) Filters shall be provided as part of the room construction and shall be installed in holding frames of leak-tight construction, and provided with means for obtaining and maintaining an effective seal of the filter into the frame.

(b) Filters shall be installed in a vertical position with horizontal air flow through the filter.

(c) Filters shall be individually removable with a minimum of loose parts and without disturbing adjacent filters or interrupting the room operation.

(d) Each single carbon filter unit shall have at least 5 internally welded spacers per 24 in. X 24 in. cell with full length welded and hemmed leading and trailing edges for complete dimensional stability and to ensure uniform carbon bed thickness throughout.

(e) The filter installation shall be arranged so that all air to be filtered will pass proportionally and uniformly through all parts of the carbon bed.

(f) Each bank of filter cells shall be fitted with a detachable test element which may be removed after a specified period of operation for analysis to indicate the extent of saturation of the installation and the expected remaining service life.

(g) Filters are to be installed downstream of, (after), particulate filters in the make-up air system for protection against dust accumulation.

(h) The filter installation shall be sized to accommodate the make-up air flow necessary to meet the process exhaust and pressurization requirements stipulated in Sections 2.4 and 3.2, and the vapor (gas) loadings stipulated above in accordance with the manufacturer's rated flow capacities and removal efficiencies.

(i)  All filters shall be in place before the supply air, make-up air, (and exhaust), fans are operated and tested.

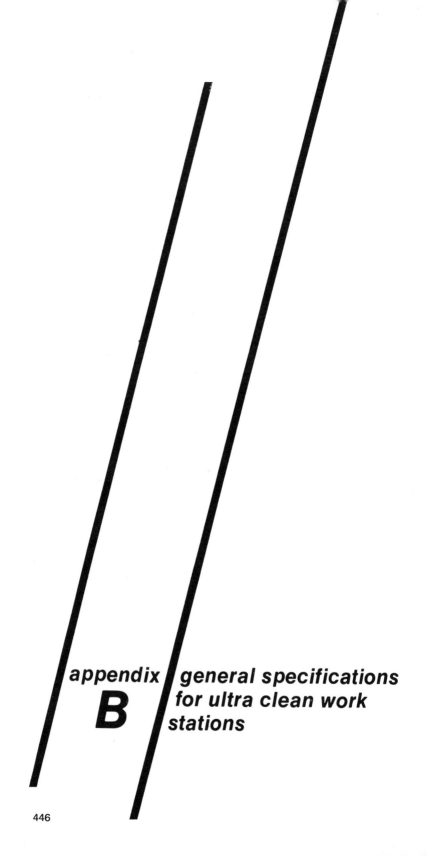

**appendix** **general specifications**
**B** **for ultra clean work stations**

# 1. GENERAL

1.1 *Compliance*–All ultra clean work stations supplied to the ———————— location ` of the ——————————Company shall comply with the general specifications listed below and where referenced shall meet the requirements of and be tested according to the procedures prescribed by the American Association for Contamination Control Tentative Standard, CS-2T, May, 1968, or latest revision. Exceptions are indicated in paragraphs 1.2, 1.5, and 14.

1.2 *Quotation Drawings*–As necessary, the ——————————— Company will provide quotation drawings with requests for quotations to: (1) indicate required dimensions, construction design preferences and any special features to be provided by supplier: and (2) to point out any additions, material preferences, or deletions to these general specifications.

1.3 *Standard Design Drawings*–Typical basic standard design arrangements and typical design dimensions and details are shown on drawing ———————————. Designs proposed by supplier shall conform to the requirements of standard design drawings.

1.4 *Supplier's Drawings*–Supplier shall provide, *for approval prior to fabrication*, drawings showing over-all dimensions and sufficient details to illustrate *construction and installation features.* Supplier shall furnish all information necessary for maintenance including electrical wiring diagrams, blower characteristics and parts, replacement information for major operating components. Such information shall be supplied to maintenance organization, after order is placed and no later than equipment delivery date in conformance with requirements of facility planning engineer or his ——————————— Company representative. As requested, Supplier shall furnish proposal intent drawings with quotations.

1.5 *Exceptions*–No exceptions to the requirements of these specifications or the quotation drawings will be permitted unless individually approved in writing by the facility planning department.

# 2. MATERIALS OF CONSTRUCTION AND WORKMANSHIP

2.1 *Quality and Workmanship*–The materials used for construction shall be of superior quality. The equipment shall be free of defects resulting from either poor materials, mediocre fabrication techniques or improper structural design.

2.11 Equipment shall be free of rough edges or sharp corners.

2.12 *All construction joints shall be sealed and smoothed* using silicone construction grade sealant.

2.2 *Materials for Work Space Surfaces*–The surfaces enclosing the work space shall be aligned with the edge of the effective HEPA filter media (see Para. 3.21 and Para. 6.24) and shall be in accordance with one or more of the following requirements, as specifically noted on the quotation drawings:

2.21 Formica, Micarta, or similar approved material, bonded to composition board; color to be similar to Formica type, *frost green*. (Not recommended for vertical flow units where wet conditions or high heat may exist.)

2.211 NOTE: Composition board base and surfacing materials other than Formica or Micarta must be specifically approved by the facility planning department.

2.22 Stainless steel, type 304 with #2B finish.

2.23 Stainless steel, type 304 with #4 finish.

2.24 Polypropylene, natural color, with all seams heat welded and ground smooth.

2.25 Cold rolled steel, PVC plastisol coated.

2.26 PVC, white color, with all seams heat welded and ground smooth.

2.27 Safety Plate Glass, 1/4 in. thick.

2.28 Transparent 1/4 in. acrylic sheets, minimum thickness.

2.3 *Materials for Surfaces Outside the Work Space*—Surfaces outside of the work space *shall* be in accordance with one or more of the following requirements as specifically noted on the quotation drawings:

2.31 Formica, Micarta, or similar approved material, bonded to composition board; color to be as specified on quotation drawings. (Not recommended for construction underneath or adjoining the work surface where wet conditions or high heat may exist.)

2.311 NOTE: Approval required as per Para. 2.211.

2.32 Stainless steel, type 304 with #4 finish.

2.33 Mild steel, finished with a baked epoxy enamel over a suitable prime coating; color to be as specified on the quotation drawings.

2.34 Exhaust and recirculating air plenums under work surfaces and recirculating air ducts are to be constructed of or lined with material identical to that used on the work table surface.

# 3. LIGHTING

3.1 *Intensity*—The lighting installation *shall* provide an average intensity at the work surface that is not less than 125 ft. candles when tested in accordance with Section 9, AACC Standard CS-2T.

3.11 NOTE: Unless quotation drawings include the statement "Provide _____ _____ watt lamps," the lamps shown on the drawings are for illustration only.

3.12 All flourescent fixtures shall have a power factor greater than 90%.

3.13 Fluorescent lamps shall be T-12 medium bi-pin standard white (or gold as specified) preheat type in 24 in. length T-12 medium bi-pin standard white (or gold as specified) rapid start type in 36 in. or 48 in. lengths, or T-12 recessed double contact standard white, or cool white, (or gold as specified) high output type in 60 in., 72 in., or 84 in. lengths.

3.2 *Obstruction to Air Flow*—When installed in the air flow path, lighting units *shall* present a minimum of obstruction to air flow, with the zone of influence limited to a maximum of 6 inches downstream from the lamp.

3.21 In vertical flow units, lamps shorter than the inside width of the work space shall be mounted on individual tube extensions from the sidewalls. Where longer tubes are used, sidewalls may be recessed from the filter plane down to the eggcrate plane for mounting of tubes; however, this construction must not adversely influence air flow in the work space nor contribute to particle count induction due to eddy current effects.

3.3 *Glare*—Conditions causing glare and reflections *shall* be minimized, and light sources shall be obscured from the direct view of the operator.

3.31 In horizontal flow units, a diffuser-type lens, gasketed in place, *shall* be provided having a smooth surface on the work space side.

3.32 In vertical flow units, except furnace loading benches, an egg-crate diffuser *of approved* material shall be provided below the lamps. The egg-crate shall have openings within the size range of 3/8 in. × 3/8 in., minimum, to 3/4 in. × 3/4 in., maximum, and shall be 1/2 in. thick. *Installation design shall be shown on the quotation drawings.*

3.33 Furnace loading units shall have an egg-crate diffuser of aluminum (see quotation drawings).

3.4 *Heat*—Lighting design *shall* minimize heat addition to the work space. Light compartment shall be ventilated outside the work space where possible.

3.5 *Access*—Lighting components shall be readily accessible for maintenance.

3.51 *Horizontal Flow Unit*—Lamps shall be replaced from the front.

3.52 *Vertical Flow Unit*—Lamps shall be replaced from the underside or front.

3.6 *Wiring Chase*—All lamp wiring shall be enclosed in raceways or chases concealed from the work space region. This raceway or chase shall not connect to any source or particle induction.

## 4. CLEANLINESS CAPABILITY

4.1 *Particle Count*—The equipment shall provide an ultra clean parallel air flow stream, without processing equipment or operator present, that is capable of sustaining a class 100 condition, (Federal Standard No. 209) in any ambient having not over 300,000 particles per cubic foot of 0.5 micron or larger particles. Particle count testing may be done on a sample basis for multiple purchases if specifically noted on the quotation drawings. This shall be demonstrated with a light scattering particle counter in accordance with the following:

    4.11 *Horizontal Flow Unit*—This level shall be obtained over the entire face of the opening of the unit in a vertical plane 4 in. from the front opening.

    4.12 *Vertical Flow Unit*—This level *shall* be obtained over the entire work area in a horizontal plane 6 in. above the work surface.

4.2 *HEPA Filter Installation Leak Test (See para. 6.13)* The work station shall meet the requirements of and be tested according to the procedures of Section 4, AACC Standard CS-2T or the following:

    4.21 *Particle Counter Leak Test*—Allows the use of a particle counter to duplicate the test methods prescribed in Section 4, AACC Standard CS-2T and modified as follows:

    4.22 *Ambient*—The work station to be leak tested shall be operated in an ambient of at least 300,000 ppcf. of 0.5 micron or larger particles.

    4.23 *Apparatus*—A particle counter having a sample rate of at least 0.1 cfm and equipped with a sampling tube having an inlet opening sized to provide isokinetic sampling (e.g., with 100 fpm air velocity in the clean-air device and 0.1 cfm sampling rate, the sample tube inlet should be approximately 7/16 inch I.D.) A particle counter having a rapid recovery rate is desirable for this test.

    4.24 *Calibration & Standardization*—Refer to section 5 (e), AACC Standard CS-2T.

    4.25 *Procedure*

        4.251 Filter and installation shall be tested at an average flow rate of 100 + 20 – 10 lineal feet per minute through the filter.

        4.252 The filter and the perimeter of the filter shall be scanned per Section 4d(5), AACC Std. CS-2T.

    4.26 *Acceptance*—The work station shall exhibit no concentration levels in excess of 100 particles per cubic foot of size equal to or larger than 0.5 micron. (See notes of Section 5 (i), AACC Standard CS-2T). *When tested for acceptance in Owner's plant, the particle counter procedure will be used.*

4.3 *Induction Leak Test (See para. 6.13)*—The work station shall meet the requirements of and be tested according to the procedures of Section 5, AACC Standard CS-2T. When tested for acceptance in Owner's plant, the particle counter procedure of that section will be used.

## 5. AIR FLOW CAPABILITY

5.1 *Air Flow Velocity—Clean Filter Condition (See Para. 6.13)*

    5.11 *Horizontal Flow Units*—The work station shall be tested in accordance with Section 6, AACC Standard CS-2T. The average velocity shall be 100 + 20 – 10 fpm. No reading shall be less than 80 fpm and no more than 130 fpm.

5.12 *Vertical Flow Units*—The work station shall be tested in accordance with Section 6, AACC Standard CS-2T except that the locations and numbers of readings shall be adjusted to suit the requirements as specified on the quotation drawings. The average velocities shall be as specified on the quotation drawings with minimum and maximum readings within 20% of the average.

5.2 *Air Flow Velocity—Loaded Filter Condition (See Para. 6.13)* The work station shall meet the requirements of Section 7, AACC Standard CS-2T, as certified by model-type test results, or as individually tested when required. (Note: for vertical flow units, numbers and locations of readings shall be as specified in para. 5.12).

5.3 *Blower*—The blower-motor assembly shall be selected and arranged to provide (see Para. 7.4 and 10.1):

5.31 Air delivery regulation to permit meeting the clean filter velocity requirements of Para. 5.1 as well as the requirements of Para. 5.2.

5.311 Where permanent split phase motors are used, an auto-transformer speed control shall be provided for each motor. Speed control shall be mounted *inside* the blower compartment for all *nonrecirculating design units*. Where *recirculating air flow* is provided, speed control shall be located within a compartment *sealed* off from the blower compartment and *ventilated* to the outside, and the motor *shall* have *no* built-in open contact thermal overload protective devices. This control *shall* permit continuous operation over an adjustable range from maximum speed to the minimum motor speed possible without stalling or overheating. At the minimum speed, the motor shall be capable of start-up without readjustment of the control. A *mechanical stop* shall be provided at this point to prevent further speed reduction. The minimum setting shall meet the requirements of 5.1, and the maximum setting shall meet the requirements of 5.2.

5.312 Splitter dampers *may* be provided on *all* blowers; however, these *do not replace* speed controls of paragraph 5.311 but may be installed in addition.

5.313 All manual dampers shall have a handle position lock to secure damper in any position, and an indicator to show damper position.

5.32 Fully stable blower operating over the entire range of filter loading and speed regulation above.

5.4 *Exhaust—Vertical Flow Units*—Based on the HEPA filter nominal face area, the average velocity requirements of Para. 5.12, and a static pressure of (−) 1 in. W.G. at the ——————— Company connection point, the exhaust plenum and duct to the connection point shall be sized to accommodate exhaust quantities as follows and shall be provided with volume control damper:

5.41 *Full Exhaust Design*—100% of the supply air through the HEPA filter.

5.42 *Recirculating Design*—A minimum of 70% of the supply air through the HEPA filter.

5.5 *Recirculating Design—Vertical Flow Units*—Where the vertical flow unit type requested on the quotation drawings indicates provision for recirculation, the following requirements *shall* be met:

5.51 *Recirculation Quantity*—Based on the HEPA filter nominal face area and the average velocity requirements of Para. 5.12, the recirculation duct work and blower, (if required), *shall* be sized to accommodate a recirculation quantity equal to 100% of the supply air through the HEPA filter, (to allow makeup air through recirculation openings, if desired).

5.511 *Volume Control*—The recirculation system shall be provided with:

    5.5111 Suitable perforations of the recirculation inlet slot at the front work surface area and adjustable dampers if necessary to insure uniform velocity across the entire width of the slot in accordance with paragraph 5.12.

    5.5112 Additional volume control dampers shall be provided as required to adjust the recirculation volume to the requirements of Para. 5.12, 5.42, or 5.51.

    5.5113 All dampers must have handle, position lock and indicator to show damper position. Dampers to be fabricated from same materials as exhaust or recirculation ducting, or work surface.

5.52 *Blower—Motor Assembly*—See Paragraphs 7.1, 7.21, 7.22, 7.41, 7.42, 7.7 and 10.11. These requirements *shall* govern the recirculation blower as the main supply blower.

5.53 *Electrical Wiring and Components in Recirculation and Supply Blower Plenums*—No electrical contacting devices or pilot lights shall be located in these plenums unless of explosion-proof design. All wiring connections located in these areas *shall* be within vapor-proof enclosures. All wiring within these areas *shall* be with flexible rubber covered cord having the proper wire size and number of conductors for the use intended and of type SJ as specified in the latest issue of the National Electrical Code. All cord penetrations through enclosure or cabinet-walls shall be sealed with tight fitting cord-grip connectors to prevent fume penetration.

5.54 *Lighting*—Ballasts required for fluorescent lighting fixtures shall conform to the NEC code and *shall* be located outside of the direct air stream and any recirculation blower plenum. All other requirements shall be as per Paragraphs 3.1 through 3.6.

5.55 *Exhausted Work Surfaces*—Work surface perforations shall be arranged to accommodate the specified exhaust quantities and air flow velocities (above the work surface) without permitting fume movement out of the work area into the recirculating slot, or the concentration of particulates or fumes in any position of the work space volume. All wells provided in the work surface shall have sides open to the exhaust plenum for at least 1/2 in. below the work surface for local exhaust control. Any well provided for hot plate or other electrical equipment wherein fume concentration may occur shall be vented to the exhaust plenum by perforating the bottom surface or sidewalls of the well just above the plenum bottom sheet.

    5.551 Quotation drawings *may* indicate specific portions of the work surface to be perforated. In such cases, these areas plus well exhaust slots plus any additional perforation necessary must be arranged to meet the provisions of paragraph 5.55.

# 6. FILTER SYSTEM

6.1 *Final Filter*—The final filter *shall* be a HEPA (High Efficiency Particulare Air) filter per Type C, Grade 2 per AACC Standard CS-1T to insure adherence to appropriate requirements of Paragraph 4.2.

    6.11 *Depth*—The final filter *shall* have a nominal depth of 6 in. unless otherwise specified in the quotation drawings.

    6.12 For horizontal flow units, HEPA Filters shall be installed with separators vertical.

    6.13 Where units are to be supplied less HEPA *filter* (as specifically indicated on quotation drawings), supplier shall not be relieved of any responsibility under this specification except for the quality of the HEPA filter itself. In any case of questionable performance for which supplier is considered responsible, the HEPA filter actually used will be demonstrated to supplier's satisfaction to be free of contribution to performance deficiencies.

6.2 *Final Filter Installation Design*—The final filter installation design must meet the requirements of Section 13, para. (b), items (1, 2, 4, and 5) of AACC Standard CS-2T and the following:

6.21 *Seal Face*—All corner joints must be sealed to present a continuous unbroken sealing face for the filter gasket. Sufficient structural support shall be provided to permit tightening up to 80% gasket compression without distortion.

6.22 *Final Filter Size*—Multiple filters *shall* be used in equipment requiring filters larger than 30 in. X 48 in. in size. Where multiple filters are used, each shall be individually clamped (see 6.23).

6.23 *Clamping*—Final filter shall be secured with clamps spaced not over 24 in. apart. Clamp bearing plates shall be provided to prevent indentation of the plywood filter frame. Any clamps 12 in. or more from face of access opening shall be provided with extension handle wrenches for easy adjustment.

6.24 *Alignment*—Equipment design *shall* provide for alignment of all work space surfaces with the edge of the effective HEPA filter media (inside edge of media sealant).

6.25 *Final Filter Protective Screen*—The protective screen for the HEPA filter *shall* be stainless steel or aluminum #16 gauge, flattened mesh, opening 5/16 in. wide, 1 in. long, diamond shape or approved equivalent (horizontal flow units only). Screen shall be located as specified on the quotation drawings.

6.26 *Final Filter Access*—The access to the final filter *shall* be as specified on the quotation drawings.

6.3 *Fresh Air Prefilter*—The prefilter(s) shall be of *non-combustible*, nonoil bearing material, Continental Air Filter Company, 1 in. thick Dycon filter media, permanent washable type "B" with UL Class 2 rating or approved equivalent.

6.4 *Prefilter Installation Design*—The prefilter installation design must meet the following requirements:

6.41 *Support Frame*—The prefilter *shall* be contained in a metal frame which provides adequate media support and permits ready access for vacuum cleaning and replacement.

6.42 *Face Area*—Sufficient face area of prefilter *shall* be provided to limit the air intake velocity to 130 feet per minute when measured 6 in. in front of the filter at any point across the face of the opening.

6.43 *Location*—Unless specifically located by dimension on the quotation drawing, the prefilter may be located at the supplier's discretion in the general vicinity shown on the quotation drawing.

6.44 *Dampers*—Opposed blade adjustable dampers shall be provided with all fresh air prefilter grilles.

## 7. ELECTRICAL STANDARDS

7.1 *Code Conformance*—All wiring and electrical components *shall* conform to Section 14, AACC Standard CS-2T.

7.2 *Wiring System*—The equipment shall be wired for operation from a 120 volt, 1 phase, 3 wire supply or a 208 volt, 3 phase, 5 wire supply, (See quotation drawing for wiring diagram) with one wire being a *green* insulated static ground wire isolated from neutral. White wires shall be used for neutrals.

7.21 The static ground shall be bonded to all metallic components to insure a common reference potential. *No* metallic component shall be left isolated (floating).

7.22 All wires shall be appropriately labeled both on the supplier's drawings and physically within the cabinet. Wire coding within the cabinet shall be correlated with the wiring diagram and shall be accomplished by the use of wire markers designating the wire at each termination or terminal block. All wires through areas exposed to corrosive atmospheres are to be Teflon covered or suitably protected, and all terminals are to be covered to prevent corrosion and/or accidental contact.

7.3 *Main Circuit Breaker*—A main circuit breaker *shall* be provided and located as shown on the quotation drawings. All internal components shall be wired and connected to this circuit breaker as shown on the standard electrical wiring drawings. Circuit breakers shall be as follows:

7.31 *Units Without Exhaust*—3 pole, 20 amp., magnetic type, trip-free.

7.32 *Units With Exhaust*—3 pole, 20 amp., magnetic type, trip-free, with 90 volt (minimum) potential trip coil on the third pole.

7.33 *Units (Requiring minimal electrical requirements)*—1 pole, 20 amp., magnetic-type, trip-free.

7.4 *Blower Motor*—The electric blower motor (s) shall be 60 cycle, permanent split phase, capacitor type, up to 3/4 hp size, with motor mounted in fan scroll direct connected. (Note: As per paragraph 5.311, no open contact thermal overload devices shall be operative when this motor type is used in recirculating air flow designs).

7.41 *Alternate*—Where necessary, for unusual conditions, and *subject to individual order approval*, totally enclosed non-ventilated split-phase general purpose type motors may be substituted. Motor installations shall be arranged for adequate cooling and de-rated in accordance with manufacturer's instructions.

7.42 *Voltage*—Fractional horsepower motors up to and including 1/2 hp shall operate on a 120 volt, single phase, three wire connection, with one wire being a *green* insulated static ground wire isolated from neutral. Motors of 3/4 hp shall operate on a 208 volt, 1 phase, 3 wire connection with one wire being a *green* insulated static ground wire isolated from neutral. For larger motors consult the facility planning department.

7.43 *Overload Protection*—A manual motor starter of the proper type and incorporating overload protection shall be provided for the blower motor, and located as shown on the quotation drawings.

7.5 *Convenience Receptacles* Convenience receptacles when required shall be Hubbell #5362, or equivalent, 20 amp., polarized with a "T" slot.

7.6 *Lighting Control*—A separate switch (or switches as per quotation drawing) *shall* be provided to control the lights independent of the blower. Location of the switch *shall* be as shown on the quotation drawings.

7.7 *Operation Pilot Light*—A red pilot light *shall* be provided to indicate blower operation. It *shall* be located as shown on the quotation drawings.

7.8 *Electrical Raceway*—When required, provide a tube with an approximate area of 6 sq. in. for use as an electrical raceway. Open ends at top and bottom for thru wiring. Provide gasketed end plates (top and bottom). Provide service cutout with blank cover approximately 12 in. below top of cabinet.

## 8. NOISE  The work station shall not create noise level conditions that exceed:

8.1 "C" scale readings (with a General Radio Type 1551 Sound Level Meter or approved equivalent) at the front edge of the work surface and 48 in. above floor level for 30 in. high work surface or 54 in. above floor level for 37 in. high work surface shown below in Column 2 when the background level at the same point with the station shut down is the corresponding "C" scale reading of Column 1:

| (1) Background Reading (Unit Shut Off) | (2) Unit Reading (Unit Operating) | (3) 24 in. Front of Unit Reading (Unit Operating) |
|---|---|---|
| 57 | 72 | 69 |
| 60 | 72.3 | 70.0 |
| 63 | 72.5 | 70.5 |
| 66 | 73.0 | 71.0 |
| 68 | 73.5 | 72.0 |
| 70 | 74.1 | 73.0 |
| 72 | 75.0 | 74.0 |

(Intermediate readings may be extrapolated.)

8.2 "C" scale readings 24 in. directly in front of station at 60 in. above floor level shown in Column 3, paragraph 8.1, for the corresponding background levels of Column 1 measured at the same location.

8.3 Speech interference levels of 58 db and 55 db at the locations of paragraphs 8.1 and 8.2 respectively, when the background "C" scale reading at the same locations is not over 72 db (based on the 600–1200, 1200–2400, and 2400–4800 Hz octave bands).

9. **VIBRATION**   The net vibration velocity on the work surface of the ultra-clean work station shall not exceed the acceptance Level "C" of Section 12 of the AACC Standard CS-2T when tested in accordance with the procedures of that section. All cabinet panels *shall* be stiffened to minimize vibration due to excitation by blower. The type and model of vibration measuring equipment used for checking shall be noted on the supplier's ultra-clean equipment drawing or test report.

## 10. MISCELLANEOUS CONSTRUCTION REQUIREMENTS

10.1 *Blower-Motor Assembly*

10.11 *Housing and Wheel*—Unless specific *corrosion resistance* requirements are shown on the quotation drawings: (a) The blower housing (s) may be painted sheet steel; and (b) The blower wheel (s) shall be fabricated of non-sparking material.

10.12 In any vertical flow unit where fresh air inlet is not located below the work surface, all closed compartments below the work surface must be ventilated.

10.2 *Access*

10.21 Access panels shall be provided for all filter and blower plenums, lighting compartments, electrical control compartments and additional spaces as specifically required on the quotation drawings.

10.22 *Quick Fasteners*—All access panels shall be secured with quick-release type fasteners, the number of which shall be a minimum consistent with adequate closure controls.

10.23 Miscellaneous equipment installed by the supplier; e.g., hot plates, gas heaters, ultrasonic generators, and transducers shall be easily accessible for adjustment and maintenance through removable panels per 10.22. Electric cords of sufficient length and suitable grounded plugs shall be provided to permit easy removal. In acid stations work surfaces shall be easily removable for access to plenums and equipment. Cord shall be Teflon covered type (see 7.22).

10.3 *Adjustable Feet*–Adjustable removable feet capable of 1 in. adjustment shall be provided (to permit equal adjustment on both sides of nominal dimensions per prints).

10.4 *Controls and Valves–for Built-In Process Equipment*–shall be mounted where possible on a canted panel located across the front of the bench below the exhaust and/or recirculating plenums. Panel should be split with switches on top half so that access to the electrical equipment does not require removal of valves, etc. Suitable cut-outs may be used instead of split panel. Panel to be recessed a minimum of 2 1/2 in. to insure that valve handles, etc. do not extend past the front of the bench. Pilot lights should be associated with all control switches except light switches. Labels of lamicoid plastic shall be provided for all controls and valves.

10.5 *Drains in Wells*–All wells (including hot-plate wells extending below bottom of exhaust plenum) shall have drains with a "pet cock" or ball valve installed in the line. All solvent drains except hot plate wells are to contain flame arrestors. Hot plate well drains will not be connected to drainage systems.

10.6 *Dimensional Tolerances*–Provisions of Section 13, (a, c, d, and e), of AACC Standard CS-2T shall apply regarding general dimensional tolerances. (Filter mounting tolerances are covered in Para. 6).

11. **INSPECTION**   Inspection of the unit at the Supplier's plant prior to shipment shall be at the discretion of the ———————— Company facility planning department. Acceptance at this time shall in no way void the requirements of paragraph 13.

12. **PACKAGING AND SHIPMENT**   All equipment shall be packed for shipment in containers that will provide physical protection to the contents during normal handling, shipping and storage in dry unheated quarters. If corrugated fiberboard containers are used they shall meet the requirements of Rule 41 of the Uniform Freight Classification.

12.1 Products that are crated for shipment shall be skidded or palletized in such manner to provide suitable material handling by forklift truck. Lumber used for crating and/or skidding and nailing practices shall meet good commercial standards.

12.2 The maximum size of any component part (uncrated) *shall* not exceed 108 in. high × 63 in. long × 36 in. deep or 72 in. high × 99 in. long × 36 in. deep.

13. **CORRECTION OF DEFECTS**   The supplier *shall* be responsible for correction of any defects causing the equipment to fail to meet these specifications that are detected within six months of shipment. In addition the supplier shall replace any blower, motor, or control that proves defective within one year of shipment.

14. **GENERAL CONDITIONS**

14.1 Any specifications, drawings, technical information or data furnished to you here under shall remain our property, shall be kept confidential, and shall be used for quotations to us on the manufacture of material covered thereby and may be used for other purposes upon such terms as may be agreen upon between us in writing.

14.2 Where the term "approved equivalent" is used, specific approval for the substitute shall be in writing from the facility planning department.

14.3 Any deviation from this specification shall be requested by the supplier in writing and shall be approved in writing by the facility planning department.

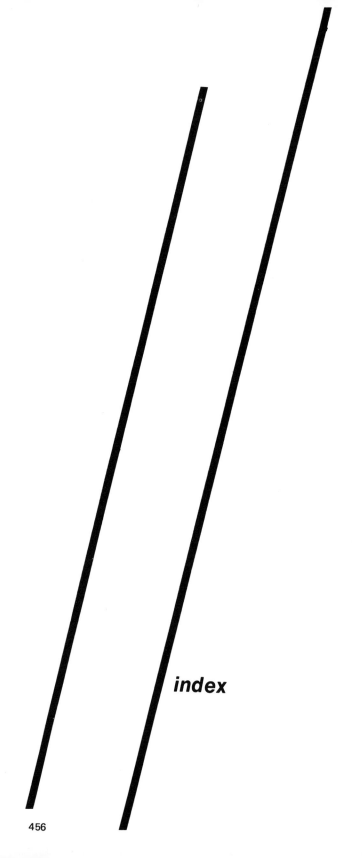

*index*

Diagnostic chemical analysis: combination of analytical instruments required, 81; defined, 53; sample collection and preparation, 81

Diagnostic methods: analytical instruments used, 55; defined, 53; device purity requirements, 55

Diatomaceous earth filters: diatoms, 214; filtration cycle, 214; water purification processes, 214

Diborane use: microelectronic usage, 316

Diffusion boats: control of process chemicals, 234

Diffusiophoresis: deposition mechanisms, 253

Dilution: factors in environment control, 254, 257

Diode: main circuit element in silicon integrated circuit (SIC), 15; one of three prominent semiconductor devices, 36

Disadvantages: surface heat treatment, 197

Dissolved gasses: raw water, 209

DOP (Dioctyl-Phthalate) Penetration Test: filter tests and standards, 262

Dopants: carrier lifetime, 229; control of process chemicals, 228; epitaxial processing, 229

Drag-out volume: rinsing methods, 199

Drybox operations: energy hazards, 365

Drying methods: air and dry nitrogen tunnel techniques, 201; alcohols and acetone, 203; centrifugal spin drying, 201; final surface cleaning of silicon wafers, 201; Freons and halocarbons, 203; ovens, 203; surface cleaning practice, 201; wiping, 201

Dust control hoods: work stations, 292

Dust holding capacity of filters: filtration, 259

Dust Holding Capacity Test: filter tests and standards, 262

Economics of clean environment facilities: clean room acquisition cost, 308; clean work station cost, 309; floor space requirements, 311; operating costs, 311

Education: accident prevention, 365

Effect of mass activity: contaminant sources, 340

Effect of particle count: contaminant sources, 338

Effect of sodium: high purity metals, 230

Effect of time duration: contaminant sources, 338

Effectiveness of ultrasonic cleaning: silicon wafers, 192

Efficient rinsing: rinsing methods, 199

Effluent requirements: liquid waste control, 372

Electrical conductivity particle counter: in liquidborne measurements, 142

Electrical monitoring of product: production monitoring, 297; types of measurement, 398

Electrochemical deposition: control of process chemicals, 227

Electrolytic cleaning: anodic cleaning, 190; cathodic cleaning, 189; surface cleaning practice, 189

Electromagnetic radiation instruments: atomic absorption spectrophotometer, 62; emission spectrograph, 60; flame emission spectrometer, 64; fourier transform spectrometer, 69; infrared spectrophotometer, 68; material behavior characteristics, 56; used in diagnostic chemical analysis, 56; ultraviolet visible absorption spectrophotometer, 66; x-ray fluorescence spectrometer, 64

Electron diffraction: in transmission electron microscope (TEM), 105

Electron diffraction patterns: thin films, 118

Electron microprobe: in sealed contact switch application, 116; monitoring plant environment, 414